Advanced Studies in Maintaining Post-harvest Quality of Fruits and Vegetables

Advanced Studies in Maintaining Post-harvest Quality of Fruits and Vegetables

Editors

Wenzhong Hu
Tian Zhong
Xiuxiu Sun

MDPI

Basel • Beijing • Wuhan • Barcelona • Belgrade • Novi Sad • Cluj • Manchester

Editors

Wenzhong Hu
School of Life Science
Zhuhai College of Science
and Technology
Zhuhai
China

Tian Zhong
Faculty of Medicine
Macau University of Science
and Technology
Macau
China

Xiuxiu Sun
United States Department of
Agriculture
Daniel K. Inouye U.S. Pacific
Basin Agricultural Research
Center
Hawaii
USA

Editorial Office
MDPI AG
Grosspeteranlage 5
4052 Basel, Switzerland

This is a reprint of articles from the Special Issue published online in the open access journal *Horticulturae* (ISSN 2311-7524) (available at: https://www.mdpi.com/journal/horticulturae/special_issues/2BLEP43X90).

For citation purposes, cite each article independently as indicated on the article page online and as indicated below:

Lastname, A.A.; Lastname, B.B. Article Title. *Journal Name* **Year**, *Volume Number*, Page Range.

ISBN 978-3-7258-2423-6 (Hbk)
ISBN 978-3-7258-2424-3 (PDF)
doi.org/10.3390/books978-3-7258-2424-3

Contents

horticulturae

MDPI

Editorial

Special Issue: 'Advanced Studies in Maintaining Post-Harvest Quality of Fruits and Vegetables'

Wenzhong Hu

College of Life Science, Zhuhai College of Science and Technology, Zhuhai 519041, China; hwz@dlnu.edu.cn

1. Introduction

The characteristics of taste, flavor, nutrition, and safety in fruits and vegetables during post-harvest, minimal processing, fresh-cutting, fermenting, and processing are critical elements in the storage and sale of products. The growing demand for fruits and vegetables has led to an increasing interest in the study of maintaining their quality, enhancing their safety, and extending their shelf life [1]. However, fruits and vegetables are prone to tissue damage, wound respiration, water loss, transpiration, ethylene production, enzymatic browning, tissue softening, and secondary metabolite production during storage, transportation, and sale along the supply chain [2,3]. Therefore, there is an urgent demand for any new preservation technologies that can maintain the quality of fruits and vegetables [4–7]. The purpose of this Special Issue, "Advanced Studies in Maintaining Post-harvest Quality of Fruits and Vegetables", is to present new preservation technologies to resolve some of these concerns, such as physiological and quality changes, the study of preservation technology and its regulation mechanism, metabolomics and microbial interactions, and any other method that might improve the quality and safety of fruits and vegetables. Articles on maintaining the quality and safety of any kind of fruits and vegetables, including post-harvest, minimally processed, fresh-cut, fermented, and processed fruits and vegetables were welcomed for this Special Issue. After a rigorous review, a total of fourteen articles (one review paper and thirteen research papers) were included in this Special Issue, proposing new preservation technology to improve the quality and shelf life of various fresh horticultural products.

2. Overview of Published Articles

2.1. Physical Preservation Technologies

The quality attributes of post-harvest fruits and vegetables include a fresh appearance, firmness, acceptable texture, characteristic flavor, vitamin content, and a sufficient shelf life to meet distribution. The main significant factors influencing the quality of fruits and vegetables are the temperature, the atmosphere, and the storage time. Modified atmosphere packaging is considered an effective technology for extending the shelf life of fruits and vegetables. Da Silva et al. (contribution 1) evaluated the impact of dipping 'Palmer' mangoes in 0.1% and 2.5% (*w/v*) sorbitol solutions and storing the fruit under a controlled atmosphere (CA) without atmosphere modification (21 kPa O_2 + 0.03 kPa CO_2) at 8 °C with 95% relative humidity (RH) or with 5 kPa O_2 + 5 kPa CO_2 at 4 °C/95% RH for 28 days. The result showed sorbitol was effective in minimizing the chilling injury symptoms and did not compromise the fruit quality, especially when it was stored at 4 °C in association with a CA containing 5 kPa O_2 + 5 kPa CO_2. It reduced lipid peroxidation and increased the activities of the superoxide dismutase (SOD) and ascorbate peroxidase (APX) enzymes in the epicarp and mesocarp, providing greater cold tolerance. Thus, this treatment represents a viable alternative for managing chilling injuries in mangoes. Erbas (contribution 2) evaluated the effects of combinations of modified atmosphere packaging and oxalic acid (OA) treatment on the quality and biochemical content changes in rocket (Eruca sativa Mill.

Citation: Hu, W. Special Issue: 'Advanced Studies in Maintaining Post-Harvest Quality of Fruits and Vegetables'. *Horticulturae* **2024**, *10*, 1039. https://doi.org/10.3390/horticulturae10101039

Received: 24 September 2024
Accepted: 29 September 2024
Published: 30 September 2024

cv. Bengi) leaves. The result showed rocket leaves could be stored at 0 °C for 8–9 days with 1 mM OA treatment. Temperature preservation is one of the most common treatments and traditional physical methods to extend the shelf life of fruits and vegetables. Zhao et al. (contribution 3) studied the effect of near-freezing temperature (NFT) on the chilling injury (CI), proline metabolism, and antioxidant capability of peach fruit during storage times. The results showed that the NFT completely inhibited the occurrence of CI in peach fruit. The NFT significantly ($p < 0.05$) enhanced the activities of the superoxide dismutase, catalase, ascorbate peroxidase, and the 1,1-diphenyl-2-picrylhydrazyl scavenging capacity. The results suggest that NFT storage can improve the chilling tolerance of peach fruit by regulating the antioxidant defense and proline metabolism, which might represent a potential novel method to store fruits and vegetables for longer times. Elataff et al. (contribution 4) evaluated the effect of different storage temperatures and MeJA on the quality of and antioxidants in post-harvest Shine Muscat table grapes. The results indicate MeJA could play a critical role as a stimulator of fruit quality as well as enhance the physicochemical parameters and antioxidant activities for extending the shelf life of grapes during storage. Hu et al. (contribution 5) investigated the effect of the combination of low temperature and different wounding intensities on the quality of fresh-cut pumpkin; the critical indexes involved in reactive oxygen species (ROS) metabolism, the vitamin C–glutathione cycle, phenylpropanoid metabolism, and membrane lipid peroxidation were monitored for pumpkin. The results showed that with the increase in the cutting injury strength, the lightness, whiteness index, respiration rate, ethylene content, lipoxygenase activity, and malondialdehyde content of fresh-cut pumpkin increased, while the hardness, sensory quality, appearance, and total soluble solid content continuously decreased. Lee et al. (contribution 6) evaluated the quality index changes in the green maturity of Jin Huang mangoes with different ethylene and post-ripening treatments and then when stored at different storage periods at 4 °C followed by 6 days at 20 °C. The result showed that the mangoes treated with 500 ppm ethylene were slow to ripen during 4 °C storage, which could be sustainable even under 20 °C storage. The treatment conditions of Jin Huang mango with ethylene ripening for 1 day and post-ripening at 20 °C for 1 day helped extend its shelf life at 20 °C, stocking and minimizing CI and anthracnose, thereby maintaining a certain quality.

2.2. Chemical Preservation Technologies

The use of edible films/coatings alone or in combination with antioxidants has beneficial effects on the quality of fresh-cut fruits and vegetables [8–11]. Edible films/coatings can offer a possibility to extend the shelf life of fresh-cut produce by providing a semi-permeable barrier to gases and water vapor and, therefore, reduce respiration, enzymatic browning, and water loss. Sarengaowa et al. (contribution 7) evaluated the quality and safety of fresh-cut potatoes with an alginate-based edible coating containing thyme essential oil (AEC-TEO) during a storage period of 16 days at 4 °C. The research showed that AEC-TEO at a 0.05% concentration was the most beneficial for maintaining the quality and inhibiting microorganisms in the fresh-cut potatoes. This represents a potential application prospect for the preservation of fresh-cut potatoes. Li et al. (contribution 8) investigated the effect of solid lipid nanoparticles containing cinnamaldehyde (SLN-CA) on post-harvest strawberry. It showed that SLN-CA treatment can effectively reduce the probability of decay and softening, maintain a high level of SOD activity in cells, and improve the sensory characteristics of strawberries and thereby their shelf life. A recently developed inhibitor of ethylene action, 1-methylcyclopropene (1-MCP), has been shown to be effective in reducing the quality deterioration and in extending the shelf life of fresh-cut fruits. Kou et al. (contribution 9) summarized the latest available information on the effects of ethylene and 1-MCP with respect to enhancing or impairing sweet potato root quality. A better understanding of the influence of ethylene and 1-MCP on root quality parameters will be useful to further explore the role and mechanisms of action of ethylene in regulating the post-harvest storage of sweet potato roots and the contributions to technological development and innovation.

Habibi et al. (contribution 10) investigated the effects of glycine betaine (GB) and/or methyl salicylate (MeSA) on the physicochemical changes, the chemical attributes of the juice, and the peel color of 'Moro' blood orange at cold quarantine storage (2 °C) for 60 days. It showed the GB and MeSA treatments offer significant advantages in preserving the physicochemical characteristics and chemical attributes of 'Moro' blood oranges during cold quarantine storage. These findings underscore the potential of GB and MeSA treatments for maintaining the quality of 'Moro' blood oranges during cold quarantine storage, with a noteworthy synergistic effect between MeSA and GB in preserving the fruit quality.

2.3. *Preharvest Treatment Technologies*

Preharvest chemical spraying can significantly improve the post-harvest quality of horticultural products. Spraying has a more pronounced effect [12]. Wang et al. (contribution 11) investigated the effects of non-bagging film agents on the contents of the mineral elements and flavonoid metabolites in apple fruits and determined the feasibility of this method. Individual spraying of non-bagging film agents can significantly increase the total contents of mineral elements in apples. Application of plant growth regulators (PGRs) in apricot orchards is a common practice with the goal of improving the yield and/or the quality of fruits at harvest. However, the question of whether such a treatment alters post-harvest properties is seldom answered. Milovic´ et al. (contribution 12) examine the impact of gibberellic acid (GA3) and 6-benzyladenine (BA) on the physical characteristics and composition of apricot fruit after prolonged cold storage (i.e., 21 days) and shelf life. At harvest, significant differences were observed between the treated and untreated fruits regarding the flesh firmness, color, ethylene production and respiration rate, and the flavonoid, carotenoid, and citric acid content, while application of BA100 changed the TA and TSS. There was no difference in the sensory properties of the treated and non-treated fruit after cold storage and in the shelf life.

2.4. *Other Technologies*

Rodríguez et al., (contribution 13) developed a methodology for evaluating the ripening pattern, internal disorders, flesh composition, and sensory quality of Hass avocados and to determine predictive quality markers to manage these. The research showed the PLS-DA model could be a powerful tool for classifying the internal quality of fruits in avocado-producing countries. Therefore, in countries with highly variable growing conditions and production practices, the use of destructive analyses and multivariate data analyses could reduce the heterogeneity and allow the shipment of fruits with better internal quality to the international market. Tziotzios et al. (contribution 14) investigated the use of a non-destructive hyperspectral imaging approach for the evaluation of kiwifruit cv. "Hayward" internal quality, focusing on physiological traits such as the soluble solid concentration (SSC), dry matter (DM), firmness, and the tannins, widely used as quality attributes. The study concluded that machine learning algorithms, especially neural networks, offer substantial accuracy, surpassing traditional methods for evaluating kiwifruit quality traits. Overall, the current study highlighted the potential of such non-destructive techniques in revolutionizing quality assessment during post-harvest by yielding rapid and reliable predictions regarding the critical quality attributes of fruits.

3. Conclusions

These articles, which are written to a very high standard with excellent citations, were selected from a large number of submissions through rigorous evaluation. This Special Issue strives to promote more in-depth communication among professional and technical personnel in the field of horticultural post-harvest preservation technology, improving post-harvest quality, ensuring product safety, and serving the healthy development of the horticultural industry.

Conflicts of Interest: The authors declare no conflicts of interest.

List of Contributions

1. da Silva, M.B.; Pedrosa, V.M.D.; Izidoro, M.; Balbuena, T.S.; Sanches, A.G.; de Almeida Teixeira, G.H. Controlled Atmosphere Storage and Sorbitol Dipping Minimize Chilling Injuries in 'Palmer' Mangoes. Horticulturae 2024, 10, 354. https://doi.org/10.3390/horticulturae10040354.
2. Erbaș, D. Effect of Oxalic acid treatments and modified atmosphere packaging on the quality attributes of rocket leaves during different storage temperatures. Horticulturae 2023, 9, 718. https://doi.org/10.3390/horticulturae9060718.
3. Zhao, H.; Meng, S.; Fu, M.; Chen, Q. Near-Freezing Temperature Storage Improves Peach Fruit Chilling Tolerance by Regulating the Antioxidant and Proline Metabolism. Horticulturae 2024, 10, 337. https://doi.org/10.3390/horticulturae10040337.
4. Elataff, E.; Elshahat, A.; Xue, Y.; Shaonan, L.; Suwen, L.; Tianyu, D.; Fang, J. Effects of Different Storage Temperatures and Methyl Jasmonate on Grape Quality and Antioxidant Activity. Horticulturae 2023, 9, 1282. https://doi.org/10.3390/horticulturae9121282.
5. Hu, W.; Guan, Y.; Wang, Y.; Yuan, N. Effect of Wounding Intensity on Edible Quality by Regulating Physiological and ROS Metabolism in Fresh-Cut Pumpkins. Horticulturae 2023, 9, 512. https://doi.org/10.3390/horticulturae9040512.
6. Lee, Y.-C.; Yu, M.-C.; Yen, C.-Y.; Tsay, J.-S.; Hou, C.-Y.; Li, P.-H.; Huang, P.-H.; Liang, Y.-S. Exploitation of Post-Ripening Treatment for Improving Cold Tolerance and Storage Period of Jin Huang Mango. Horticulturae 2024, 10, 103. https://doi.org/10.3390/horticulturae10010103.
7. Sarengaowa; Feng, K.; Li, Y.; Long, Y.; Hu, W. Effect of Alginate-Based Edible Coating Containing Thyme Essential Oil on Quality and Microbial Safety of Fresh-Cut Potatoes. Horticulturae 2023, 9, 543. https://doi.org/10.3390/horticulturae9050543.
8. Li, S.; Chen, J.; Liu, Y.; Zheng, Q.; Tan, W.; Feng, X.; Feng, K.; Hu, W. Application of Cinnamaldehyde Solid Lipid Nanoparticles in Strawberry Preservation. Horticulturae 2023, 9, 607. https://doi.org/10.3390/horticulturae9050607.
9. Kou, J.; Zang, X.; Li, M.; Li, W.; Zhang, H.; Chen, Y.; Zhu, G. Effects of Ethylene and 1-Methylcyclopropene on the Quality of Sweet Potato Roots during Storage: A Review. Horticulturae 2023, 9, 667. https://doi.org/10.3390/horticulturae9060667.
10. Habibi, F.; Sarkhosh, A.; Guillén, F.; Serrano, M.; Valero, D. Changes in Physicochemical Characteristics, Peel Color, and Juice Attributes of 'Moro' Blood Orange Fruit Treated with Glycine Betaine and Methyl Salicylate during Cold Quarantine Storage. Horticulturae 2023, 9, 1103. https://doi.org/10.3390/horticulturae9101103.
11. Wang, F.; Wu, X.; Ding, Y.; Liu, X.; Wang, X.; Gao, Y.; Tian, J.; Li, X. Study of the Effects of Spraying Non-Bagging Film Agent on the Contents of Mineral Elements and Flavonoid Metabolites in Apples. Horticulturae 2024, 10, 198. https://doi.org/10.3390/horticulturae10030198.
12. Milovic´, M.; Kevrešan, Ž.; Mastilovic´, J.; Kovacˇ, R.; Kalajdžic´, J.; Magazin, N.; Bajic´, A.; Milic´, B.; Barac´, G.; Keserovic´, Z. Could an Early Treatment with GA and BA Impact Prolonged Cold Storage and Shelf Life of Apricot? Horticulturae 2022, 8, 1220. https://doi.org/10.3390/horticulturae8121220.
13. Rodríguez, P.; Soto, I.; Villamizar, J.; Rebolledo, A. Fatty Acids and Minerals as Markers Useful to Classify Hass Avocado Quality: Ripening Patterns, Internal Disorders, and Sensory Quality. Horticulturae 2023, 9, 460. https://doi.org/10.3390/horticulturae9040460.
14. Tziotzios, G.; Pantazi, X.E.; Paraskevas, C.; Tsitsopoulos, C.; Valasiadis, D.; Nasiopoulou, E.; Michailidis, M.; Molassiotis, A. Non-Destructive Quality Estimation Using a Machine Learning-Based Spectroscopic Approach in Kiwifruits. Horticulturae 2024, 10, 251. https://doi.org/10.3390/horticulturae10030251.

References

1. De Corato, U. Improving the shelf-life and quality of fresh and minimally-processed fruits and vegetables for a modern food industry: A comprehensive critical review from the traditional technologies into the most prom-ising advancements. *Crit. Rev. Food Sci. Nutr.* **2019**, *60*, 940–975. [CrossRef]
2. Francis, G.A.; Gallone, A.; Nychas, G.J.; Sofos, J.N.; Colelli, G.; Amodio, M.L.; Spano, G. Factors affecting quality and safety of fresh-cut produce. *Crit. Rev. Food Sci. Nutr.* **2012**, *52*, 595–610. [CrossRef] [PubMed]
3. Guan, Y.; Ji, Y.; Yang, X.; Pang, L.; Cheng, J.; Lu, X.; Zheng, J.; Yin, L.; Hu, W. Antioxidant activity and microbial safety of fresh-cut red cabbage stored in different packaging films. *LWT* **2023**, *175*, 114478. [CrossRef]
4. Zhao, Q.; Zhao, Q.; Li, J.; Yi, L. Antibacterial activity and action mechanism of bacteriocin paracin wx7 as a selective biopreservative against vancomycin-resistant Enterococcus faecalis in fresh-cut lettuce. *Foods* **2024**, *13*, 1448. [CrossRef] [PubMed]
5. Azam, I.; ul Ain, Q.; Sidiq MMuzzaffar, S. Lycopene extract as an antibrowning agent for edible coating of fresh cut apples. *J. Food Sci. Technol.* **2024**, *61*, 1758–1766. [CrossRef] [PubMed]
6. Zhang, J.; Zhao, J.; Zuo, X.; You, W.; Ru, X.; Xu, F.; Jin, P.; Zheng, Y.; Cao, S. Glutamate application maintains quality and antioxidant capacity of fresh-cut carrots by modulating GABA shunt, phenylpropanoid and ROS metabolism. *Food Chem.* **2024**, *443*, 138545. [CrossRef]
7. You, W.; Wang, C.; Zhang, J.; Ru, X.; Xu, F.; Wu, Z.; Jin, P.; Zheng, Y.; Cao, S. Exogenous chlorogenic acid inhibits quality deterioration in fresh-cut potato slices. *Food Chem.* **2024**, *446*, 138866. [CrossRef] [PubMed]
8. Firdous, N.; Moradinezhad, F.; Farooq, F.; Dorostkar, M. Advances in formulation, functionality, and application of edible coatings on fresh produce and fresh-cut products: A review. *Food Chem.* **2023**, *407*, 135186. [CrossRef] [PubMed]
9. Naqash, F.; Masoodi, F.A.; Ayob, O.; Parvez, S. Effect of active pectin edible coatings on the safety and quality of fresh-cut apple. *Int. J. Food Sci. Technol.* **2022**, *57*, 57–66. [CrossRef]
10. Xin, Y.; Yang, C.; Zhang, J.; Xiong, L. Application of whey protein-based emulsion coating treatment in fresh-cut apple preservation. *Foods* **2023**, *12*, 1140. [CrossRef] [PubMed]
11. Duong, N.; Uthairatanakij, A.; Laohakunjit, N.; Jitareerat, P.; Kaisangsri, N. Cross-linked alginate edible coatings incorporated with hexyl acetate: Film characteristics and its application on fresh-cut rose apple. *Food Biosci.* **2023**, *52*, 102410. [CrossRef]
12. Xie, P.D.; Yang, Y.Y.; Gong, D.; Li, Y.; Wang, Y.; Li, Y.C.; Prusky, D.; Bi, Y. Spraying L-phenylalanine during fruit development alleviates chilling injury in harvested muskmelons by regulating membrane lipid metabolism. *Postharvest Biol. Technol.* **2024**, *211*, 112858. [CrossRef]

horticulturae

MDPI

Article

Controlled Atmosphere Storage and Sorbitol Dipping Minimize Chilling Injuries in 'Palmer' Mangoes

Maryelle Barros da Silva [1], Vanessa Maria Dantas Pedrosa [1], Maiqui Izidoro [1], Tiago Santana Balbuena [2], Alex Guimarães Sanches [1] and Gustavo Henrique de Almeida Teixeira [1,3,*]

[1] Faculdade de Ciências Agrárias e Veterinárias (FCAV), Campus de Jaboticabal, Departamento de Ciências da Produção Agrícola, Universidade Estadual Paulista (UNESP), Via de Acesso Prof. Paulo Donato Castellane s/n, Jaboticabal 14884-900, SP, Brazil; maiqui.izidoro@unesp.br (M.I.)

[2] Faculdade de Ciências Agrárias e Veterinárias (FCAV), Campus de Jaboticabal, Departamento de Biotecnologia Agropecuária e Ambiental, Universidade Estadual Paulista (UNESP), Via de Acesso Prof. Paulo Donato Castellane s/n, Jaboticabal 14884-900, SP, Brazil

[3] Kimberly Research and Extension Center, University of Idaho (U of I), 3806 N 3600 E, Kimberly, ID 83341-5082, USA

* Correspondence: gteixeira@uidaho.edu

Abstract: Our previous studies have shown that 'Palmer' mangoes immersed in solutions containing 2.5% sorbitol and stored under a controlled atmosphere (CA) at 8 °C for 30 days had fewer symptoms of a chilling injury. However, there is no information regarding the effectiveness of sorbitol treatment in other atmospheres and/or in combination with lower temperatures. Thus, the objective of this study was to assess the impact of dipping 'Palmer' mangoes in 0.1% and 2.5% (w/v) sorbitol solutions and storing the fruit under a CA without atmosphere modification (21 kPa O_2 + 0.03 kPa CO_2) at 8 °C/95% relative humidity (RH) or with 5 kPa O_2 + 5 kPa CO_2 at 4 °C/95% RH for 28 days. The fruits were evaluated periodically for chilling injuries, quality, and oxidative metabolism. A chilling injury (CI) was correlated with increased fresh weight loss (FWL) and changes in the color of the epicarp (L_{peel}, $h°_{peel}$, and C_{peel}) and mesocarp (L^*_{pulp}). Lipid peroxidation (LP_{pulp} and LP_{peel}) and the hydrogen peroxide content (H_2O_{2peel} and H_2O_{2pulp}) were associated with the development of a CI, particularly after being transferred to ambient. The treatment with 2.5% sorbitol was more effective in minimizing the chilling injury symptoms and did not compromise the fruit quality, especially when it was stored at 4 °C in association with a CA containing 5 kPa O_2 + 5 kPa CO_2. This treatment reduced lipid peroxidation and increased the activities of the superoxide dismutase (SOD) and ascorbate peroxidase (APX) enzymes in the epicarp and mesocarp, providing greater cold tolerance. The use of 2.5% sorbitol has been identified as the most efficacious approach for mitigating the adverse impacts of chilling injuries, preserving the fruit quality, and enhancing oxidative metabolism, even at lower temperatures. Thus, this treatment represents a viable alternative for managing chilling injuries in mangoes.

Keywords: *Mangifera indica* L.; polyols; oxidative metabolism; SOD; APX

Citation: da Silva, M.B.; Pedrosa, V.M.D.; Izidoro, M.; Balbuena, T.S.; Sanches, A.G.; de Almeida Teixeira, G.H. Controlled Atmosphere Storage and Sorbitol Dipping Minimize Chilling Injuries in 'Palmer' Mangoes. *Horticulturae* **2024**, *10*, 354. https://doi.org/10.3390/horticulturae10040354

Academic Editor: Maria Dulce Carlos Antunes

Received: 29 January 2024
Revised: 26 March 2024
Accepted: 28 March 2024
Published: 3 April 2024

1. Introduction

As observed in other horticultural products, cold storage is the primary post-harvest technology used to extend the shelf life of mangoes because low temperatures reduce their metabolic activity [1]. Mangoes stored at temperatures between 8 and 13 °C have a post-harvest shelf life of up to 21–30 days, depending on the cultivar and fruit maturity [2,3]. However, prolonged storage at temperatures below 13 °C can lead to the development of a physiological disorder known as a chilling injury (CI), resulting in qualitative and quantitative losses [4,5].

The symptoms of a CI in mangoes are most evident in the epicarp and are characterized by the presence of dark, sunken spots resembling burns, which can hinder the marketing of fresh fruit, as this defect may render the product unacceptable according to market standards. The mesocarp is also affected, as the fruit may exhibit irregular ripening [5,6].

The mechanisms related to the development of CIs in plants are related to changes in the lipid bilayer of the plasma membrane, i.e., transitioning from a more flexible fluid state to a rigid gel phase due to exposure to low temperatures. This change in rigidity leads to the loss of functions and the rupture of cell membranes, resulting in cell death [7]. Several studies have reported changes in the rigidity of cell membranes in mangoes with CIs [8–11].

Exposure to low temperatures also induces the overproduction of reactive oxygen species (ROS), such as hydrogen peroxide (H_2O_2), superoxide radicals ($O_2^{\bullet -}$), and hydroxyl radicals ($^{\bullet}OH$), that react with various molecules, leading to lipid membrane peroxidation [12]. In mangoes, ROS overproduction has been correlated with a higher incidence of CIs [5,13–15] and the manifestation of their various symptoms [16–19].

Sorbitol ($C_6H_{14}O_6$) is a water-soluble polyol that naturally occurs in various fruits [20,21] and plays an important role in the osmotic adjustment of the cytoplasm under stress conditions, potentially stabilizing membranes [22]. It can increase cold stress tolerance [23] by binding to water and lowering the dielectric constant, even at temperatures above freezing [24]. Sorbitol acts as an osmoprotective agent that has been linked to CI tolerance [25], contributing to membrane stabilization and reducing structural damage [26]. It can also act as an antioxidant, affecting the activity of enzymes, such as superoxide dismutase (SOD), peroxidase (POD), and catalase (CAT), as well as scavenging ROS [27].

In this context, Sanches et al. [15] have reported that 'Palmer' mangoes stored at 8 °C for 30 days, when treated with 2.5% sorbitol, showed reduced H_2O_2 accumulation and polyphenol oxidase (PPO) activity, higher membrane integrity (malondialdehyde—MDA), and increased activity of the enzymes SOD, CAT, and ascorbate peroxidase (APX), both in the epicarp and the mesocarp compared to those in standard storage or storage after treatment with the polyols propylene glycol and glycerol. Consequently, sorbitol immersion has been considered a feasible approach to alleviating CI during refrigerated storage.

Under standard cold quarantine treatment conditions (1 °C for 14 days), immersion in a solution containing 0.1% sorbitol was the most effective in mitigating Cis. This effect was associated with lower MDA and H_2O_2 concentrations and PPO activity levels, as well as higher SOD, CAT, and APX enzyme activity and ascorbic acid levels, especially in the pericarp [28]. The combination of immersion of 'Palmer' mangoes in solutions containing 2.5% sorbitol with controlled atmosphere (CA) storage reduced CIs through the activation of non-enzymatic (ascorbic acid and total phenolic compounds) and enzymatic (SOD, CAT, and APX) mechanisms in fruits kept at 8 °C for 30 days [28]. Despite these results, there is no information available regarding the effectiveness of sorbitol treatment in other storage atmospheres and/or in combination with lower temperatures.

Thus, the objective of this study was to evaluate the effect of dipping 'Palmer' mangoes in 0.1 and 2.5% (*w*/*v*) sorbitol solutions and storing them under a CA with ambient atmospheric conditions (21 kPa O_2 + 0.03 kPa CO_2) at 8 °C/95% relative humidity (RH) or under 5 kPa O_2 + 5 kPa CO_2 at 4 °C/95% RH for 28 days. The development of chilling injuries, quality modification, and oxidative metabolism were evaluated periodically.

2. Materials and Methods

2.1. Plant Material

'Palmer' mangoes were obtained from commercial orchards located in Taquaritinga (21°25′57.72″ latitude south, 48°32′50.46″ longitude west), São Paulo, Brazil (Experiment I), and Belém do São Francisco (8°32′1.89″ latitude south, 38°58′53.97″ longitude west), Pernambuco, Brazil (Experiment II). The fruits were harvested at physiological maturity and selected manually, and then classified according to size, soundness, the absence of mechanical damage, and pest and disease lesions. Fruit physiological maturity was evaluated based on the dry matter (DM) content of 20 fruits from each experiment. The DM content was 13.1% ± 1% and 13.0% ± 1% in the fruits from Experiments I and II, respectively.

2.2. Experiment I: Without Gas Modification (Air)

After selection, the mangoes were washed with soap, rinsed with running water, and dried. Following that, they were dipped in one of the following solutions at 5 °C for 60 min: i. distilled water (control), ii. 0.1% (*w*/*v*) food-grade sorbitol, or iii. 2.5% (*w*/*v*) food-grade sorbitol (Sigma-Aldrich, St. Louis, MO, USA) [15]. Subsequently, the mangoes were stored at 8.0 ± 1.0 °C and 95 ± 0.5% RH in a controlled atmosphere chamber (Venezia PCM 1000 model; Fruit Control Equipment, Milan, Italy), with ambient atmosphere conditions maintained (21 kPa O_2 + 0.03 kPa CO_2) for 28 days. The fruits were evaluated every 7 days. During each evaluation, the chilling injury was immediately evaluated, and some of the fruits were moved to ambient conditions (~24 ± 2.0 °C and 75 ± 2.0% RH) to evaluate them for signs of chilling injuries whenever the fruit became ripe (from 5 to 10 days). This study was set according to a completely randomized design (CRD) in a factorial arrangement of 3 (treatments: control, 0.1% sorbitol, and 2.5% sorbitol) × 5 (storage periods: 0, 7, 14, 21, and 28 days) with three replicates.

2.3. Experiment II: CA—Modification of Atmosphere Gases

The fruits for this experiment were washed, rinsed with water, and dried as in the other study. Following this, they were i. dipped in distilled water and stored in CA with 21 kPa O_2 + 0.03 kPa CO_2 (negative control), ii. dipped in distilled water and stored in CA with 5 kPa O_2 + 5 kPa CO_2 (positive control), iii. dipped in 0.1% (*w*/*v*) sorbitol and stored in CA with 5 kPa O_2 + 5 kPa CO_2, or iv. dipped in 2.5% (*w*/*v*) sorbitol and stored in CA with 5 kPa O_2 + 5 kPa CO_2. The fruits were stored at 4.0 ± 1.0 °C and 95 ± 0.5% RH in a controlled atmosphere chamber (Venezia PCM 1000 model; Fruit Control Equipment, Milan, Italy), and control of the oxygen (O_2), carbon dioxide (CO_2), and ethylene (C_2H_4) levels was performed using SWINGLOS® software, GAC 5000 (Fruit Control Equipment, Milan, Italy). These conditions were maintained for 28 days, and evaluations were conducted every 7 days. At each evaluation, some of the fruits were moved to ambient conditions (~24 ± 2.0 °C and 75 ± 2.0% RH) to be evaluated for signs of chilling injuries whenever the fruit became ripe (from 4 to 7 days). This experiment was set according to a completely randomized design (CRD) in a factorial arrangement of 4 (treatments: negative control, positive control, 0.1% sorbitol, and 2.5% sorbitol) × 5 (storage periods: 0, 7, 14, 21, and 28 days) with three replications.

2.4. Evaluations

In both experiments, the fruits were evaluated using the same method.

2.4.1. Chilling Injury Development

To evaluate chilling injury (CI) development, the fruits had their pericarp lesions visually rated according to the process of Miguel et al. [29] with modifications. Fruits with no visible symptoms (CI = 0%) were rated as 1, mild symptoms (CI = 0–25%) as 2, moderate symptoms (CI = 25–50%) as 3, and severe symptoms (CI $\geq$ 50%) as 4.

2.4.2. Fresh Weight Loss

Fresh weight loss (FWL) was determined by weighing the mangoes on an analytical balance (AS 2000 model; Mars, Brazil), and the accumulated weight loss was calculated according to the method of Sanches et al. [15] for each experiment and each evaluation day. FWL is expressed as a percentage (%).

2.4.3. Firmness

Fruit firmness was measured in the mid portion between the stem end and the remains of stigma and style of each mango without the epicarp (skin). Firmness was measured using a penetrometer (Effegi Fruit Tester, Forlì, Italy) with an 8 mm tip, and the results are expressed in Newtons (N), as has been previously described by Watkins and Harman [30].

2.4.4. Color

The epicarp (skin) and mesocarp (flesh) colors were measured in the mid portion between the stem end and the remains of the stigma and style. The skin color was also measured on opposite sides of each fruit on the blush and green areas of the protruding side. A reflectometer (CR-400; Minolta, Osaka, Japan) was used to obtain the L*, a*, and b* values, and chromaticity (C*) and hue angle (h°) were calculated according to the method of McGuire [31].

2.4.5. Physicochemical Evaluation

The soluble solid content (SSC) of the fruits was measured using a digital refractometer (Alpha; Atago Co., Ltd., Tokyo, Japan), and the results are expressed as percentages [32]. Titratable acidity (TA) was measured via titration using 0.1% phenolphthalein as an indicator, and the results are expressed in g kg^{-1} citric acid [31]. The ratio (SSC/TA) was calculated according to the method of the AOAC [31], and the pH was determined using a pH meter (Orion 3 Star; Thermo Scientific, Waltham, MA, USA).

2.5. Oxidative Metabolism

2.5.1. Lipid Peroxidation

For this analysis, samples of peel and pulp (0.5 g) were homogenized with 2.5 mL of a solution containing 0.1% (*w*/*v*) trichloroacetic acid (TCA) and 20% (*w*/*v*) polyvinylpyrrolidone (PVP). After centrifugation (ST16-R; Thermo Scientific, Waltham, MA, USA) at 12,298× *g* and 4 °C for 15 min, 250 µL of the supernatant was mixed with 1 mL 20% TCA (*w*/*v*) and 0.5% thiobarbituric acid (TBA) and incubated in a water bath at 95 °C for 30 min. After this period, the samples were cooled in ice for 10 min and centrifuged at 12,298× *g* for 15 min at 4 °C. Lipid peroxidation was calculated according to Gratão et al. [33] and is expressed in µmol of malondialdehyde (MDA) per kg of fresh weight.

2.5.2. Hydrogen Peroxide (H_2O_2)

The hydrogen peroxide (H_2O_2) content was determined in 1.0 g peel and pulp samples according to the process of Alexieva et al. [34]. The samples were homogenized in 0.1% (*w*/*v*) trichloroacetic acid at 4 °C and centrifuged at 12,298× *g* for 20 min at 4 °C. Following this, the supernatant (200 µL) was mixed with 200 µL of 100 mM potassium phosphate buffer (pH 7.5) and 800 µL of 1 M potassium iodide (KI). The samples were incubated in ice for 1 h in the dark. The hydrogen peroxide content was measured at 390 nm and expressed in mol H_2O_2 per kg of fresh weight.

2.5.3. Superoxide Dismutase (SOD) and Ascorbate Peroxidase (APX) Extraction

These two enzymes were extracted according to the process of Yang et al. [35]. The peel and pulp samples (1.0 g) were homogenized in 100 mM potassium phosphate buffer (pH 7.0), 1 mM ethylenediaminetetraacetic acid (EDTA), and 20% (*w*/*v*) polyvinylpolypyrrolidone (PVP). The homogenates were filtered using a fine nylon mesh, centrifuged at 12,298× *g* for 25 min, and immediately frozen at −18 °C. The protein levels of the homogenates were determined [36].

2.5.4. Superoxide Dismutase (SOD) Activity

SOD (SOD, EC 1.15.1.1) activity was measured using the method described by Giannopolitis and Ries [37]. The enzymatic extract (50 µL) was mixed with 1.0 mL of 50 mM sodium phosphate buffer (pH 7.8), 19.5 mM methionine, 150 µL of NBT, and 300 µL of riboflavin. The incubation took place under light, and after 15 min, the absorbance was measured at 560 nm (Beckman spectrophotometer, DU-640; GMI: Mequon, WI, USA). The SOD activity is defined and expressed in U min^{-1} kg^{-1} 10^6 protein [38].

2.5.5. Ascorbate Peroxidase (APX) Activity

APX (APX, EC 1.11.1.1) activity was measured according to the process of Nakano and Asada [39]. The enzymatic extract (50 µL) was mixed with 800 µL of 50 mM potassium phosphate buffer (pH 7.0), 0.1 mM EDTA, 100 µL of H_2O_2 (0.03 M), and 50 µL of ascorbic acid (0.015 M). Ascorbic acid oxidation was measured at 290 nm (Beckman spectrophotometer, DU-640; USA), and the APX results are expressed in mol H_2O_2 min^{-1} kg^{-1} protein.

2.5.6. Polyphenol Oxidase (PPO) Extraction and Activity Measurements

To extract the polyphenol oxidase (PPO) enzyme (EC 1.14.18.1), 1.0 g of peel and pulp was homogenized with 50 mM potassium phosphate buffer (pH 7.0) and 1% (*w*/*v*) polyvinylpyrrolidone (PVP) at 4 °C. The homogenate was centrifuged (12,298× *g* for 10 min at 4 °C), and the supernatant was immediately frozen [40]. PPO activity was measured by mixing 100 µL of enzymatic extract with 1.85 mL of 100 mM potassium phosphate buffer (pH 6.0). Catechol (100 mM) was used as a substrate, and after 30 min of incubation in a water bath (30 °C for 30 min), the reactions were terminated (800 µL of 2 N perchloric acid) and the absorbance measured at 395 nm (Beckman spectrophotometer, DU-640; USA). The activity is expressed in U min^{-1} kg^{-1} 10^6 protein [41].

2.6. *Statistical Analysis*

2.6.1. Univariate

Analysis of variance (ANOVA) was performed on the data. R software version 3.0 (R Core Team, 2020; Auckland, New Zealand) was employed to obtain the means and compare them using Tukey's test at 0.05%.

2.6.2. Multivariate

Multivariate analysis was performed to reduce the variables capable of interpretation, summarizing much of the variability among them. Numerical matrices were constructed using chilling injury (CI) symptoms, physicochemical variables, and oxidative metabolism in the peel and pulp of the fruits from the different storage periods in Experiment I (0, 0 + 10, 7, 7 + 5, 14, 14 + 6, 21, 21 + 5, 28, 28 + 5 days) and Experiment II (0, 0 + 7, 7, 7 + 6, 14, 14 + 4, 21, 21 + 4, 28, and 28 + 4 days). Principal component analysis (PCA) was performed by extracting the principal components via correlation matrix using R software (R Core Team, 2020; Auckland, New Zealand), and the first two principal components (PC1 and PC2) were used.

3. Results

The results of univariate analysis can be observed in the Supplementary Materials. Due to the large number of variables evaluated, we chose to focus on the results from multivariate analysis. Thus, the score plots and loading plots (biplots) of the PCA were obtained through correlation matrices based on the analyzed variables (Tables 1 and 2). The factor loadings used in both the experiments were based on the first two principal components (PC1 and PC2), correlating chilling injuries with the other analyzed variables.

Table 1. Variable codes for quality attributes used in the principal component analysis of Experiments I and II.

Variables	Codes
Firmness	Firmness
pH	pH
Soluble solids content	SSC
Titratable acidity	TA
Ratio SSC/TA	SSC/TA
Fresh weight loss	FWL
Chilling injury	CI
Luminosity peel	L^*_{peel}

Table 1. *Cont.*

Variables	Codes
Luminosity pulp	L^*_{pulp}
Hue angle peel	h°_{peel}
Hue angle pulp	h°_{pulp}
Chromaticity peel	C^*_{peel}
Chromaticity pulp	C^*_{pulp}

Table 2. Variable codes for oxidative metabolism variables used in the principal component analysis of Experiments I and II.

Variables	Codes
Lipid peroxidation peel	LP_{peel}
Lipid peroxidation pulp	LP_{pulp}
Hydrogen peroxide peel	H_2O_{2peel}
Hydrogen peroxide pulp	H_2O_{2pulp}
Superoxide dismutase peel	SOD_{peel}
Superoxide dismutase pulp	SOD_{pulp}
Ascorbate peroxidase peel	APX_{peel}
Ascorbate peroxidase pulp	APX_{pulp}
Polyphenol oxidase peel	PPO_{peel}
Polyphenol oxidase pulp	PPO_{pulp}

3.1. *Experiment I: Without Gas Modification (Air)*

3.1.1. Chilling Injury and Physicochemical Variables

During cold storage, chilling injury (CI) development was not affected by the sorbitol treatments (Table 3). At the end of cold storage, the mangoes showed mild-to-moderate CI symptoms (Figure 1). On the other hand, after being transferred to ambient conditions, the 2.5% sorbitol treatment reduced CI development (Table 4) and the mangoes showed a better visual quality (Figures S1 and S2). If the score 3 (moderate symptoms—CI = 25–50%) is used as a shelf-life threshold, mangoes from the control treatment had a shelf-life of 21 + 5 days and 28 + 5 days when treated with 0.1 and 2.5% sorbitol (Figure 2).

Table 3. Effect of dipping 'Palmer' mangoes (Experiment I) in sorbitol (0.1 and 2.5%) and storage at 8.0 ± 1.0 °C and 95 ± 0.5% RH without atmosphere modification (21 kPa O_2 + 0.03 kPa CO_2) for 28 days on parameters fresh weight loss (FWL), chilling injury (CI), soluble solids content (SSC), titratable acidity (TA), ratio (SSC/TA), and pH.

Main Effect	FWL (%)	CI (1–4)	SSC (%)	TA (g kg^{1})	SSC/TA	pH
Treatments (A)						
Control	0.86 ^{a}	1.30 ^{a}	6.49 ^{a}	0.280 ^{a}	24.39 ^{a}	3.83 ^{a}
Sorbitol 0.1%	0.86 ^{a}	1.15 ^{a}	6.39 ^{a}	0.276 ^{a}	25.91 ^{a}	3.79 ^{a}
Sorbitol 2.5%	0.72 ^{b}	1.10 ^{a}	6.47 ^{a}	0.254 ^{a}	23.70 ^{a}	3.88 ^{a}
F test	7.63 **	1.86 ^{ns}	0.44 ^{ns}	2.91 ^{ns}	1.96 ^{ns}	1.54 ^{ns}
Days (B)						
0	0.00 ^{e}	1.00 ^{c}	6.38 ^{ab}	0.255 ^{bc}	25.07 ^{b}	3.84 ^{a}
7	0.46 ^{d}	1.11 ^{bc}	6.21 ^{b}	0.267 ^{abc}	23.74 ^{b}	3.87 ^{a}
14	0.82 ^{c}	1.11 ^{bc}	6.38 ^{ab}	0.305 ^{a}	21.70 ^{b}	3.75 ^{a}
21	1.23 ^{b}	1.25 ^{ab}	6.54 ^{ab}	0.292 ^{ab}	23.22 ^{b}	3.80 ^{a}
28	1.60 ^{a}	1.50 ^{a}	6.73 ^{a}	0.231 ^{c}	29.60 ^{a}	3.93 ^{a}
F test	220.93 **	4.07 **	3.59 *	7.74 **	8.33 **	2.00 ^{ns}
Interaction						
F test	0.75 ^{ns}	1.50 ^{ns}	0.50 ^{ns}	3.12 **	2.86 *	0.70 ^{ns}

Means followed by the same letter within each column do not differ statistically from each other using Tukey's test. Non-significant interaction (^{ns}), significant interaction at $p < 0.05$ (*), and significant interaction at $p < 0.01$ (**).

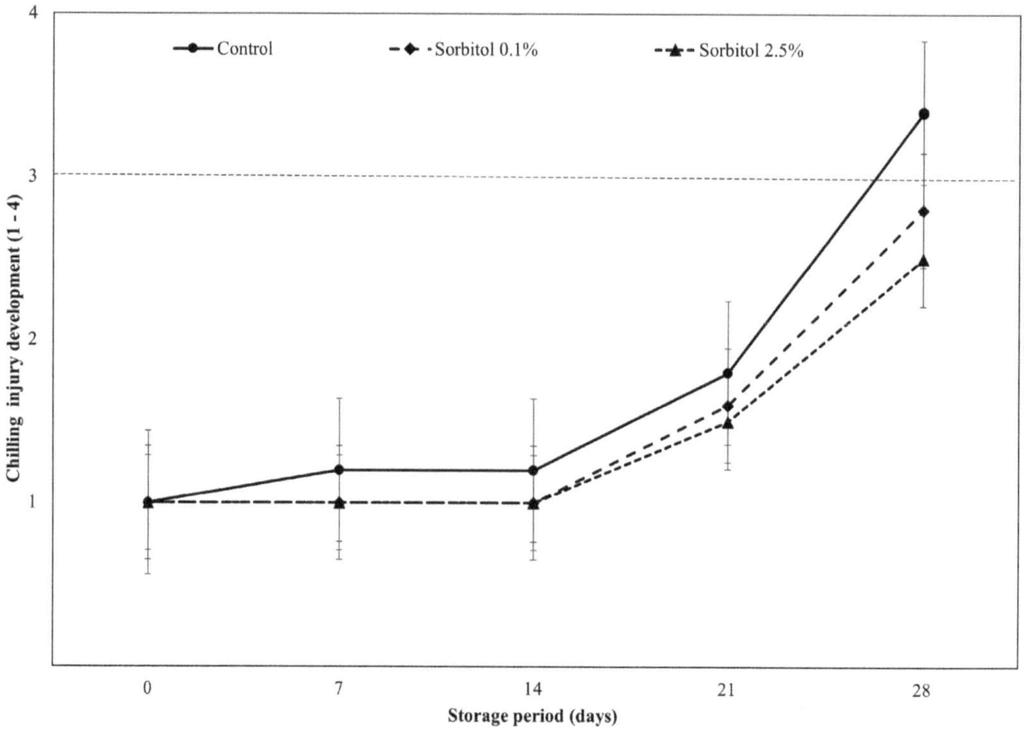

Figure 1. Interaction between treatments and storage period for chilling injury (CI) in 'Palmer' mangoes (Experiment I) stored at 8.0 ± 1.0 °C and 95 ± 0.5% RH without atmosphere modification (21 kPa O_2 + 0.03 kPa CO_2) for 28 days, and then transferred to ambient conditions (24 ± 2.0 °C and 75 ± 2.0% RH).

Table 4. Effect of dipping 'Palmer' mangoes (Experiment I) in sorbitol (0.1 and 2.5%) and storage at 8.0 ± 1.0 °C and 95 ± 0.5% RH without atmosphere modification (21 kPa O_2 + 0.03 kPa CO_2) for 28 days on the parameters of chilling injury (CI), soluble solids content (SSC), titratable acidity (TA), SSC/TA ratio, and pH after being transferred to ambient conditions (24 ± 2.0 °C and 75 ± 2.0% RH) from 5 to 10 days.

Main Effects	CI (1–4)	SSC (%)	TA (g kg^{-1})	SSC/TA	pH
Treatments (A)					
Control	1.80 ^{a}	14.13 ^{a}	0.252 ^{a}	60.53 ^{b}	4.00 ^{a}
Sorbitol 0.1%	1.70 ^{a}	14.05 ^{a}	0.209 ^{b}	73.21 ^{a}	4.04 ^{a}
Sorbitol 2.5%	1.65 ^{a}	14.29 ^{a}	0.234 ^{ab}	66.64 ^{ab}	4.06 ^{a}
Test F	0.30 ^{ns}	0.27 ^{ns}	4.52 *	7.77 **	0.38 ^{ns}
Days (B)					
0 + 10	1.25 ^{b}	15.19 ^{a}	0.245 ^{ab}	62.56 ^{a}	3.94 ^{ab}
7 + 5	1.33 ^{b}	13.98 ^{ab}	0.197 ^{b}	70.46 ^{a}	4.19 ^{a}
14 + 6	1.75 ^{b}	14.06 ^{ab}	0.209 ^{b}	67.30 ^{a}	4.12 ^{a}
21 + 5	1.41 ^{b}	14.35 ^{ab}	0.214 ^{b}	67.05 ^{a}	4.09 ^{ab}
28 + 5	2.83 ^{a}	13.19 ^{b}	0.292 ^{a}	45.17 ^{b}	3.83 ^{b}
Test F	12.95 **	5.44 **	8.40 **	3.17 ^{ns}	4.49 **
Interaction					
Test F	2.46 **	1.69 ^{ns}	3.77 **	3.81 **	1.5 ^{ns}

Means followed by the same letter within each column do not differ statistically from each other using Tukey's test. Non-significant interaction (^{ns}), significant interaction at $p < 0.05$ (*), and significant interaction at $p < 0.01$ (**).

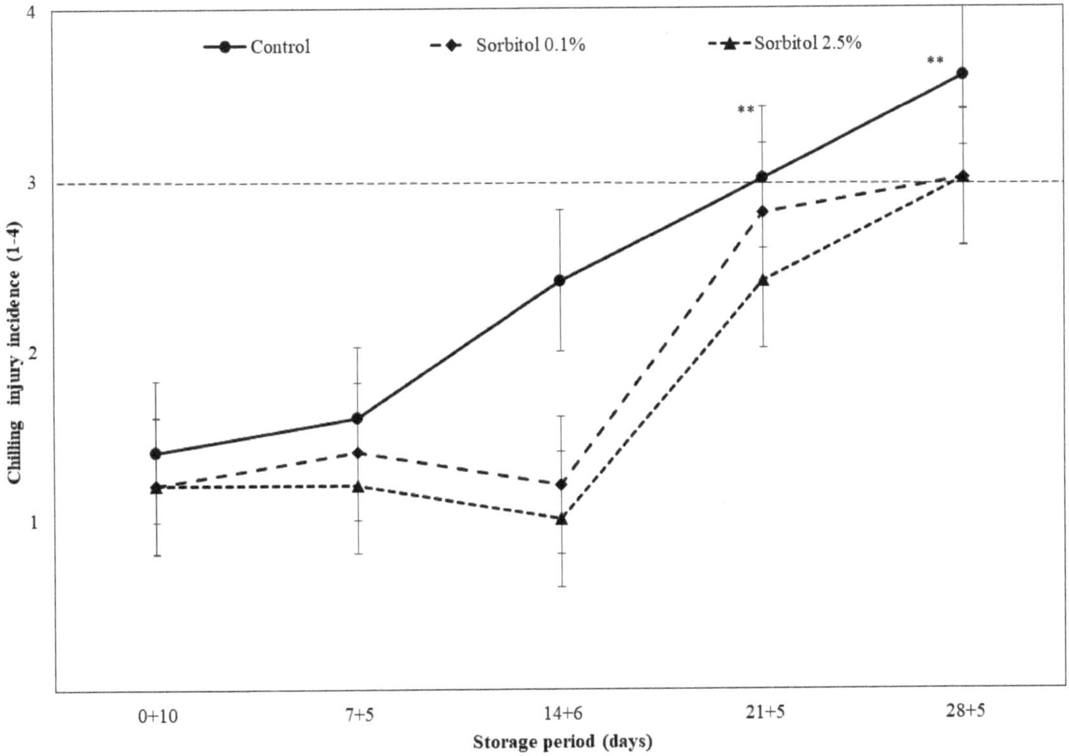

Figure 2. Interaction between treatments and storage period for (Experiment I) stored at 8.0 ± 1.0 °C and 95 ± 0.5% RH without atmosphere modification (21 kPa O_2 + 0.03 kPa CO_2) for 28 days, and then transferred to ambient conditions (24 ± 2.0 °C and 75 ± 2.0% RH) from 5 to 10 days. Means followed by double asterisks (**) are significantly different using Tukey's test at $p < 0.01$.

The variable related to chilling injury lesions (CI vector) was positioned in the lower positive quadrant of PC1, grouping with fresh weight loss (FWL), the soluble solid content (SSC), and pulp luminosity (L_{pulp}), as shown in Figure 3A. CIs do not correlate with firmness, peel and pulp chromaticity (C^*_{peel} and C^*_{pulp}), the pulp hue angle (h°_{pulp}), the ratio (SSC/TA), or the pH. The data from the treatments evaluated on the last storage days (21 and 28 days) tended to be closer to the CI vector, while those from the control treatment, followed by immersion in 0.1% sorbitol and the 2.5% sorbitol treatment, were distanced more from this vector (Figure 3A).

Upon moving the fruits to ambient conditions, the CI vector was positioned in the upper-left quadrant of PC2, grouping with pulp luminosity (L_{pulp}), titratable acidity (TA), and firmness, and was not associated with the SSC levels, pH, SSC/TA, peel chromaticity (C^*_{peel}), or peel hue angle (h°_{peel}), as shown in Figure 3A. At the beginning of the storage period, CIs were associated with the control treatment and with the 0.1% and 2.5% sorbitol treatments at the end of storage (21 and 28 days) (Figure 3B).

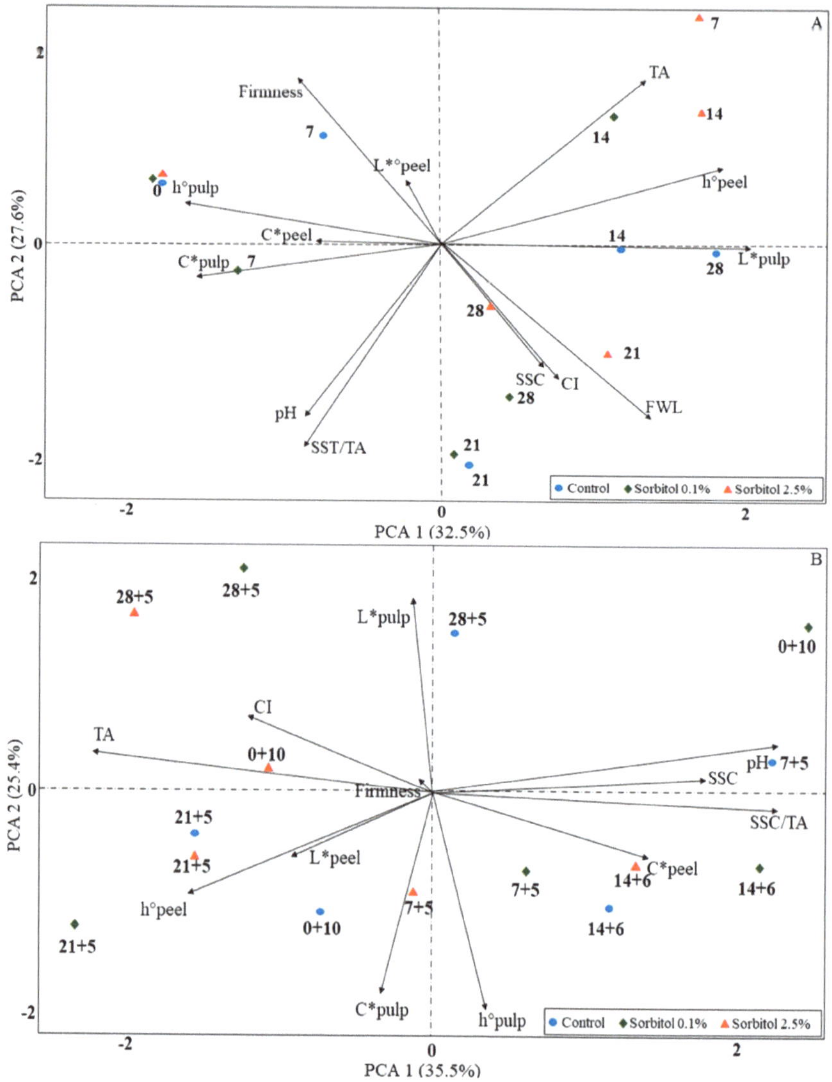

Figure 3. Biplot of the first principal component (PC1) and the second principal component (PC2) of 'Palmer' mango samples (Experiment I) stored at 8.0 ± 1.0 °C and 95 ± 0.5% RH without atmosphere modification (21 kPa O_2 + 0.03 kPa CO_2) for 28 days (**A**) and transferred to ambient conditions (24 ± 2.0 °C and 75 ± 2.0% RH) from 5 to 10 days (**B**). Chilling injury (CI), fresh weight loss (FWL), luminosity (L*), hue angle (h°), chromaticity (C*), firmness, soluble solids content (SSC), titratable acidity (TA), ratio (SSC/TA), and pH. Treatments include control, 0.1% sorbitol, and 2.5% sorbitol.

3.1.2. Cold Damage and Oxidative Metabolism

The sorbitol treatments affected the oxidative metabolism during cold storage (Table 5) and after moving to ambient conditions (Table 6). The CI vector correlated with the hydrogen peroxide content in the peel (H_2O_{2peel}), especially for the samples evaluated at the end of storage (21 and 28 days). The 0.1% and 2.5% sorbitol treatments were distanced from the CI vector, associating with the vectors of the superoxide dismutase enzyme in the peel and pulp (SOD_{peel} and SOD_{pulp}) and the ascorbate peroxidase in the peel (APX_{peel}). On the other hand, the control treatment correlated with the vectors related to the hydrogen peroxide content in the pulp (H_2O_{2pulp}) and the PPO activity in the peel (PPO_{peel}), as shown in Figure 4A.

Table 5. Effect of dipping 'Palmer' mangoes (Experiment I) in sorbitol (0.1 and 2.5%) and storage at 8.0 ± 1.0 °C and 95 ± 0.5% RH without atmosphere modification (21 kPa O_2 + 0.03 kPa CO_2) for 28 days on the parameters of lipid peroxidation (LP), hydrogen peroxide content (H_2O_2), and the activity of the enzymes superoxide dismutase (SOD), ascorbate peroxidase (APX), and polyphenol oxidase (PPO).

Main Effects	LP Peel	LP Pulp	H_2O_2 Peel	H_2O_2 Pulp	SOD Peel	SOD Pulp	APX Peel	APX Pulp	PPO Peel	PPO Pulp
Treatments (A)										
Control	2.72 ^{a}	2.28 ^{a}	76.87 ^{a}	57.59 ^{a}	129.06 ^{a}	204.12 ^{a}	64.65 ^{c}	123.60 ^{b}	133.13 ^{a}	180.13 ^{a}
Sorbitol 0.1%	2.76 ^{a}	2.28 ^{a}	75.04 ^{a}	51.68 ^{b}	120.66 ^{ab}	218.50 ^{a}	83.82 ^{b}	127.87 ^{ab}	107.08 ^{b}	161.14 ^{b}
Sorbitol 2.5%	2.51 ^{a}	2.05 ^{b}	70.57 ^{a}	48.52 ^{b}	129.06 ^{a}	227.54 ^{a}	96.21 ^{a}	134.76 ^{a}	100.53 ^{b}	119.17 ^{c}
Test F	2.27 ^{ns}	6.27 **	2.66 ^{ns}	9.41 **	4.63 *	2.80 ^{ns}	70.62 **	4.90 *	51.85 **	56.95 **
Days (B)										
0	3.06 ^{a}	2.34 ^{a}	8.80 ^{d}	91.94 ^{b}	105.24 ^{a}	101.46 ^{d}	41.19 ^{d}	227.41 ^{a}	184.02 ^{a}	81.63 ^{c}
7	2.33 ^{b}	2.21 ^{a}	53.14 ^{c}	100.93 ^{a}	106.24 ^{a}	203.85 ^{c}	96.77 ^{ab}	97.37 ^{c}	90.32 ^{cd}	151.36 ^{b}
14	2.56 ^{b}	1.87 ^{b}	89.08 ^{b}	14.94 ^{d}	125.34 ^{a}	296.07 ^{b}	102.15 ^{a}	84.06 ^{cd}	107.83 ^{b}	307.77 ^{a}
21	2.78 ^{ab}	2.27 ^{a}	108.91 ^{a}	37.63 ^{c}	130.58 ^{a}	362.51 ^{a}	77.83 ^{c}	154.71 ^{b}	102.46 ^{bc}	133.20 ^{b}
28	2.59 ^{b}	2.33 ^{a}	110.86 ^{a}	17.48 ^{d}	105.24 ^{a}	119.71 ^{d}	89.85 ^{b}	80.18 ^{d}	83.27 ^{d}	93.45 ^{c}
Test F	5.67 **	8.20 **	284.82 **	450.63 **	2.55 ^{ns}	152.09 **	99.21 **	364.92 **	172.00 **	289.64 **
Interaction										
Test F	2.04 ^{ns}	1.62 ^{ns}	0.99 ^{ns}	1.40 ^{ns}	0.94 ^{ns}	0.37 ^{ns}	14.67 **	1.32 ^{ns}	5.59 **	24.47 **

Means followed by the same letter within each column do not differ statistically from each other by Tukey's test. Non-significant interaction (^{ns}), significant interaction at $p < 0.05$ (*), and significant interaction at $p < 0.01$ (**).

Table 6. Effect of dipping 'Palmer' mangoes (Experiment I) in sorbitol (0.1 and 2.5%) and storage at 8.0 ± 1.0 °C and 95 ± 0.5% RH without atmosphere modification (21 kPa O_2 + 0.03 kPa CO_2) for 28 days on the parameters of lipid peroxidation (LP), hydrogen peroxide content (H_2O_2), and the activity of the enzymes superoxide dismutase (SOD), ascorbate peroxidase (APX), and polyphenol oxidase (PPO) after transferring the mangoes to ambient conditions (24 ± 2.0 °C and 75 ± 2.0% RH) from 5 to 10 days.

Main Effects	LP Peel	LP Pulp	H_2O_2 Peel	H_2O_2 Pulp	SOD Peel	SOD Pulp	APX Peel	APX Pulp	PPO Peel	PPO Pulp
Treatments (A)										
Control	4.93 ^{a}	4.23 ^{a}	104.03 ^{a}	41.43 ^{a}	89.19 ^{b}	181.85 ^{c}	80.55 ^{c}	276.41 ^{c}	212.56 ^{a}	269.28 ^{a}
Sorbitol 0.1%	4.57 ^{ab}	4.32 ^{a}	99.59 ^{ab}	37.36 ^{ab}	92.29 ^{b}	192.95 ^{b}	90.17 ^{b}	320.59 ^{b}	183.17 ^{b}	233.92 ^{b}
Sorbitol 2.5%	4.21 ^{b}	4.01 ^{a}	95.55 ^{b}	35.71 ^{b}	113.96 ^{a}	202.52 ^{a}	110.69 ^{a}	353.81 ^{a}	170.95 ^{c}	197.12 ^{c}
Test F	4.88 *	1.79 ^{ns}	7.29 **	5.15 *	45.45 **	15.43 **	159.49 **	254.76 **	52.74 **	37.95 **
Days (B)										
0 + 10	4.64 ^{ab}	4.08 ^{ab}	89.85 ^{c}	63.87 ^{a}	57.91 ^{d}	257.26 ^{a}	90.37 ^{b}	374.25 ^{a}	189.69 ^{b}	150.12 ^{c}
7 + 5	5.10 ^{a}	3.61 ^{b}	108.41 ^{a}	25.25 ^{c}	162.49 ^{a}	119.01 ^{c}	111.85 ^{b}	311.69 ^{c}	236.111 ^{a}	284.72 ^{a}
14 + 6	4.67 ^{ab}	4.56 ^{a}	90.78 ^{c}	23.75 ^{c}	77.87 ^{c}	120.19 ^{c}	87.39 ^{c}	263.10 ^{e}	203.12 ^{b}	300.59 ^{a}
21 + 5	4.27 ^{ab}	4.51 ^{a}	99.11 ^{b}	40.92 ^{b}	122.27 ^{b}	200.14 ^{b}	129.06 ^{a}	345.08 ^{b}	154.53 ^{c}	208.38 ^{b}
28 + 5	4.18 ^{b}	4.16 ^{ab}	110.47 ^{a}	37.01 ^{b}	71.91 ^{c}	265.26 ^{a}	63.68 ^{d}	290.73 ^{d}	161.53 ^{c}	223.38 ^{b}
Test F	3.11 *	6.14 **	22.41 **	92.83 **	265.85 **	436.84 **	260.14 **	194.87 **	76.03 **	64.76 **
Interaction										
Test F	1.65 ^{ns}	0.33 ^{ns}	0.12 ^{ns}	0.39 ^{ns}	15.52 **	0.81 ^{ns}	14.59 **	48.50 **	12.07 **	2.05 ^{ns}

Means followed by the same letter within each column do not differ statistically from each other by Tukey's test. Non-significant interaction (^{ns}), significant interaction at $p < 0.05$ (*), and significant interaction at $p < 0.01$ (**).

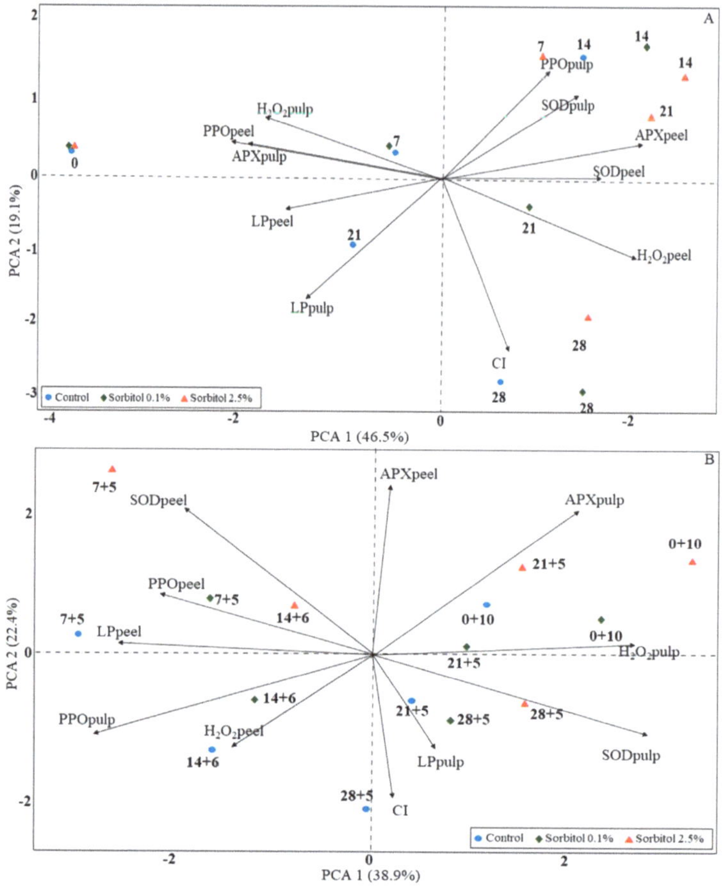

Figure 4. Biplot of the first principal component (PC1) and the second principal component (PC2) of 'Palmer' mango samples (Experiment I) stored at 8.0 ± 1.0 °C and 95 ± 0.5% RH without atmosphere modification (21 kPa O_2 + 0.03 kPa CO_2) for 28 days (**A**) and transferred to ambient conditions (24 ± 2.0 °C and 75 ± 2.0% RH) from 5 to 10 days (**B**). Lipid peroxidation (LP), hydrogen peroxide (H_2O_2), superoxide dismutase (SOD), ascorbate peroxidase (APX), and polyphenol oxidase (PPO). Treatments: control, 0.1% sorbitol, and 2.5% sorbitol.

Upon being transferred to ambient conditions, the vector related to chilling injuries (Cis) was positioned in the lower positive quadrant of PC1, grouping with the vectors of lipid peroxidation (LP_{pulp}) and superoxide dismutase (SOD_{pulp}), and even more with the data from the control treatment on evaluation days 21 + 5 and 28 + 5, which were closer to the CI vector (Figure 4B).

3.2. *Experiment II: CA—Modification of Atmospheric Gases*

3.2.1. Cold Damage and Physicochemical Variables

The association of sorbitol treatments and CA resulted in fewer chilling injuries (CIs) developing (Table 7) and a better fruit quality during cold storage compared to the control treatment (Figure 5). However, the CI symptoms were more severe in this condition with respect to Experiment I (Tables 3 and 4). Upon moving the fruits to ambient conditions, the 0.1% and 2.5% sorbitol treatments reduced CI development (Table 8), and the mangoes showed a better visual quality (Figures S3 and S4). Again, if score 3 (moderate symptoms—CI = 25–50%) is used as the shelf-life threshold, the only treatment that reached this level of CI was the control and CA treatments (Figure 6). Overall, the shelf-life was 21 + 4 and 28 + 4 days for all the treatments (Figure 6).

Table 7. Effect of dipping 'Palmer' mangoes (Experiment II) in sorbitol (0.1 and 2.5%) and storage at 4.0 ± 1.0 °C and 95 ± 0.5% RH under controlled atmosphere with modified gases (5 kPa O_2 + 5 kPa CO_2) for 28 days on the parameters of fresh weight loss (FWL), chilling injury (CI), soluble solids content (SSC), titratable acidity (TA), the ratio (SSC/TA), and pH.

Main Effect	FWL (%)	CI (1–4)	SSC (%)	TA (g kg^1)	SSC/TA	pH
Treatments (A)						
Control	0.86 ^{a}	1.70 ^{a}	8.91 ^{a}	0.435 ^{ab}	20.65 ^{a}	3.63 ^{a}
CA	0.77 ^{a}	1.60 ^{ab}	8.74 ^{a}	0.427 ^{ab}	20.88 ^{ab}	3.70 ^{a}
Sorbitol 0.1% + CA	0.76 ^{a}	1.40 ^{b}	8.38 ^{a}	0.476 ^{a}	18.37 ^{b}	3.72 ^{a}
Sorbitol 2.5% + CA	0.73 ^{a}	1.35 ^{b}	8.66 ^{a}	0.418 ^{b}	21.16 ^{ab}	3.71 ^{a}
Test F	1.43 ^{ns}	5.24 **	2.43 ^{ns}	3.15 *	3.06 *	1.14 ^{ns}
Days (B)						
0	0.00 ^{e}	1.00 ^{c}	6.90 ^{d}	0.340 ^{c}	21.70 ^{ab}	3.81 ^{a}
7	0.39 ^{d}	1.00 ^{c}	7.88 ^{c}	0.521 ^{a}	15.37 ^{c}	3.57 ^{b}
14	0.76 ^{c}	1.00 ^{c}	8.42 ^{c}	0.452 ^{b}	18.81 ^{bc}	3.66 ^{ab}
21	1.12 ^{b}	1.68 ^{b}	9.74 ^{b}	0.439 ^{b}	22.41 ^{ab}	3.73 ^{ab}
28	1.62 ^{a}	2.87 ^{a}	10.42 ^{a}	0.441 ^{b}	24.28 ^{a}	3.68 ^{ab}
Test F	149.57 **	102.72 **	80.90 **	15.60 **	13.84 **	4.23 **
Interaction						
Test F	0.82 ^{ns}	2.64 **	2.85 **	1.78 ^{ns}	0.85 ^{ns}	0.43 ^{ns}

Means followed by the same letter within each column do not differ statistically from each other using Tukey's test. Non-significant interaction (^{ns}), significant interaction at $p < 0.05$ (*), and significant interaction at $p < 0.01$ (**).

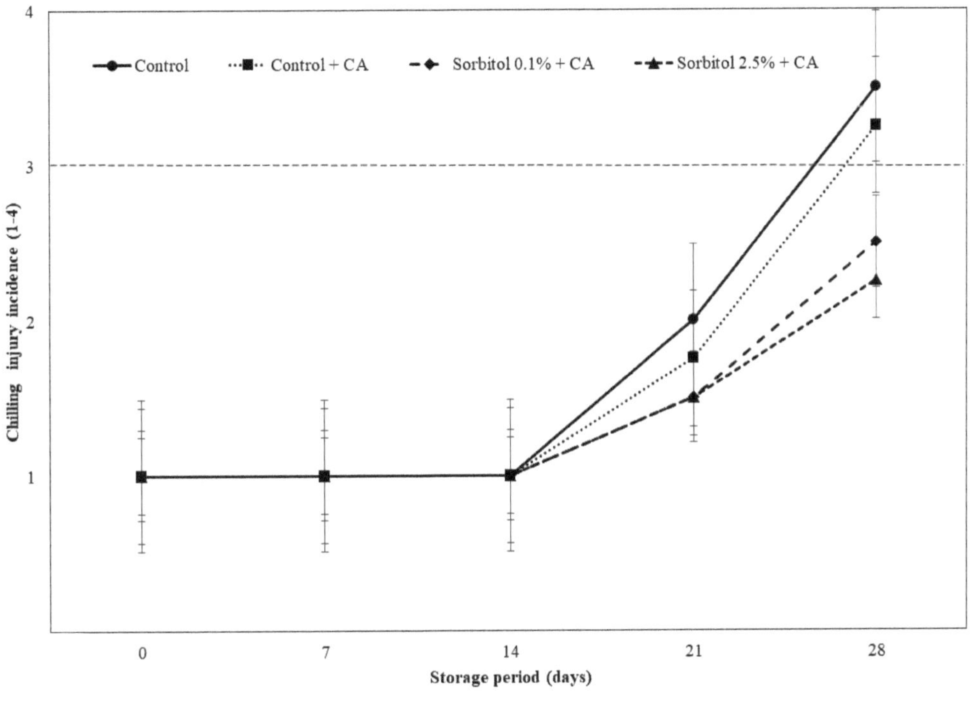

Figure 5. Interaction between treatments and storage period for chilling injury (CI) in 'Palmer' mangoes (Experiment II) stored at 4.0 ± 1.0 °C and 95 ± 0.5% RH under controlled atmosphere with gas modification (5 kPa O_2 + 5 kPa CO_2) for 28 days, and then transferred to ambient conditions (24 ± 2.0 °C and 75 ± 2.0% RH).

Table 8. Effect of dipping 'Palmer' mangoes (Experiment II) in sorbitol (0.1 and 2.5%) and storage at 4.0 ± 1.0 °C and 95 ± 0.5% RH under controlled atmosphere with modified gases (5 kPa O_2 + 5 kPa CO_2) for 28 days on the parameters of chilling injury (CI), soluble solids content (SSC), titratable acidity (TA), ratio (SSC/TA), and pH after being transferred to ambient conditions (24 ± 2.0 °C and 75 ± 2.0% RH) from 4 to 7 days.

Main Effects	CI (1–4)	SSC (%)	TA (g kg^{-1})	SSC/TA	pH
Treatments (A)					
Control	2.10 ^{a}	11.86 ^{a}	0.351 ^{a}	34.07 ^{a}	3.99 ^{a}
CA	1.90 ^{ab}	11.75 ^{a}	0.336 ^{ab}	35.28 ^{a}	4.06 ^{a}
Sorbitol 0.1% + CA	1.55 ^{b}	11.50 ^{a}	0.330 ^{b}	34.87 ^{a}	4.08 ^{a}
Sorbitol 2.5% + CA	1.50 ^{b}	10.61 ^{b}	0.350 ^{a}	30.52 ^{b}	4.13 ^{a}
Test F	5.41 **	6.63 **	4.96 **	7.65 **	1.29 ^{ns}
Days (B)					
0 + 7	1.12 ^{c}	11.40 ^{ab}	0.338 ^{bc}	33.85 ^{b}	4.02 ^{a}
7 + 6	1.12 ^{c}	12.21 ^{a}	0.324 ^{c}	37.81 ^{a}	4.19 ^{a}
14 + 4	1.56 ^{c}	11.41 $^{a b}$	0.327 ^{c}	35.04 ^{ab}	4.02 ^{a}
21 + 4	2.18 ^{b}	10.83 ^{b}	0.368 ^{a}	29.44 ^{c}	4.07 ^{a}
28 + 4	2.81 ^{a}	11.30 ^{ab}	0.353 ^{ab}	32.30 ^{bc}	4.03 ^{a}
Test F	28.09 **	4.11 **	12.77 **	12.59 **	1.72 ^{ns}
Interaction					
Test F	0.86 ^{ns}	2.68 **	3.74 **	1.75 ^{ns}	1.34 ^{ns}

Means followed by the same letter within each column do not differ statistically from each other using Tukey's test. Non-significant interaction (^{ns}), and significant interaction at $p < 0.01$ (**).

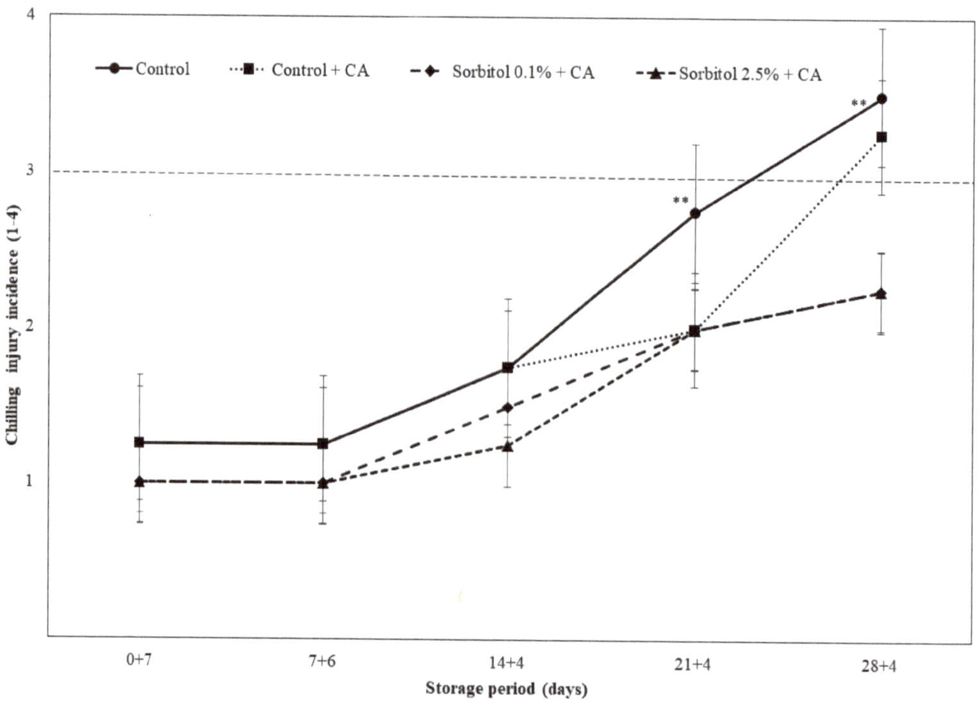

Figure 6. Interaction between treatments and storage period for chilling injury (CI) in 'Palmer' mangoes (Experiment II) stored at 4.0 ± 1.0 °C and 95 ± 0.5% RH under controlled atmosphere with gas modification (5 kPa O_2 + 5 kPa CO_2) for 28 days, and then transferred to ambient conditions (24 ± 2.0 °C and 75 ± 2.0% RH) from 4 to 7 days. Means followed by double asterisks (**) are significantly different using Tukey's test at $p<0.01$.

The CI vector was in the lower-right quadrant of PC1 and correlated with the FWL, SSC, C^*_{peel}, SSC/TA, and hue angle ($h°_{peel}$) vectors, grouping with the vectors of the control and CA treatments at both 21 and 28 days of storage (Figure 7A). The vectors of the variables of firmness, L_{pulp}, pH, $h°_{pulp}$, and C^*_{pulp} were in PC2 and did not correlate with CIs. In the ambient conditions, the CI vector positioned itself in the upper quadrant of PC1 and correlated with L_{peel}, L_{pulp}, $h°_{peel}$, and C^*_{peel}, with the vectors of the control and CA treatments more grouped with this vector, especially on the last evaluation days (14 + 4, 21 + 4, and 28 + 4). On the other hand, the data from the 2.5% sorbitol treatment grouped in the lower quadrant of PC1 (Figure 7B).

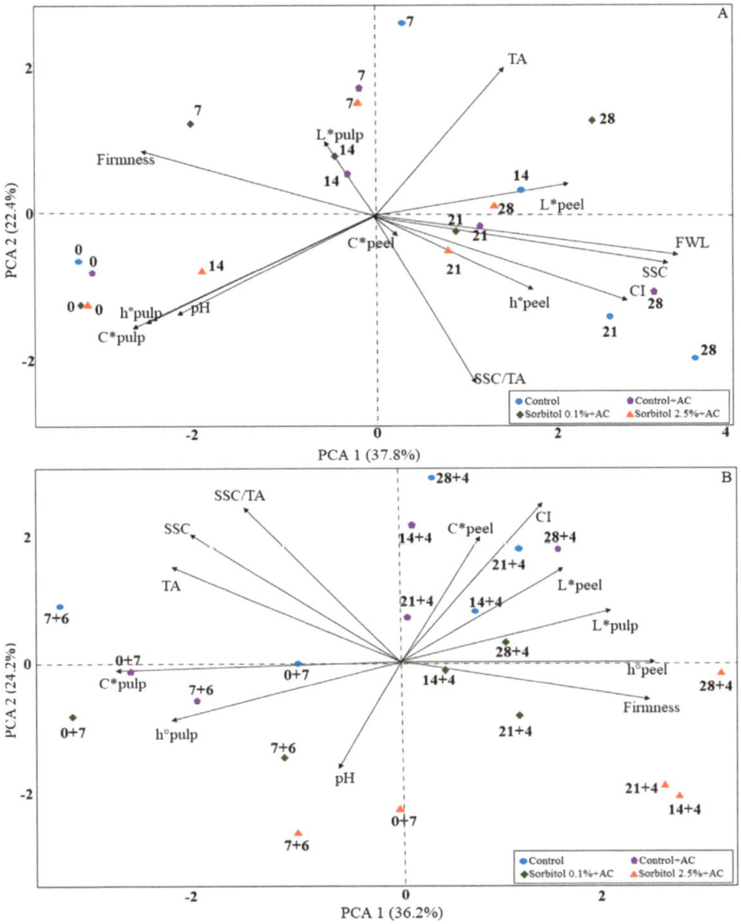

Figure 7. Biplot of the first principal component (PC1) and the second principal component (PC2) of 'Palmer' mango samples (Experiment II) stored at 4 ± 1.0 °C and 95 ± 0.5% RH under controlled atmosphere (5 kPa O_2 + 5 kPa CO_2) for 28 days (**A**) and transferred to ambient conditions (24 ± 2.0 °C and 75 ± 2.0% RH) from 4 to 7 days (**B**). Chilling injury (CI), fresh weight loss (FWL), luminosity (L*), hue angle (h°), chromaticity (C*), firmness, soluble solids content (SSC), titratable acidity (TA), ratio (SSC/TA), and pH. Treatments: negative control, positive CA, 0.1% sorbitol + CA, and 2.5% sorbitol + CA.

3.2.2. Cold Damage and Oxidative Metabolism

The sorbitol treatments affected the oxidative metabolism during cold storage (Table 9) and after the move to ambient conditions (Table 10). The CI vector was positioned in the upper-right quadrant of PC1, correlating with LP_{pulp} and grouping with the data from the negative control treatment on all the evaluated days as well as the CA and 0.1% sorbitol treatments on day 28 (Figure 8A). On the other hand, the vectors related to the activity of the ascorbate peroxidase enzyme (APX_{peel} and APX_{pulp}) and superoxide dismutase (SOD_{pulp}) were located on the opposite side of CI, highlighting the 0.1% and 2.5% sorbitol treatments (7 and 14 days), which grouped with the vectors of these enzymes (Figure 8A).

Table 9. Effect of dipping 'Palmer' mangoes (Experiment II) in sorbitol (0.1 and 2.5%) and storage at 4.0 ± 1.0 °C and 95 ± 0.5% RH under controlled atmosphere with modified gases (5 kPa O_2 + 5 kPa CO_2) for 28 days on the parameters of lipid peroxidation (LP), hydrogen peroxide content (H_2O_2), and the activity of the enzymes superoxide dismutase (SOD), ascorbate peroxidase (APX), and polyphenol oxidase (PPO).

Main Effects	LP Peel	LP Pulp	H_2O_2 Peel	H_2O_2 Pulp	SOD Peel	SOD Pulp	APX Peel	APX Pulp	PPO Peel	PPO Pulp
Treatments (A)										
Control	2.92 a	2.27 a	58.45 a	68.71 a	168.40 d	363.31 c	86.98 c	193.83 c	84.94 a	149.18 a
CA	2.76 a	1.97 a	56.05 ab	53.46 b	182.48 c	378.48 c	99.39 b	232.09 b	73.16 b	131.22 b
Sorbitol 0.1% + CA	2.76 a	1.91 a	52.08 b	48.18 b	200.17 b	403.71 b	102.88 b	238.93 b	67.59 c	100.98 c
Sorbitol 2.5% + CA	2.72 a	1.88 a	50.21 b	38.46 c	207.86 a	424.14 a	112.19 a	267.53 a	61.38 d	91.34 d
Test F	0.53 ns	2.74 ns	5.24 **	54.63 **	176.66 **	27.05 **	24.05 **	157.39 **	50.98 **	275.54 **
Days (B)										
0	2.10 c	1.82 b	69.99 a	11.85 c	178.66 b	646.47 a	160.55 a	2131.24 b	134.80 a	127.04 b
7	2.45 bc	1.99 ab	46.25 c	54.71 b	148.81 d	420.82 b	79.08 d	268.81 a	39.30 d	119.52 c
14	2.66 b	2.43 a	46.77 c	59.99 b	112.11 e	320.52 c	124.69 b	261.82 a	50.98 c	142.11 a
21	3.42 a	1.76 b	46.36 c	80.53 a	349.38 a	333.08 c	102.28 c	216.51 c	47.32 c	94.19 e
28	3.33 a	2.05 ab	61.56 b	53.94 b	159.67 c	241.17 d	35.20 e	186.10 d	86.43 b	108.05 d
Test F	18.12 **	4.71 **	36.18 **	170.95 **	3837.48 **	722.86 **	394.41 **	156.76 **	637.60 **	102.51 **
Interaction										
Test F	0.59 ns	0.64 ns	1.22 ns	6.59 **	65.60 **	6.12 **	4.11 **	29.51 **	13.01 **	30.02 **

Means followed by the same letter within each column do not differ statistically from each other using Tukey's test. Non-significant interaction (ns), and significant interaction at $p < 0.01$ (**).

Table 10. Effect of dipping 'Palmer' mangoes (Experiment II) in sorbitol (0.1 and 2.5%) and storage at 4.0 ± 1.0 °C and 95 ± 0.5% RH under controlled atmosphere with modified gases (5 kPa O_2 + 5 kPa CO_2) for 28 days on the parameters of lipid peroxidation (LP), hydrogen peroxide content (H_2O_2), and the activity of the enzymes superoxide dismutase (SOD), ascorbate peroxidase (APX), and polyphenol oxidase (PPO) after being transferred to the environment (24 ± 2.0 °C and 75 ± 2.0% RH) from 4 to 7 days.

Main Effects	LP Peel	LP Pulp	H_2O_2 Peel	H_2O_2 Pulp	SOD Peel	SOD Pulp	APX Peel	APX Pulp	PPO Peel	PPO Pulp
Treatments (A)										
Control	4.11 a	3.41 a	82.29 a	53.85 a	104.39 d	125.64 d	56.16 c	162.15 d	190.58 a	119.27 a
CA	4.01 a	3.32 a	71.72 b	50.94 a	131.48 c	131.77 c	65.07 b	174.05 c	167.62 b	104.32 b
Sorbitol 0.1% + CA	3.81 a	3.30 a	66.37 c	42.66 b	145.41 b	141.46 b	67.27 b	181.72 b	160.28 c	99.93 b
Sorbitol 2.5% + CA	3.75 a	2.95 a	58.92 c	36.86 b	161.92 a	155.30 a	76.92 a	190.04 a	143.45 d	90.31 c
Test F	1.04 ns	1.84 ns	32.23 **	18.74 **	241.88 **	86.97 **	45.45 **	44.85 **	149.68 **	50.56 **
Days (B)										
0 + 7	3.43 b	3.15 ab	71.87 ab	47.60 b	69.92 e	202.64 a	67.60 c	185.59 b	140.00 d	57.69 d
7 + 6	3.58 b	3.80 a	71.01 ab	27.92 c	95.68 c	103.92 d	86.84 a	166.61 c	180.65 c	33.91 e
14 + 4	4.42 a	2.90 b	77.66 a	59.88 a	241.43 a	147.17 b	75.51 b	147.69 d	82.92 e	111.31 c
21 + 4	3.42 b	2.90 b	62.66 c	64.25 a	185.57 b	103.12 d	51.55 d	178.61 c	232.32 a	128.54 b
28 + 4	4.74 a	3.48 ab	65.93 b c	27.90 c	86.39 d	135.87 c	50.33 d	206.15 a	191.52 b	185.91 a
Test F	11.35 **	5.37 **	8.91 **	67.64 **	1796.24 **	690.96 **	121.81 **	120.85 **	1005.68 **	1003.14 **
Interaction										
Test F	0.14 ns	0.2 ns	1.04 ns	2.45 *	63.20 **	9.55 **	3.38 **	3.95 **	26.77 **	3.45 **

Means followed by the same letter within each column do not differ statistically from each other using Tukey's test. Non-significant interaction (ns), significant interaction at $p < 0.05$ (*), and significant interaction at $p < 0.01$ (**).

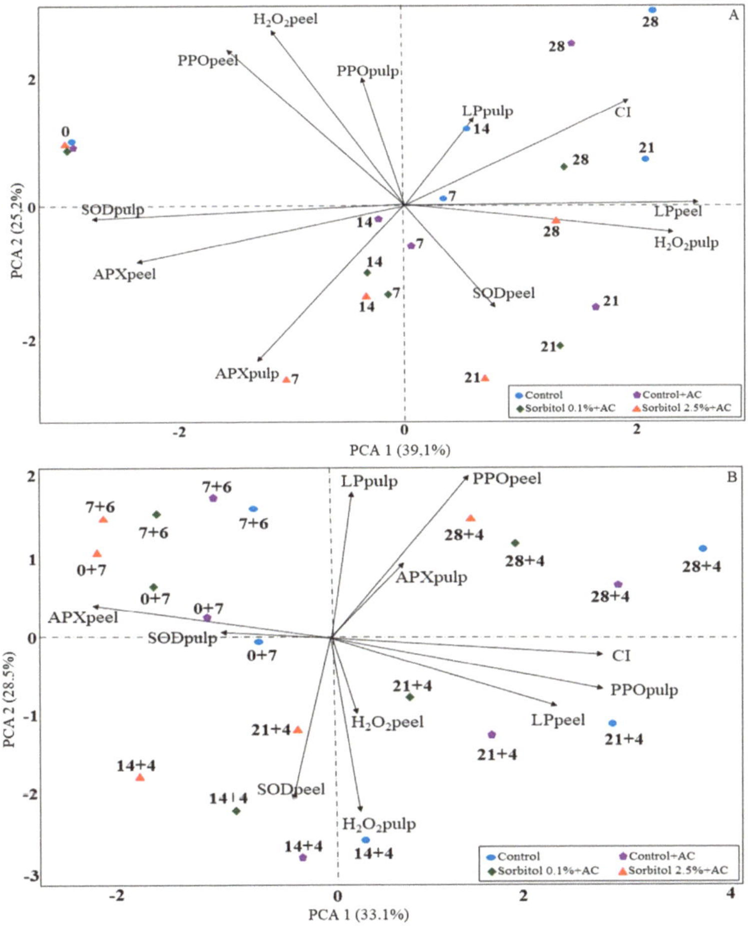

Figure 8. Biplot of the first principal component (PC1) and the second principal component (PC2) of 'Palmer' mango samples (Experiment II) stored at 4 ± 1.0 °C and 95 ± 0.5% RH under controlled atmosphere (5 kPa O_2 + 5 kPa CO_2) for 28 days (**A**) and transferred to ambient conditions (24 ± 2.0 °C and 75 ± 2.0% RH) from 4 to 7 days (**B**). Lipid peroxidation (LP), hydrogen peroxide (H_2O_2), superoxide dismutase (SOD), ascorbate peroxidase (APX), and polyphenol oxidase (PPO). Treatments: negative control, CA, 0.1% sorbitol + CA, and 2.5% sorbitol + CA.

When the fruits were transferred to ambient conditions, the vector of chilling injury (CI) was positioned in the lower-right quadrant of PC1, correlating with the activity of the PPO enzyme in the pulp (PPO_{pulp}), LP_{peel}, H_2O_{2peel}, and H_2O_{2pulp} and grouping with the data from the negative control and CA treatments on days 21 + 4 and 28 + 4, which presented more pronounced chilling injury lesions at the end of storage compared to the 2.5% sorbitol treatment (Figure 8B).

4. Discussion

4.1. Cold Damage and Physicochemical Variables

In both the experiments, the incidence of chilling injuries (CIs) correlated with fresh weight loss (FWL), especially in the control treatments on the last evaluation days. The same was reported by Sanches et al. [15], who stored mangoes at 8 °C. However, those authors reported that a 75% RH cold room provided a fresh weight loss of 18%, and even under those conditions, immersion in solutions containing polyols (propylene glycol, glycerol, and sorbitol) significantly reduced the fresh weight loss to 17% [15]. This may suggest an effect of the modified atmosphere even at reduced sorbitol concentrations (0.1%).

To verify the effectiveness of sorbitol immersion by isolating the control of fresh weight loss on chilling injury development, we stored the fruit at 8.0 ± 1.0 °C and 95 ± 0.5% RH without gas modification (21 kPa O_2 + 0.03 kPa CO_2) for 28 days in CA chambers. This minimized the fresh weight loss (0.78–0.89%), and the sorbitol treatments had no significant effect. The same occurred during the storage of the fruits at 4.0 ± 1.0 °C and 95 ± 0.5% RH under a CA containing 5 kPa O_2 + 5 kPa CO_2 for 28 days, i.e., the fresh weight loss was reduced to 0.65–0.88%. Thus, the effect of the immersion in the sorbitol-containing solutions was not related to the reduction in fresh weight loss, as there was no significant effect compared to the control treatments, which showed more severe chilling injury symptoms regardless of the CA conditions.

The worsening of the chilling injury symptoms when the fruits were moved to ambient conditions was evidenced by the darkening of the epicarp, which is usually observed after removal from cold storage and fruits become ripe [42]. For this reason, the chilling injuries correlated with variables such as L_{peel}, $h°_{peel}$, C^*_{peel}, and $h°_{pulp}$ (Tables S1–S4). However, the chilling injuries were not restricted to damage to the epicarp and led to irregular ripening, affecting the color parameters of the pulp, the SSC, the TA, and the ratio. The discoloration of the epicarp (peel) and irregular ripening are commonly observed in mangoes stored at low temperatures [43].

Sanches et al. [44] have reported that a combination of immersion in solutions containing 2.5% sorbitol with storage in CA (5 kPa O_2 + 5 kPa CO_2) has delayed mango ripening and controlled chilling injury development. We observed that the mangoes subjected to the 0.1% and 2.5% sorbitol treatments were firmer even after being transferred to the ambient conditions, which may have affected the ripening process. In both of our experiments, dipping the fruit in 2.5% sorbitol reduced chilling injury incidence, and the fruits remained firmer. Thus, maintaining firmness may be related to the control of chilling injuries. Salazar-Salas et al. [45] observed that 'Keitt' mangoes dipped in hot water (HWT) showed a lower incidence of chilling injuries and had better firmness. Thus, firmness may be indicative of maintaining the structure of the cell wall and plasma membrane and be associated with reduced chilling injuries.

4.2. Cold Damage and Oxidative Metabolism

Chilling injury development was related to the accumulation of malondialdehyde (MDA) (LP_{pulp} and LP_{peel}), which is considered a marker of lipid membrane peroxidation [46]. In PCA, it was observed that chilling injury correlated with hydrogen peroxide content (H_2O_{2peel}), LP_{pulp}, LP_{pulp}, LP_{peel}, H_2O_{2peel}, H_2O_{2pulp}, and LP_{peel}, especially in the control treatment on the last evaluation days (21, 21 + 4, 28, and 28 + 4 days), as represented by increases in the MDA and H_2O_2 contents.

Dipping the mangoes in sorbitol (0.1% and 2.5%) resulted in a higher activity level of superoxide dismutase (SOD), which is considered the first enzyme to be activated in the elimination of ROS via O_2 dismutation [13]. Thus, it can be inferred that the sorbitol influenced the control of oxidative stress by increasing the SOD activity level in the fruit, similar to other polyols [47].

Other enzymes, such as APX, CAT, and POD, are also considered important in initiating plant defenses against oxidative stress. In this sense, higher activity levels of APX enzymes can reduce hydrogen peroxide accumulation, which was observed in the fruit treated with the sorbitol (0.1% and 2.5%) compared to the H_2O_{2peel} and H_2O_{2pulp} vectors of the control fruit, which showed significant increases in H_2O_2 and more severe chilling injury symptoms. Pomegranates (*Punica granatum* L.) with chilling injury symptoms had a higher accumulation of H_2O_2, which was lower in the fruits treated with arginine, also resulting in increased activity levels of the SOD, CAT, and APX enzymes [48]. Thus, the activation of the enzymatic defense system (SOD, APX, and CAT) is fundamental for reducing H_2O_2 accumulation in plants under stress conditions [49] and, consequently, minimizes chilling injuries in fruits stored at low temperatures, which was possible with the immersion in 2.5% sorbitol.

The most visible symptom of chilling injuries in mangoes is the presence of dark, sunken spots, similar to burns, associated with a higher polyphenol oxidase enzyme (PPO) activity level [50]. In both the experiments, the PPO activity correlated with the highest incidence of chilling injuries, especially in the control treatment of the experiment without gas modification (21 kPa O_2 + 0.03 kPa CO_2) after the fruits were transferred to the ambient conditions in the last days of storage (21 + 4, 28 + 4). However, the treatments with sorbitol, especially the 2.5% treatment, significantly reduced the PPO activity. This effect was evident when the CA (5 kPa O_2 + 5 kPa CO_2) and 2.5% sorbitol treatment were combined, even at lower temperatures (4.0 °C). This demonstrates a synergistic effect between CA and sorbitol treatment in activating oxidative metabolism to control chilling injuries. However, this combination only minimizes the chilling injuries, and further studies should be conducted to explain which mechanisms are involved combining sorbitol and CA.

5. Conclusions

The immersion of 'Palmer' mangoes in sorbitol-containing solutions demonstrated efficacy in mitigating chilling injuries under varied storage conditions, which has not been extensively explored in the existing literature, such as differing gas compositions (21 kPa O_2 + 0.03 kPa CO_2 and 5 kPa O_2 + 5 kPa CO_2) and temperatures (8 °C and 4 °C). When the mangoes were stored under controlled atmosphere (CA) conditions with 95% RH, it was possible to mitigate the impact of fresh weight loss and chilling injury symptoms, enabling the preservation of fruit quality for up to 21 + 5 days at 8 °C and to 21 + 4 days at 4 °C.

The effectiveness of controlling chilling injury was attributed to the attenuation of lipid peroxidation in the cell membranes and the activation of the antioxidant enzymes SOD and APX in the epicarp and mesocarp, particularly in the mangoes treated with sorbitol. Notably, the application of 2.5% sorbitol emerged as the most effective approach to mitigating chilling injuries, sustaining the fruit quality, and enhancing oxidative metabolism even at lower temperatures. Thus, this treatment option has established itself as a viable strategy for chilling injury management in mangoes.

Supplementary Materials: The following supporting information can be downloaded at: https://www.mdpi.com/article/10.3390/horticulturae10040354/s1, Figure S1: Epicarp chilling injury (CI) development in 'Palmer' mangoes (Experiment I) stored at 8.0 ± 1.0 °C and 95 ± 0.5% RH without atmosphere modification (21 kPa O_2 + 0.03 kPa CO_2) for 28 days and then transferred to ambient conditions (24 ± 2.0 °C and 75 ± 2.0% RH) for up to 10 to 5 days; Figure S2: Mesocarp chilling injury (CI) development in 'Palmer' mangoes (Experiment I) stored at 8.0 ± 1.0 °C and 95 ± 0.5% RH without atmosphere modification (21 kPa O_2 + 0.03 kPa CO_2) for 28 days and then transferred to ambient conditions (24 ± 2.0 °C and 75 ± 2.0% RH) for up to 10 to 5 days; Figure S3: Epicarp chilling injury (CI) development in 'Palmer' mangoes (Experiment II) stored at 4.0 ± 1.0°C and 95 ± 0.5% RH under controlled atmosphere with gas modification (5 kPa O_2 + 5 kPa CO_2) for 28 days and then transferred to ambient conditions (24 ± 2.0 °C and 75 ± 2.0% RH) for up to 7 to 4 days; Figure S4: Mesocarp chilling injury (CI) development in 'Palmer' mangoes (Experiment II) stored at 4.0 ± 1.0 °C and 95 ± 0.5% RH under controlled atmosphere with gas modification (5 kPa O_2 + 5 kPa CO_2) for 28 days and then transferred to ambient conditions (24 ± 2.0°C and 75 ± 2.0% RH) for up to 7 to 4 days. Table S1: Effect of dipping 'Palmer' mangoes (Experiment I) in sorbitol (0.1 and 2.5%) and storage at 8.0 ± 1.0 °C and 95 ± 0.5% RH RH without atmosphere modification (21 kPa O_2 + 0.03 kPa CO_2) for 28 days on firmness, luminosity, hue angle, and chromaticity parameters; Table S2: Effect of dipping 'Palmer' mangoes (Experiment I) in sorbitol (0.1 and 2.5%) and storage at 8.0 ± 1.0 °C and 95 ± 0.5% RH RH without atmosphere modification (21 kPa O_2 + 0.03 kPa CO_2) for 28 days on the parameters of firmness, luminosity, hue angle, and chromaticity after transfer to ambient conditions (24 ± 2.0 °C and 75 ± 2.0% RH) for up to 10 to 7 days. Table S3: Effect of dipping 'Palmer' mangoes (Experiment II) in sorbitol (0.1 and 2.5%) and storage at 4.0 ± 1.0 °C and 95 ± 0.5% RH under controlled atmosphere with modification of gases (5 kPa O_2 + 5 kPa CO_2) for 28 days on firmness, luminosity, hue angle, and chromaticity. Table S4: Effect of dipping 'Palmer' mangoes (Experiment II) in sorbitol (0.1 and 2.5%) and storage at 4.0 ± 1.0 °C and 95 ± 0.5% RH under controlled atmosphere with modified gases (5 kPa O_2 + 5 kPa CO_2) for 28 days on the parameters of firmness, luminosity, hue angle, and chromaticity after transfer to ambient conditions (24 ± 2.0 °C and 75 ± 2.0% RH) for up to 7 to 4 days.

Author Contributions: Methodology, M.B.d.S. and G.H.d.A.T.; validation, M.B.d.S.; formal analysis, M.B.d.S., V.M.D.P., M.I. and A.G.S.; investigation, M.B.d.S., V.M.D.P. and M.I.; data curation, M.B.d.S.; writing—original draft preparation, M.B.d.S., T.S.B., A.G.S. and G.H.d.A.T.; supervision, T.S.B. and G.H.d.A.T.; project administration, G.H.d.A.T.; funding acquisition, G.H.d.A.T. All authors have read and agreed to the published version of the manuscript.

Funding: This research was funded by the Fundaçao de Amparo à Pesquisa do Estado de São Paulo (FAPESP) grant number [2019/15116-0] and in part by the Coordenação de Aperfeiçoamento de Pessoal de Nível Superior—(CAPES) Brazil—finance code [001].

Data Availability Statement: Data are contained within the article and supplementary materials.

Acknowledgments: The authors would like to thank Ogata Citrus, represented by Ademar Ogata, and Agropecuária Roriz Dantas (AGRODAN) for providing the fruit.

Conflicts of Interest: The authors declare no conflict of interest.

References

1. Vega-Alvarez, M.; Salazar-Salas, N.Y.; López-Ângulo, G.; Pineda-Hidalgo, K.V.; López-López, M.E.; Vega-García, M.O.; Delgado-Vargas, F.; López-Valenzuela, J.A. Metabolomic changes in mango fruit peel associated with chilling injury tolerance induced by quarantine hot water treatment. *Postharvest Biol. Technol.* **2020**, *169*, 111299. [CrossRef]
2. Singh, Z.; Singh, R.K.; Sane, V.A.; Nath, P. Mango-postharvest biology and biotechnology. *Crit. Rev. Plant Sci.* **2013**, *32*, 217–236. [CrossRef]
3. Penchaiya, P.; Tijskens, L.M.M.; Uthairatanakij, A.; Srilaong, V.; Tansakul, A.; Kanlayanarat, S. Modelling quality and maturity of 'Namdokmai Sithong' mango and their variation during storage. *Postharvest Biol. Technol.* **2020**, *159*, 111000. [CrossRef]
4. Zaharah, S.S.; Singh, Z. Postharvest nitric oxide fumigation alleviates chilling injury, delays fruit ripening and maintains quality in cold-stored 'Kensington Pride' mango. *Postharvest Biol. Technol.* **2011**, *60*, 202–210. [CrossRef]
5. Zhang, Z.; Zhu, Q.; Hu, M.; Gao, Z.; An, F.; Li, M.; Jiang, Y. Low-temperature conditioning induces chilling tolerance in stored mango fruit. *Food Chem.* **2017**, *219*, 76–84. [CrossRef] [PubMed]
6. Chaplin, G.R.; Cole, S.R.; Landrin, M.; Nuevo, P.A.; Lam, P.F.; Graham, G. Chilling injury and storage of mango (*Mangifera indica* L.) held under low temperatures. *Acta Hort.* **1991**, *291*, 461–471. [CrossRef]
7. Lyons, J.M.; Raison, J.K.; Steponkus, P.L. The plant membrane in response to low temperature: An overview. In *Low Temperature Stress in Crop Plants: The Role of the Membrane*; Lyons, J.M., Graham, D., Raison, J.K., Eds.; Academic Press: New York, NY, USA, 1979; pp. 1–24.
8. Kane, O.; Marcellin, P.; Mazliak, P. Incidence of ripening and chilling injury on the oxidative activities and fatty acid compositions of the mitochondria from mango fruit. *Plant Physiol.* **1978**, *61*, 634–638. [CrossRef] [PubMed]
9. Hidalgo, M.; Cruz, J.; Parkin, K.L.; Garcia, H.S. Refrigerated storage and chilling injury development of Manila mangoes (*Mangifera indica* L.). *Acta Hort.* **1997**, *455*, 718–725. [CrossRef]
10. Agillon, A.B.; Lizada, M.C.C. Responses of 'Carabao' mango (*Mangifera indica*) fruit to chilling stress. *Acta Hort.* **2012**, *877*, 467–474. [CrossRef]
11. Kumpoun, W.; Nishizawa, T.; Motomura, Y.; Puthmee, T.; Aikawa, T. Histological and biochemical traits of chilling-injured pulp tissues as affected by cold storage of mango fruit. *HortScience* **2017**, *52*, 1380–1384. [CrossRef]
12. Shewfelt, R.L.; Rosario, B.A. The role of lipid peroxidation in storage disorders of fresh fruit and vegetables. *HortScience* **2000**, *35*, 575–579. [CrossRef]
13. Wang, B.; Zhu, S. Pre-storage cold acclimation maintained quality of cold-stored cucumber through differentially and orderly activating ROS scavengers. *Postharvest Biol. Technol.* **2017**, *129*, 1–8. [CrossRef]
14. Khaliq, G.; Mohamed, M.T.M.; Ghazali, H.M.; Ding, P.; Ali, A. Influence of gum arabic coating enriched with calcium chloride on physiological, biochemical and quality responses of mango (*Mangifera indica* L.) fruit stored under low temperature stress. *Postharvest Biol. Technol.* **2016**, *111*, 362–369. [CrossRef]
15. Sanches, A.G.; Pedrosa, V.M.D.; Checchio, M.V.; Fernandes, T.F.S.; Guevara, J.E.M.; Gratão, P.L.; Teixeira, G.H.A. Polyols alleviate chilling injury in 'Palmer' mangoes during cold storage. *Food Control* **2021**, *129*, 108248. [CrossRef]
16. Lyons, J.M.; Raison, J.K. Oxidative activity of mitochondria isolated from plant tissues sensitive and resistant to chilling injury. *Plant Physiol.* **1970**, *45*, 386–389. [CrossRef]
17. Raison, J.K.; Lyons, J.M. Chilling injury: A plea for uniform terminology. *Plant Cell Environ.* **1986**, *9*, 685–686. [CrossRef]

18. Raison, J.K.; Orr, G.R. Proposals for a better understanding of the molecular basis of chilling injury. *Chill. Inj. Hortic. Crops* **1990**, *145*, 164.
19. Sevillano, L.; Sanchez-Ballesta, M.T.; Romojaro, F.; Flores, F.B. Physiological, hormonal and molecular mechanisms regulating chilling injury in horticultural species. Postharvest technologies applied to reduce its impact. *J. Sci. Food Agric.* **2009**, *89*, 555–573. [CrossRef]
20. Rice, T.; Zanini, E.; Arendt, E.K.; Coffey, A.A. Review of polyols–biotechnological production, food applications, regulation, labeling and health effects. *Crit. Rev. Food Sci. Nutr.* **2020**, *60*, 2034–2051. [CrossRef]
21. Fang, T.; Cai, Y.; Yang, Q.; Ogutu, C.O.; Liao, L.; Han, Y. Analysis of sorbitol content variation in wild and cultivated apples. *J. Sci. Food Agric.* **2020**, *100*, 139–144. [CrossRef]
22. Kanayama, Y. Physiological roles of polyols in horticultural crops. *J. Jpn. Soc. Hortic. Sci.* **2009**, *78*, 158–168. [CrossRef]
23. Ito, A.; Sugiura, T.; Sakamoto, D.; Moriguchi, T. Effects of dormancy progression and low-temperature response on changes in the sorbitol concentration in xylem sap of Japanese pear during winter season. *Tree Physiol.* **2013**, *33*, 398–408. [CrossRef] [PubMed]
24. Shirgire, S.D.; Talware, R.B.; Kadam, S.S.; Kumbharkhane, A.C. Dielectric relaxation of d-sorbital-water mixture using a time domain reflectometry technique. *J. Mol. Liq.* **2012**, *169*, 33–36. [CrossRef]
25. Busttamante, C.; Monti, L.; Gabilondo, J.; Scossa, F.; Valentini, G.; Budde, C.; Lara, M.; Fernie, A.; Drincovish, M. Differential metabolic rearrangements after cold storage are correlated with chilling injury resistance of peach fruit. *Front. Plant Sci.* **2016**, *7*, 1478. [CrossRef] [PubMed]
26. Thomashow, M.F. Plant cold acclimation: Freezing tolerance genes and regulatory mechanisms. *Annu. Rev. Plant Physiol.* **1999**, *50*, 571–599. [CrossRef] [PubMed]
27. Yang, G.; Li, L.Y.; Huang, M.L.; Yang, F.C.; Zhang, F.K.; Xu, R.C.; Yan, D.Y. Progresses in study on sorbitol effect on plants resistance. *Soils* **2018**, *50*, 446–454. [CrossRef]
28. Sanches, A.G.; Silva, M.B.; Fernandes, T.F.S.; Pedrosa, V.M.D.; Wong, M.C.C.; Gratão PLTeixeira, G.H.A. Quarantine cold treatment hardness of 'Palmer' mangoes treated with polyols. *J. Sci. Food Agric.* **2022**, *102*, 6112–6122. [CrossRef] [PubMed]
29. Miguel, A.C.A.; Durigan, J.F.; Morgado, C.M.A.; Oliveira, R.F. Lesão por frio em manga cv Palmer. *Rev. Bras. Frutc.* **2011**, *33*, 255–260. [CrossRef]
30. Watkins, C.; Harman, J. Use of penetrometer to measure flesh firmness of fruit. *Orchardist N. Z.* **1981**, *54*, 14–16.
31. McGuire, R.G. Reporting of objective color measurements. *HortScience* **1992**, *27*, 1254–1255. [CrossRef]
32. AOAC. *Official Methods of Analysis of the Association of Official Analytical Chemistry*, 20th ed.; Patricia Cunniff: Washington, DC, USA, 2016.
33. Gratão, P.L.; Monteiro, C.C.; Carvalho, R.F.; Teozotto, T.; Piotto, F.A.; Peres, L.E.P.; Azevedo, R.A. Biochemical dissection of diageotropica and Never ripe tomato mutants to Cd-stressful conditions. *Plant Physiol. Biochem.* **2012**, *56*, 79–96. [CrossRef] [PubMed]
34. Alexieva, V.; Sergiev, I.; Mapelli, S.; Karanov, E. The effect of drought and ultraviolet radiation on growth and stress markers in pea and wheat. *Plant Cell Environ.* **2001**, *24*, 1337–1344. [CrossRef]
35. Yang, Z.; Zheng, Y.; Cao, S. Effect of high oxygen atmosphere on quality, antioxidant enzymes, and DPPH-radical scanvenging activity of Chinese bayberry fruit. *J. Agric. Food Chem.* **2009**, *57*, 176–181. [CrossRef] [PubMed]
36. Bradford, M.M. A rapid and sensitive method for the quantification of microgram quantities of protein utilizing the principle of protein-dye binding. *Anal. Biochem.* **1976**, *722*, 248–254. [CrossRef]
37. Giannopolitis, C.N.; Ries, S.K. Superoxide dismutase. I. Occurrence in higher plants. *Plant Physiol.* **1977**, *59*, 309–314. [CrossRef] [PubMed]
38. Beauchamp, C.; Fridovich, I. Superoxide Dismutase: Improved Assays and an Assay Applicable to Acrylamide Gels. *Anal. Biochem.* **1971**, *44*, 276–287. [CrossRef] [PubMed]
39. Nakano, Y.; Asada, K. Hydrogen peroxide is scavenged by ascorbate specific peroxidases in spinach chloroplast. *Plant Cell Physiol.* **1981**, *22*, 867–880. [CrossRef]
40. Sojo, M.M.; Nuñez-Delicado, E.; García-Carmona, F. Partial purification of a banana polyphenol oxidase using triton X-114 and PEG 8000 for removal of polyphenols. *J. Agric. Food Chem.* **1998**, *46*, 4924–4930. [CrossRef]
41. Wissemann, K.W.; Lee, C.Y. Polyphenoloxidase activity during grape maturation and wine production. *Am. J. Enol. Vitic.* **1980**, *31*, 206–211. [CrossRef]
42. Lima, F.T.D.; Martins, L.P.; Toscano, B.L.M.D.; Costa, E.S.; Souza, W.F.C.; Lucena, F.A.; Silva Filho, J.F.; Sousa, S. Cold damage affects the quality of noni fruit (*Morinda citrifolia* L.). *Biocatal. Agric. Biotechnol.* **2023**, *47*, 102566. [CrossRef]
43. Narayana, C.K.; Rao, D.V.S.; Roy, S.K.; Siddiq, M. Mango Production, Postharvest Physiology and Storage. In *Tropical and Subtropical Fruit: Postharvest Physiology, Processing and Packaging*; Wiley: Hoboken, NJ, USA, 2012; pp. 259–276. [CrossRef]
44. Sanches, A.G.; Silva, M.B.; Wong, M.C.C.; Oliveira, A.R.G.; Pedrosa, V.M.D.; Fernandes, T.F.S.; Gratão, P.L.; Teixeira, G.H.A. Sorbitol immersion controls chilling injury in CA stored 'Palmer' mangoes. *Postharvest Biol. Technol.* **2022**, *185*, 111800. [CrossRef]
45. Salazar-Salas, N.Y.; Chairez-Veja, D.A.; Vega-Alvarez, M.; González-Nuñez, D.G.; Pineda-Hidalgo, K.V.; Chávez-Ontiveros, J.; Delgado-Vargas, F.; Lopez-Valenzuela, J.A. Proteomic changes in mango fruit peel associated with chilling injury tolerance induced by quarantine hot water treatment. *Postharvest Biol. Technol.* **2022**, *186*, 111838. [CrossRef]
46. Wei, C.; Ma, L.; Cheng, Y.; Guan, Y.; Guan, J. Exogenous ethylene alleviates chilling injury of 'Huangguan' pear by enhancing the proline content and antioxidant activity. *Sci. Hortic.* **2019**, *257*, 108671. [CrossRef]

47. Saeed, F.; Chaudhry, U.K.; Raza, A.; Charagh, S.; Bakhsh, A.; Bohra, A.; Ali, S.; Chitikineni, A.; Saeed, Y.; Visser, R.G.F.; et al. Developing future heat-resilient vegetable crops. *Funct. Integr. Genom.* **2023**, *23*, 47. [CrossRef] [PubMed]
48. Babalar, M.; Pirzad, F.; Sarcheshmeh, M.A.A.; Talaei, A.; Lessani, H. Arginine treatment attenuates chilling injury of pomegranate fruit during cold storage by enhancing antioxidant system activity. *Postharvest Biol. Technol.* **2018**, *137*, 31–37. [CrossRef]
49. Dumanović, J.; Nepovimova, E.; Natić, M.; Kuča, K.; Jaćević, V. The significance of reactive oxygen species and antioxidant defense system in plants: A concise overview. *Front. Plant. Sci.* **2020**, *11*, 1–13. [CrossRef]
50. Song, C.; Wang, K.; Xiao, X.; Liu, Q.; Yang, M.; Li, X.; Feng, Y.; Li, S.; Shi, L.; Chen, W.; et al. Membrane lipid metabolism influences chilling injury during cold storage of peach fruit. *Food Res. Int.* **2022**, *157*, 111249. [CrossRef]

horticulturae

MDPI

Article

Near-Freezing Temperature Storage Improves Peach Fruit Chilling Tolerance by Regulating the Antioxidant and Proline Metabolism

Handong Zhao [1,2,*], Shuqi Meng [2], Maorun Fu [2,*] and Qingmin Chen [3]

[1] School of Food and Health, Beijing Technology and Business University, Beijing 100048, China
[2] College of Food Science and Engineering, Qilu University of Technology (Shandong Academy of Sciences), Jinan 250353, China; mengshuqi2580@163.com
[3] Department of Food Science and Engineering, Shandong Agriculture and Engineering University, Jinan 250100, China; qmchen1207@163.com
* Correspondence: qlspzhd@163.com (H.Z.); skyfmr@163.com (M.F.)

Abstract: Chilling injury (CI) in peach fruit (*Prunus persica* cv. Yuhualu) is generally caused by long-time low temperature (5 °C or 0 °C) storage. However, peach fruit stored at near-freezing temperature (NFT in this research is −1 °C), defined as within 0.5 °C above the biological freezing point of biological tissue, does not exhibit CI symptoms. The effect of NFT on the CI, proline metabolism, and antioxidant capability of peach fruit during storage was studied and compared with 5 °C and 0 °C storage as controls. The results exhibit that NFT completely inhibited the occurrence of CI in peach fruit. NFT significantly ($p < 0.05$) enhanced the activities of superoxide dismutase, catalase, ascorbate peroxidase, and 1,1-diphenyl-2-picrylhydrazyl scavenging capacity. Moreover, the increase of malondialdehyde, ion leakage, and H_2O_2 accumulation were inhibited remarkably by NFT, and decreases in the contents of phenolics and ascorbic acid were slowed significantly in peach fruit stored at NFT ($p < 0.05$). Additionally, NFT storage enhanced proline accumulation by modulating the activity of proline metabolizing enzymes. In conclusion, the above results suggest that NFT storage can improve the chilling tolerance of peach fruit by regulating the antioxidant defense and proline metabolism, which might represent a potential novel method to store fruits and vegetables for longer storage times.

Keywords: near-freezing temperature storage; peach fruit; chilling injury; antioxidant defense system; proline metabolism

Citation: Zhao, H.; Meng, S.; Fu, M.; Chen, Q. Near-Freezing Temperature Storage Improves Peach Fruit Chilling Tolerance by Regulating the Antioxidant and Proline Metabolism. *Horticulturae* **2024**, *10*, 337. https://doi.org/10.3390/horticulturae10040337

Academic Editor: Michailidis Michail

Received: 29 February 2024
Revised: 23 March 2024
Accepted: 27 March 2024
Published: 28 March 2024

1. Introduction

Peach fruit (*Prunus persica* cv. Yuhualu) is bright in color, rich in aroma and nutrients, sweet and sour, and well loved by consumers [1]. However, postharvest challenges arise due to the climacteric nature of peaches, leading to softening and decay when stored at room temperature. Although low-temperature refrigeration ranging from 0 °C to 8 °C is a widely employed strategy to extend peach storage, it often triggers chilling injury (CI). CI generally results in flesh browning, fibrillation, and aroma loss. Underlying these manifestations are disruptions in plant cell metabolism, including membrane damage, reactive oxygen disorder, and alterations in cell wall materials [2–4].

Membrane damage, caused by disturbance of oxidative stress metabolism, is a critical reason for CI [5]. The antioxidant defense system can reduce CI damage in plants by reducing damage to the plasma membrane caused by free radicals. Reactive oxygen species are essential redox signaling factors involved in developmental and physiological processes, and induce the expression of defense genes and adaptive response. Nitric oxide treatment can extenuate mitochondrial swelling, maintain mitochondrial membrane potential and membrane fluidity, and delay the decrease in activities of the cytochrome pathway and cyanide-insensitive pathway of the mitochondrial respiratory chain in peach fruit [6]. Similarly, low temperature generally reduces the activities of catalase (CAT) and superoxide dismutase (SOD) in postharvest cherry fruit [7].

Various abiotic and biotic stresses generally lead to proline accumulation [8]. Both exogenous ethylene and glycine betaine treatments can delay plant chilling tolerance by promoting reactive oxygen species metabolism and proline accumulation [9]. Chitosan regulates proline metabolism and enhances the chilling tolerance of rice [10]. Generally, the accumulation of proline can maintain the structure and function of cells, which may improve the chilling tolerance of plants [11].

Near-freezing temperature (NFT) refers to the temperature range from the freezing point of the organism to 0 °C; the cells of the organism will not suffer freezing damage under this temperature range. NFT storage technology can inhibit the respiration and microbial metabolism of postharvest fruits while maintaining their quality [12,13]. Recent studies have suggested that NFT can significantly restrain CI occurrence and prolong the storage period of nectarines [14,15]. While this phenomenon shows that the chilling tolerance of the fruit is significantly enhanced at this temperature, the specific mechanism has not yet been elucidated.

This research delves into the repercussions of NFT storage on CI, quality, antioxidant properties, and the metabolism of reactive oxygen species and proline in peach fruits. The investigation seeks to unravel potential correlations between antioxidant and proline metabolism and chilling tolerance during NFT storage, presenting valuable insights for the preservation of fresh vegetables and fruits.

2. Materials and Methods

2.1. Plant Materials and Fruit Treatment

For this experiment, peach fruits (*Prunus persica* cv. Yuhualu), with commercial maturity (soluble solids content 8~9° brix), color, and size and absence of diseases and mechanical damage were collected and immediately transferred to the laboratory.

Following established protocols outlined in our previous research [14], the fruits underwent a precooling process at 5 °C ± 0.1 °C for 12 h. Subsequently, they were randomly divided into three groups: 5 °C (±0.1 °C), 0 °C (±0.1 °C), and Near-Freezing Temperature (NFT, −1 °C ± 0.1 °C). The storage duration was 28 days and the storage conditions included a relative humidity of 90%. Each treatment was replicated three times, with each replicate consisting of 100 peach fruits. Sampling was conducted at 0, 7, 14, 21, and 28 days, and fifteen fruits were sampled from each batch. Tissue slices of approximately 1 cm thick were rapidly frozen in liquid nitrogen and stored at −80 °C for subsequent analysis.

2.2. Confirmation of Near-Freezing Point Temperature of Peach Fruit

After calibration by a 0 °C of water and ice, the probe of the high-precision temperature tester (Elitech RC-4HC, JiangChuang, Xuzhou, China) was used for testing the freezing point temperature [14]. According to the freezing curve (Supplementary Material Figure S1), the super-cooling point was −2.3 °C and the freezing point was −1.1 °C. Thus, to prevent freezing damage, the near-freezing storage temperature was set at −1 °C and the temperature fluctuation was controlled within ±0.1 °C.

2.3. Measurement of Incidence and Incidence Index of CI in Peach Fruit

Subjective assessment of CI development was performed with 15 fruits from each replicate according to the previous method [15]. The CI degree was investigated by comparing the internal browning area of peaches transferred to room temperature for three days after storage [16]. The degree of internal browning was calculated using a 0 to 4 rating, as follows: 0, no browning; 1, browning area less than 20%; 2, browning area 20~40%; 3, browning area 40~60%; and 4, browning area greater than 60%. The CI index = $\sum$ (browning degree × number of fruits with browning in each grade)/(4 × total number of fruit) × 100%.

2.4. Determination of Ion Leakage and Contents of Malonaldehyde and H_2O_2 in Peach Fruits

Ion leakage was determined using 15 pieces of pulp (3 mm thickness × 8 mm diameter) from 15 fruits [17]. The pulp was allowed to stand in 30 mL of double-distilled water for 30 min. After shaking, the electrical conductivity of the mixed solution was detected by an electrical conductivity meter (DDS-12B, Beijing, China), then the mixed sample was warmed in boiling water for 15 min. After cooling, the final conductivity of the mixed sample was recorded. The relative conductivity is generally used to represent the amount of ion leakage (conductivity of mixed solution/final conductivity) × 100%.

To test the malonaldehyde (MDA) content, tissue (10 g) was placed in a system of thiobarbituric acid as previously reported [18]. After chromogenic reaction, the absorbance was recorded at 532 nm and 600 nm and the accumulation was described in μmol kg^{-1}. Moreover, flesh tissue (2.0 g) was used for measuring the H_2O_2 content as previously reported [19], and the H_2O_2 content was recorded as mmol kg^{-1}.

2.5. Determination of Proline Level and Related Enzyme Activities in Peach Fruits

The proline content was measured by adding sulfosalicylic acid to ground flesh in an ice bath, followed by boiling and centrifugation [20]. The absorbance of the supernatant was tested at 520 nm, with the proline content expressed in mg kg^{-1}.

We modified a previous method for measuring the enzyme activities of pyrroline-5-carboxylate reductase (P5CR) and Δ^1-pyrroline-5-carboxylate synthetase (P5CS) [21]. Fruit tissue (1.0 g) was mixed with formulated extraction buffer (pH 7.5, 2 mmol L^{-1} phenylmethyl sulfonyl fluoride, 10 mol L^{-1} $MgC1_2$, 0.5 mol L^{-1} Tris-HCl, 2% Polyvinyl pyrrolidone). The resulting homogenate was cooled in an ice bath, followed by centrifugation at 20,000× *g* for 20 min at 4 °C. The residue was repeatedly extracted once, and the collected supernatant from the centrifugation was combined to form the enzyme extract.

The reaction solution, including 10 mmol L^{-1} ATP, 20 mmol L^{-1} $MgCl_2$, 50 mmol L^{-1} L-glutamate, and 100 mmol L^{-1} hydroxylamine hydrochloride, was mixed with 1 mL of the preceding enzyme extract. The reaction termination buffer, containing 5% $FeCl_3$ and 12% trichloroacetic acid in 5 mol L^{-1} HCl, was added to this system, then the mixture was reacted in a water bath at 37 °C for 15 min. After centrifugation, the supernatant was used to test the absorbance at 535 nm and the blank control without ATP was used as a reference. The unit of P5CS (U) is defined as the quantity of enzyme required to produce 1 μmol of γ-glutamine every 60 s, and results are described in U g^{-1} protein.

The activity of P5CR was determined by the previous method [22]. Reaction solution buffer, including 0.1 mmol L^{-1} NADPH, 0.56 mmol L^{-1} P5C, was reacted with enzyme solution (20 μL). The oxidation of 1 μmol of NADPH per minute is defined as one unit activity of P5CR.

We modified the proline dehydrogenase (PRODH) activity determination method proposed previously [23]. Fruit tissue (3.0 g) was blended at low temperature with potassium phosphate buffer (100 mmol L^{-1} 5 mL pH 7.4) containing 1.0% (*w*/*v*) PVPP, 1.0 mmol L^{-1} EDTA, 5.0 mmol L^{-1} $MgCl_2$, 10 mmol L^{-1} β-mercaptoethanol, and 60 mmol L^{-1} KCl. After centrifugation (10,000× *g*, 15 min, 5 °C), the enzymatic activity was estimated using the obtained supernatant.

The reaction mixture (2.5 mL pH 10.3) contained 1.6 mL 0.15 mol L^{-1} Na_2CO_3-$NaHCO_3$ buffer, 0.2 mL 0.1 mol L^{-1} L-proline, and 0.2 mL 0.9 mmol L^{-1} 2,6-dichlorophenol indiophenol. After 5 min warming at 30 °C, 0.5 mL reaction mixture was mixed with 0.2 mL enzyme extract and 9 mg mL^{-1} phenazine methyl sulfate reagent. The absorbance change at 600 nm was marked immediately. The enzyme activity unit (U) was clarified as 0.01 ΔA600 nm g^{-1} min^{-1}.

2.6. Measurement of Enzyme Activities of Antioxidant in Fruit

Fresh pulp (10 g) was used for determining the SOD activity of peach according to a previous method [24]. The amount of enzyme required to inhibit the tetrazolium photoredox reaction by 50% was defined as one enzyme unit (U). Following a previous method [25], 2.0 g pulp was used to prepare the crude enzymatic extract of CAT and ascorbate peroxidase (APX). Following reported methods, we utilized the enzymatic extract to calculate the estimated activities of CAT and APX [8]. The enzyme reaction system of peroxidase (POD) contained 2.5 mL of 0.2 mol L^{-1} phosphate buffer (pH 5.0), 150 μL of 0.08% H_2O_2, 0.1 mL of 0.1% guaiacol, and 10 μL enzyme solution. After adding the hydrogen peroxide for 5 min, the absorbance at 460 nm was recorded. Each enzyme activity unit was defined as every 0.01 change in absorbance [26].

2.7. Determination of Phenolics, Ascorbic Acid, and Free Radical Scavenging Capacity

The level of phenolics in the fruits was measured by the Folin–Ciocalteu method [27], with the phenolics content expressed in gallic acid equivalents and the unit expressed in g kg^{-1}. In addition, flesh tissue (1.0 g) was used to estimate the ascorbic acid content in peach fruits following previously reported methods [14]. The volume of the titrated solution was recorded, with the ascorbic acid content expressed as mg kg^{-1}.

1,1-Diphenyl-2-picrylhydrazyl (DPPH) is generally utilized to evaluate the antioxidant capacity of tissue [28]. Fruit extract (10 μL) was thoroughly mixed with distilled water (90 μL) and DPPH methanol (3.9 mL 0.0250 g L^{-1}), then the mixture was placed in the dark for 30 min. The absorbance was tested at 515 nm without DPPH. The calculation formula of the clearance rate was as follows: (%) DPPH clearance = $(A_0 - A_1)/A_0 \times 100\%$; in the above formula, A_0 represents the Abs control and A_1 represents the Abs sample.

2.8. Statistical Analysis

The experiment used a completely randomized design. Analysis of variance was performed among different groups, and Duncan's multiple interval test was used to compare the mean values with a significance level of 0.05 (SPSS 11.0 Inc., Chicago, IL, USA). Final values represent the mean ± SE (standard error) of three replicates. Figures were prepared using Origin Program 2016 (OriginLab Co., Northampton, MA, USA).

3. Results

3.1. NFT Storage Enhances the Chilling Tolerance of Peach Fruit

Both the CI incidence and CI incidence index of peach fruits stored at 5 °C and 0 °C showed an upward trend with the prolongation of storage time. However, the NFT group did not have any CI symptoms during the whole storage time (Figure 1A,B). On Day 7, for the 5 °C group the CI incidence was 12.61% and the CI incidence index was 5.31%. On Day 28, for the 0 °C group the CI incidence and CI incidence index were 20.93% and 10.52%, respectively. The above results, as exhibited in Figure 1, indicate that NFT was able to reduce the occurrence of CI. Figure 1C shows that the degree of browning inside the peach fruits in the 5 °C and 0 °C groups increased sharply, while the peach fruits at NFT kept their original luster and internal color during the whole storage time. This phenomenon suggests that NFT storage was able to remarkably alleviate the internal browning of peach fruits.

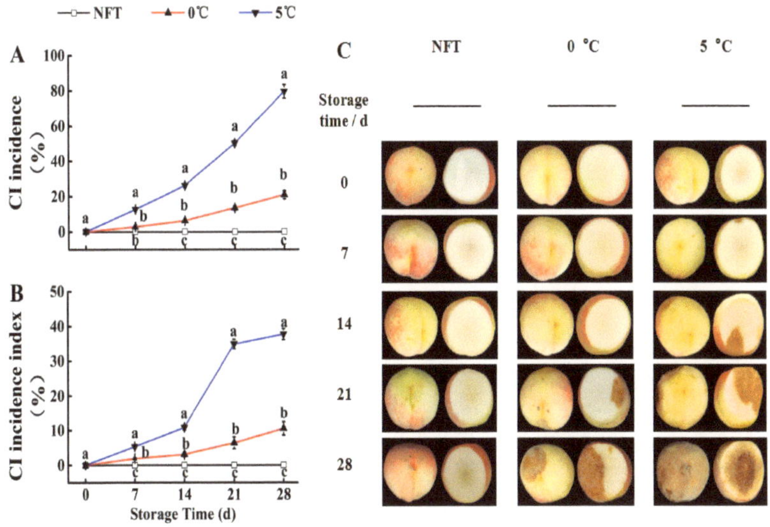

Figure 1. Effects of different temperature treatments on CI incidence and CI incidence index of peach fruits. CI incidence (**A**), CI incidence index (**B**), and photographs of stored peach fruits (**C**). Data are expressed as the mean ± SE (n = 3). Vertical bars represent the standard error of the mean ($p < 0.05$). Duncan's test letters represent the difference among the different temperatures within the same days.

3.2. NFT Can Reduce Ion Leakage and Contents of MDA and H_2O_2

Figure 2A shows that the ion leakage gradually emerged as an incremental trend in all groups with the extension of storage time. At 5 °C, the ion leakage increased rapidly by 28.40% within 28 days. This increase in ion leakage was significantly suppressed when decreasing the storage temperature. The ion leakage level of peach fruits stored at NFT was significantly ($p < 0.05$) lower than those of the 0 °C and 5 °C groups.

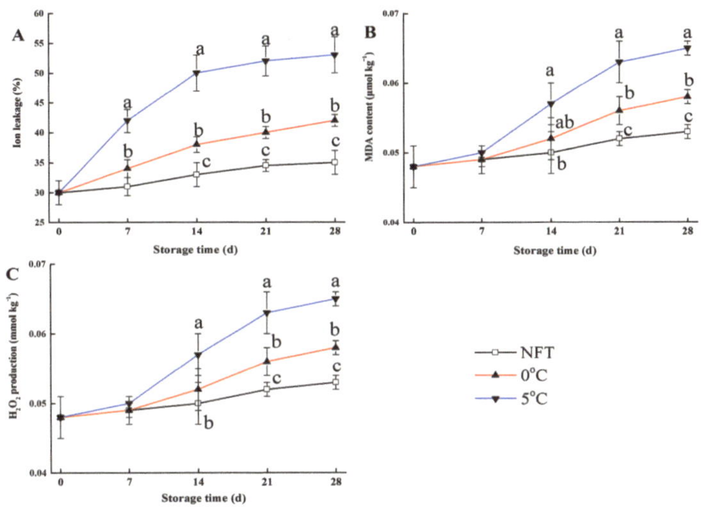

Figure 2. Effects of NFT storage on ion leakage (**A**), MDA content (**B**), and H_2O_2 (**C**) content of peach fruits. Data are expressed as the mean ± SE (n = 3). Vertical bars represent the standard error of the mean ($p < 0.05$). Duncan's test letters represent the difference among the different temperatures within the same days.

During the first of 14 days of storage, NFT inhibited the increase of MDA content, with a level of 0.048 μmol kg^{-1}. The MDA content of the 5 °C group increased rapidly after Day 7, an abnormal situation that may have been caused by severe fruit rot. Moreover, the MDA content of the NFT and 0 °C groups decreased slowly during the storage period, and the MDA value of the NFT group increased to 0.591 μmol kg^{-1} until the end of storage, which is remarkably lower than the 0 °C and 5 °C groups (Figure 2B).

As Figure 2C shows, the highest H_2O_2 contents appeared on Day 28 in the 0 °C and 5 °C groups, respectively. Compared with the 0 °C and 5 °C groups, the content of H_2O_2 in the NFT increased mildly and remained at a lower level ($p < 0.05$) in the whole storage time.

3.3. *Effect of NFT Storage on Proline Metabolism in Peach Fruit*

Compared with the beginning, the proline level of the 5 °C storage group decreased by 7% after 14 days of storage (Figure 3A). During the first 21 days of storage, the proline value of the 0 °C storage group was consistently lower than that of the 5 °C and NFT groups. On Day 28 of storage, the proline content of the NFT group was similar to that of the 0 °C group, and was 16% higher than that of the 5 °C group.

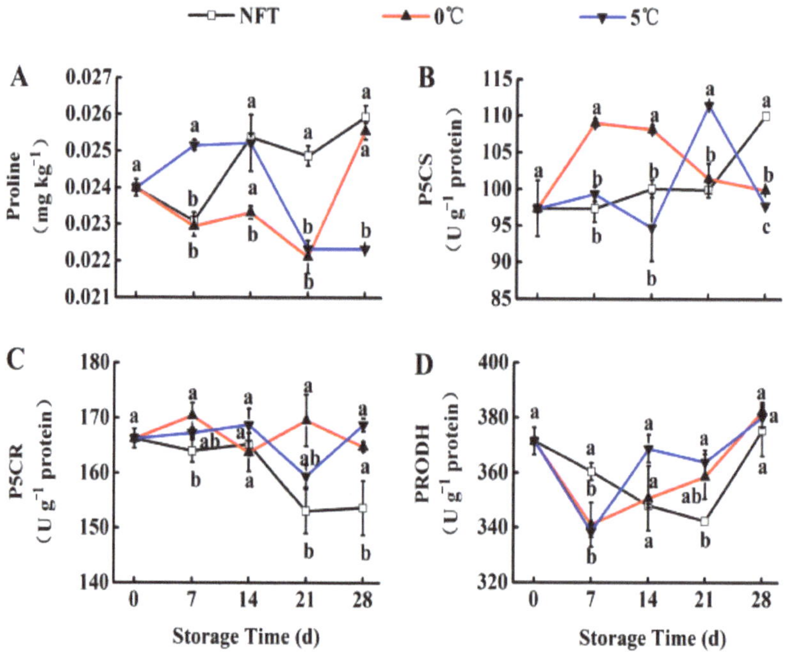

Figure 3. Effects of NFT storage on proline content and related enzyme activities in peach fruits. Proline content (**A**) and activities of P5CS (**B**), P5CR (**C**), and PRODH (**D**). Data are expressed as the mean ± SE (n = 3). Vertical bars represent the standard error of the mean ($p < 0.05$). Duncan's test letters represent the difference among the different temperatures within the same days.

The P5CS activity of fruits stored at NFT showed a gradual increase throughout the storage period. The P5CS activity of the NFT group was significantly higher ($p < 0.05$) on Day 28 than that of the 5 °C and 0 °C groups (Figure 3B). In the whole storage time, there were some fluctuations in the activity of P5CR in the 0 °C and 5 °C treatment groups, while the overall change was not significant (Figure 3C). The P5CR activity of the NFT group was always lower than that of the other two groups during the whole storage process.

During storage, the activity in the different temperature groups first decreased and then increased. On Day 7, the activity of PRODH in the 0 °C and 5 °C groups decreased rapidly to the lowest level and then gradually increased. PRODH activity in the NFT group decreased laxly and arrived at the lowest value on Day 21, maintaining a lower level than the 0 °C and 5 °C groups (Figure 3D).

3.4. NFT Enhances the Antioxidant Defense System

The SOD activities at different storage temperatures showed signs of a mild retreat at the beginning, then SOD activities increased and peaked in all groups on Day 21 (Figure 4A). Compared with the 0 °C and 5 °C groups, the SOD activity of the NFT group was the highest, and showed an upward trend during storage. These results indicate that NFT storage was able to effectively improve the SOD activity of peach fruits.

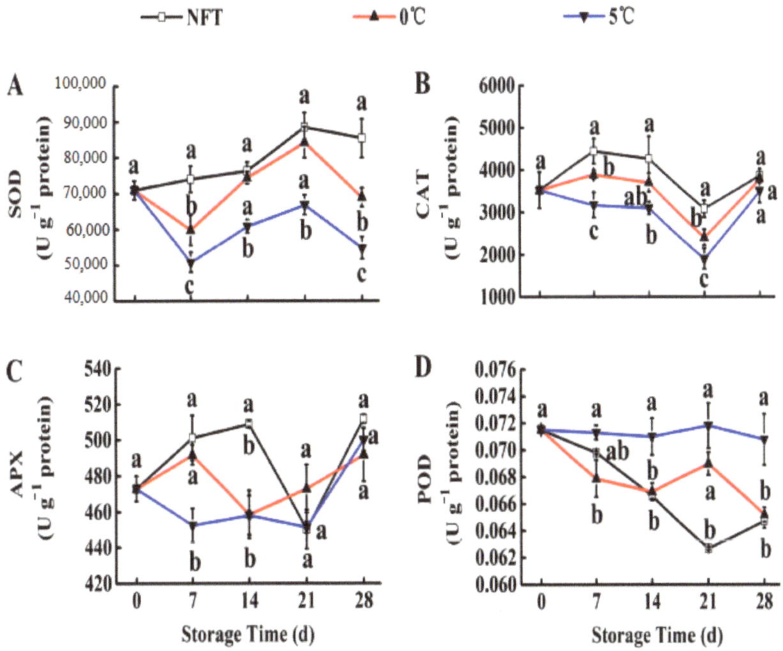

Figure 4. Effects of NFT storage on reactive oxygen species metabolism in peach fruit. Measurement of SOD (**A**), CAT (**B**), APX (**C**), and POD (**D**) activities. Data are expressed as the mean ± SE (n = 3). Vertical bars represent the standard error of the mean ($p < 0.05$). Duncan's test letters represent the difference among the different temperatures within the same days.

The activities of CAT in all group exhibited a slight fluctuation at the end of storage. CAT activity in the 0 °C and NFT groups decreased to the lowest value at Day 21, then increased again. NFT treatment delayed the decrease in CAT activity and maintained a higher level than in the other groups during the whole storage period ($p < 0.05$, Figure 4B).

Figure 4C shows that APX activity exhibited an increase at beginning of storage, followed by a decrease; moreover, the APX activity in the NFT storage group was higher than that of the 0 °C and 5 °C groups, and was 1.1-fold higher on Day 14. In the whole storage time except for Day 21, the APX activity of peach fruits stored at NFT was higher than in the 0 °C and 5 °C groups.

According to Figure 4D, the POD activity of the peach fruits in the 5 °C group remained unchanged throughout the storage time and was remarkably loftier ($p < 0.05$) than that of the other two groups. The POD activity of peach fruits stored at NFT gradually decreased, and was lower than that of the 0 °C and 5 °C groups on Day 27.

3.5. Effects of NFT Storage on Phenolics Content, Ascorbic Acid Content, and DPPH Scavenging Capacity of Peach Fruit

According to Figure 5A, the contents of peach phenolics at different temperatures showed a decreasing trend at first and then increasing trend during storage. Among the different groups, the contents of phenolics at NFT was significantly higher than that of the 5 °C and 0 °C groups ($p < 0.05$). During storage, the ascorbic acid level exhibited a declining trend at all temperatures, with that of the 5 °C group being significantly ($p < 0.05$) lower than those of the other two groups. There was no significant difference in ascorbic acid level between the 0 °C and NFT groups ($p > 0.05$), indicating that both the 0 °C and NFT treatments inhibited the decrease in ascorbic acid content (Figure 5B).

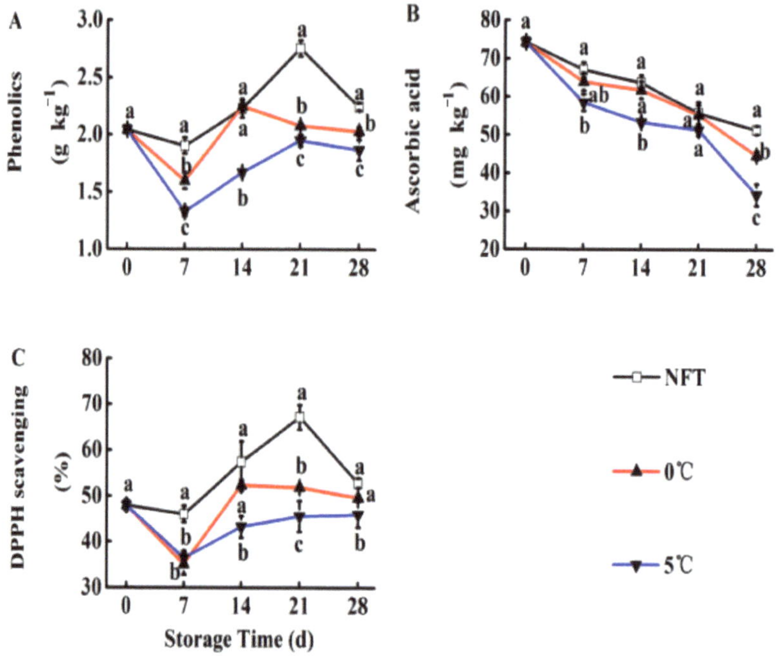

Figure 5. Effects of NFT storage on phenolics content (**A**), ascorbic acid content (**B**), and DPPH scavenging capacity (**C**) of peach fruits. Data are expressed as the mean ± SE (n = 3). Vertical bars represent the standard error of the mean ($p < 0.05$). Duncan's test letters represent the difference among the different temperatures within the same days.

DPPH scavenging and phenolics exhibited similar trends (Figure 5C). The DPPH scavenging capacity in the 5 °C and 0 °C groups showed a decreasing trend at first, followed by an increase during storage. Moreover, the scavenging rates on Day 28 and Day 0 were almost the same. The free radical scavenging rate of the NFT group was considerably higher ($p < 0.05$) than those of the 5 °C and 0 °C groups during the whole storage time.

4. Discussion

The symptoms of CI in peach fruits are browning and softening, which seriously affect fruit quality and commercial value. In recent years, methods such as treatment with heat, 1-Methylcyclopropene, glycine betaine, melatonin, and other methods have been used to inhibit CI [29–32]. In our study, both the CI incidence and CI index of peach fruits at NFT did not exhibit growth during the whole storage time, indicating that NFT completely inhibited the occurrence of CI. Ion leakage is an important physiological indicator for judging membrane damage, and can accurately reflect its development [33]. The increase of ion leakage in the NFT group was very slow, indicating that the fluidity of the cell membrane was maintained, meaning that the cell membrane remained almost undamaged.

Similarly, low temperature stress can enhance membrane peroxidation and reactive oxygen metabolism in plant cells; the accumulation of MDA and H_2O_2 disrupts the plasma membrane and affects cell membrane fluidity, leading to the development of CI [18,34]. In our study, the results showed that NFT storage significantly decreased both MDA and H_2O_2 contents, thereby reducing the toxic effects of free radicals on the cell membrane and maintaining cellular homeostasis. These results indicate that NFT can enhance chilling tolerance in fruit.

Proline regulates cell osmotic balance and protects proteins, helping to stabilize the cell structure when plants are under abiotic stress [35,36]. At the same time, proline can scavenge free radicals and protect the antioxidant enzyme system [37]. The synthesis of proline in plants is catalyzed by P5CS and P5CR. For example, P5CS plays a critical role in the synthesis of proline when plants suffer stress [38]. As the key enzyme in proline synthesis, the activity of P5CS was reinforced by NFT storage in peach fruits while the activity of P5CR was weakened, indicating that NFT accelerated proline synthesis. Proline metabolizing enzymes can coordinate with PRODH to degrade the accumulated proline in plants, which might result in the breakdown of osmotic balance and cause damage [39]. NFT reduced the activity of PRODH and inhibited the degradation of proline in peach fruits. The above results suggest that NFT storage can improve proline accumulation by regulating the activities of enzymes involved in proline metabolism, ultimately enhancing the chilling tolerance of fruit (Figure 6).

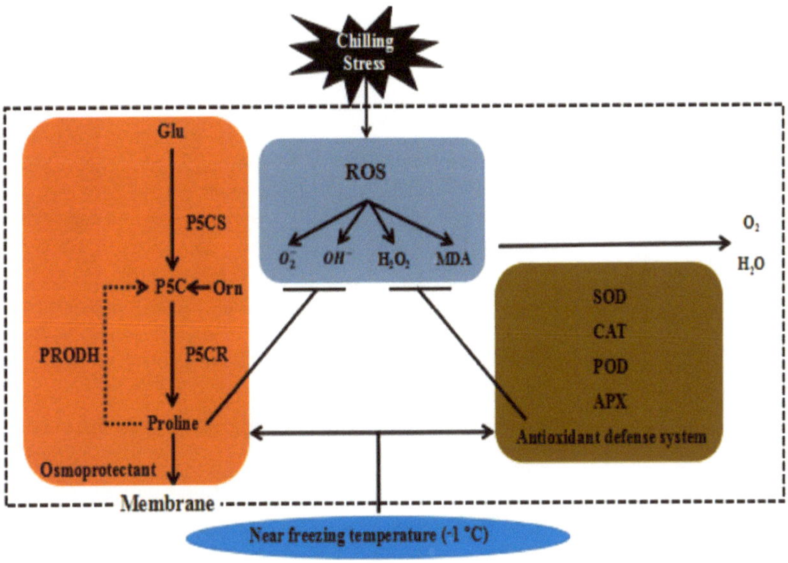

Figure 6. Schematic presentation of NFT inhibiting the occurrence of CI.

Excessive concentrations of reactive oxygen species can lead directly to plant senescence and even death; the antioxidant defense system, including CAT, SOD, APX, and POD, is critical in regulating the purging of reactive oxygen [40]. When plants suffer from abiotic stress, the metabolic balance of reactive oxygen species is broken and a large number of free radicals is generated in the cell, which damages the cell membrane and eventually induces CI. SOD, as a key enzyme in preventing oxidative stress, catalyzes the formation of H_2O_2 and O_2, after which H_2O_2 is converted into H_2O and O_2 under the action of CAT and POD, retarding the toxic effect of reactive oxygen species [41]. As an enzyme that catalyzes the breakdown of peroxides, CAT reduces the toxic effects of H_2O_2 on metabolic tissues [42]. In addition, APX catalyzes the redox reaction between ascorbic acid and H_2O_2 so that ascorbic acid is oxidized to form hydroascorbic acid, with H_2O_2 decomposed in the process of maintaining the balance of free radical metabolism in cells [43]. Furthermore, both phenolics and ascorbic acid can scavenge free radicals and improve antioxidant capacity [44]. Zhao et al. found that NFT can effectively promote the accumulation of phenolics in peach fruits and slow the decline in ascorbic acid content [8]. The changes in the activities of CAT, SOD, and APX in our research indicate that NFT storage helped to promote scavenging of reactive oxygen radicals, prevent the accumulation of high oxygen concentrations, maintain the balance of reactive oxygen species, and re-establish cellular redox homeostasis, thereby enhancing the chilling tolerance of peach fruits (Figure 6).

5. Conclusions

In conclusion, this study highlights intriguing findings regarding the efficacy of NFT storage in mitigating chilling injury in peaches, especially when compared to storage at higher temperatures. Notably, NFT storage exhibited consistent preservation of elevated levels of antioxidant enzyme activities, ascorbic acid, phenolics, and DPPH scavenging capacity. In addition, it effectively mitigated the accumulation of malondialdehyde and inhibited the activity of proline dehydrogenase (PRODH). Finally, NFT storage facilitated increased proline production and bolstered the overall antioxidant defense system.

Supplementary Materials: The following supporting information can be downloaded at: https://www.mdpi.com/article/10.3390/horticulturae10040337/s1, Figure S1. Freezing point curve of peach fruit.The temperature initially decreased to the super-cooling point of peach fruit during the freezing process. Owing to exothermic phenomena, the temperature curve raised and temporarily formed distinct plateaus, which correspond to the biological freezing point of peach fruit. According to the curve, the super-cooling point and freezing point of fruits were –2.3 °C and –1.1 °C, respectively. Moreover, temperature fluctuation was confirmed within a small region (± 0.1 °C) to avoid freezing damage.

Author Contributions: H.Z.: funding acquisition, writing and editing, investigation, parameter testing. S.M.: methodology establishment, parameter testing, funding acquisition. M.F.: experiment administration, validation. Q.C.: writing—review. All authors have read and agreed to the published version of the manuscript.

Funding: This work was supported by the Youth Fund of the Natural Science Foundation of Shandong Province (ZR2020QC245), the talent research projects of Qilu University of Technology in 2023 (2023RCKY234), the Major Pilot Innovative Project of the Integration of Science and Production (2022JBZ01-08), and the College Students' Innovation and Entrepreneurship Project of Shandong Province (No. S202210431026).

Data Availability Statement: The original date presented in the research are included in the article. Further inquiries can be directed to the corresponding authors.

Conflicts of Interest: The authors declare no conflicts of interest.

References

1. Ye, J.B.; Wang, G.Y.; Tan, J.P.; Zheng, J.R.; Zhang, X.; Xu, F.; Cheng, S.Y.; Chen, Z.X.; Zhang, W.W.; Liao, Y.L. Identification of candidate genes involved in anthocyanin accumulation using illmuina-based RNA-seq in peach skin. *Sci. Hortic.* **2019**, *250*, 184–198. [CrossRef]
2. Bustamante, C.A.; Brotman, Y.; Montia, L.L.; Gabilondoc, J.; Buddec, C.O.; Laraa, M.V.; Fernied, A.R.; Drincovicha, M.F. Differential lipidome remodeling during postharvest of peach varieties with different susceptibility to chilling injury. *Physiol. Plantarum* **2018**, *163*, 2–17. [CrossRef] [PubMed]
3. Chen, M.S.; Guo, H.M.; Chen, S.Q.; Li, T.T.; Li, M.Q.; Rashid, A.; Xu, C.J.; Wang, K. Methyl jasmonate promotes phospholipid remodeling and jasmonic acid signaling to alleviate chilling injury in peach fruit. *J. Agric. Food Chem.* **2019**, *67*, 9958–9966. [CrossRef] [PubMed]
4. Lin, Y.X.; Lin, H.T.; Fan, Z.Q.; Wang, H.; Lin, M.S.; Chen, Y.H.; Hung, Y.C.; Lin, Y.F. Inhibitory effect of propyl gallate on pulp breakdown of longan fruit and its relationship with ROS metabolism. *Postharvest Biol. Technol.* **2020**, *168*, 111272. [CrossRef]
5. Marangoni, A.G.; Palma, T.; Stanley, D.W. Membrane effects in postharvest physiology. *Postharvest Biol. Technol.* **1996**, *7*, 193–217. [CrossRef]

6. Wang, C.; Huang, D.; Tian, W.; Zhu, S. Nitric oxide alleviates mitochondrial oxidative damage and maintains mitochondrial functions in peach fruit during cold storage. *Sci. Hortic.* **2021**, *287*, 110249. [CrossRef]
7. Yadu, B.; Chandrakar, V.; Meena, R.K.; Keshavkant, S. Glycinebetaine reduces oxidative injury and enhances fluoride stress tolerance via improving antioxidant enzymes, proline and genomic template stability in *Cajanus cajan* L. *S. Afr. J. Bot.* **2017**, *111*, 68–75. [CrossRef]
8. Wei, T.L.; Wang, Z.X.; He, Y.F.; Xue, S.; Zhang, S.Q.; Pei, M.S.; Liu, H.N.; Yu, Y.H.; Guo, D.L. Proline synthesis and catabolism-related genes synergistically regulate proline accumulation in response to abiotic stresses in grapevines. *Sci. Hortic.* **2022**, *305*, 111373. [CrossRef]
9. Ignatenko, A.; Talanova, V.; Repkina, N.; Titov, A. Exogenous salicylic acid treatment induces cold tolerance in wheat through promotion of antioxidant enzyme activity and proline accumulation. *Acta Physiol. Plant.* **2019**, *41*, 80. [CrossRef]
10. Zhang, Y.; Fan, L.Q.; Zhao, M.Y.; Chen, Q.M.; Qin, Z.; Fang, Z.H.; Fujiwara, T. Chitooligosaccharide plays essential roles in regulating proline metabolism and cold stress tolerance in rice seedlings. *Acta Physiol. Plant.* **2019**, *41*, 77. [CrossRef]
11. Niu, Y.; Ye, L.; Wang, Y.; Shi, Y.; Luo, A. Salicylic acid mitigates *'Hayward'* kiwifruit chilling injury by regualting hormone and proline metabolism, as wella s maintaining cellular structure. *Food Biosci.* **2024**, *57*, 103573. [CrossRef]
12. Okamoto, T.; Nakamura, T.; Zhang, J.T.; Aoyama, A.; Chen, F.S.; Fujinaga, T.; Shoji, T.; Hamakawa, H.; Sakai, H.; Manabe, T.; et al. Successful sub-zero non-freezing preservation of rat lungs at −2 °C utilizing a new supercooling technology. *J. Heart Lung Transpl.* **2018**, *27*, 1150–1157. [CrossRef] [PubMed]
13. Cai, L.Y.; Liu, C.H.; Ying, T.J. Changes in quality of low-moisture conditioned pine nut (*Pinus gerardiana*) under near freezing temperature storage. *CyTA-J. Food* **2013**, *11*, 216–222. [CrossRef]
14. Zhao, H.D.; Shu, C.; Fan, X.G.; Cao, J.K.; Jiang, W.B. Near-freezing temperature storage prolongs storage period and improves quality and antioxidant capacity of nectarines. *Sci. Hortic.* **2018**, *228*, 196–203. [CrossRef]
15. Zhao, H.D.; Jiao, W.X.; Cui, K.B.; Fan, X.G.; Shu, C.; Zhang, W.L.; Cao, J.K.; Jiang, W.B. Near-freezing temperature storage enhances chilling tolerance in nectarine fruit through its regulation of soluble sugars and energy metabolism. *Food Chem.* **2019**, *289*, 426–435. [CrossRef] [PubMed]
16. Jin, P.; Zhu, H.; Wang, J.; Chen, J.J.; Wang, X.L.; Zheng, Y.H. Effect of methyl jasmonate on energy metabolism in peach fruit during chilling stress. *J. Sci. Food Agri.* **2013**, *93*, 1827–1832. [CrossRef] [PubMed]
17. Valenzuela, J.L.; Manzano, S.; Palma, F.; Carvajal, F.; Garrido, D.; Jamilena, M. Oxidative stress associated with chilling injury in immature fruit: Postharvest technological and biotechnological solutions. *Int. J. Mol. Sci.* **2017**, *18*, 1467. [CrossRef] [PubMed]
18. Patterson, B.D.; Macrae, E.A.; Ferguson, I.B. Estimation of hydrogen peroxide in plant extracts using titanium (IV). *Anal. Biochem.* **1984**, *139*, 487–492. [CrossRef] [PubMed]
19. Bates, L.S.; Waldren, R.P.; Teare, I.D. Rapid determination of free proline for water-stress studies. *Plant Soil* **1973**, *39*, 205–207. [CrossRef]
20. Hayzer, D.J.; Leisinger, T. The gene-enzyme relationships of proline biosynthesis in escherichia coli. *J. Gen. Appl. Microbiol.* **1980**, *118*, 287–293. [CrossRef]
21. Murahama, M.; Yoshida, T.; Hayashi, F.; Ichino, T.; Sanada, Y.; Wada, K. Purification and characterization of Δ^1-pyrroline-5-carboxylate reductase isoenzymes, indicating differential distribution in spinach (*Spinacia oleracea* L.) leaves. *Plant Cell Physiol.* **2001**, *42*, 742–750. [CrossRef] [PubMed]
22. de Freitas, P.A.; de Souza Miranda, R.; Marques, E.C.; Prisco, J.T.; Gomes-Filho, E. Salt tolerance induced by exogenous proline in maize is related to low oxidative damage and favorable ionic homeostasis. *J. Plant Growth Regul.* **2018**, *37*, 911–924. [CrossRef]
23. Wang, Y.S.; Tian, S.P.; Xu, Y. Effects of high oxygen concentration on pro- and anti-oxidant enzymes in peach fruits during postharvest periods. *Food Chem.* **2005**, *91*, 99–104. [CrossRef]
24. Dhindsa, R.S.; Plumb-Dhindsa, P.; Thorpe, T.A. Leaf senescence: Correlated with increased levels of membrane permeability and lipid peroxidation, and decreased levels of superoxide dismutase and catalase. *J. Exp. Bot.* **1981**, *32*, 93–101. [CrossRef]
25. Zhang, Q.M.; Zhu, L.S.; Wang, J.; Xie, H.; Wang, J.H.; Han, Y.N.; Yang, J.H. Oxidative stress and lipid peroxidation in the earthworm Eisenia fetida induced by low doses of fomesafen. *Environ. Sci. Pollut. Res.* **2013**, *20*, 201–208. [CrossRef] [PubMed]
26. Li, C.Y.; Li, J.; Yan, S.L.; Wang, Q.Z. Effects of ascorbic acid treatment on enzymatic browning and quality characteristics of fresh lotus rhizome juice. *J. Chin. Inst. Food Sci. Technol.* **2021**, *21*, 151–158.
27. Alothman, M.; Kaur, B.; Fazilah, A.; Bhat, R.; Karim, A.A. Ozone-induced changes of antioxidant capacity of fresh-cut tropical fruits. *Innov. Food Sci. Emerg.* **2010**, *11*, 666–671. [CrossRef]
28. Huan, C.; Han, S.; Jiang, L.; An, X.J.; Yu, M.L.; Xu, Y.; Ma, R.J.; Yu, Z.F. Postharvest hot air and hot water treatments affect the antioxidant system in peach fruit during refrigerated storage. *Postharvest Biol. Technol.* **2017**, *126*, 1–14. [CrossRef]
29. Gao, H.; Lu, Z.M.; Yang, Y.; Wang, D.N.; Yang, T.; Cai, M.M.; Cao, W. Melatonin treatment reduces chilling injury in peach fruit through its regulation of membrane fatty acid contents and phenolic metabolism. *Food Chem.* **2018**, *245*, 659–666. [CrossRef]
30. Wang, L.; Shan, T.M.; Xie, B.; Ling, C.; Shao, S.; Jin, P.; Zheng, Y.H. Glycie betaine reduces chilling injury in peach fruit by enhancing phenolic and sugar metabolisms. *Food Chem.* **2019**, *272*, 530–538. [CrossRef]
31. Choi, H.R.; Jeong, M.J.; Baek, M.W.; Choi, J.H.; Lee, H.C.; Jeong, C.S.; Tilahun, S. Transcriptome analysis of pre-storage 1-MCP and high CO_2-treated 'Madoka' peach fruit explains the reduction in chilling injury and improvement of storage period by delaying ripening. *Int. J. Mol. Sci.* **2021**, *22*, 4437. [CrossRef] [PubMed]

32. Nie, X.; Zhao, Y.; Li, Y.; Lv, H. 1-Methylcyclopropylene and heat treatment alleviate chilling injury in purple sweet potato by regulating ROS homeostasis. *Sci. Hortic.* **2024**, *324*, 112606. [CrossRef]
33. Hasanuzzaman, M.; Bhuyan, M.H.; Zulfiqar, F.; Raza, A.; Mohsin, S.M.; Mahmud, J.A.; Fujita, M.; Fotopoulos, V. Reactive oxygen species and antioxidant defense in plants under abiotic stress: Revisiting the crucial role of a universal defense regulator. *Antioxidants* **2020**, *9*, 681. [CrossRef]
34. Zhang, G.; Yan, Z.M.; Wang, Y.H.; Feng, Y.N.; Yuan, Y.N. Exogenous proline improve the growth and yield of lettuce with low potassium content. *Sci. Hortic.* **2020**, *271*, 109469. [CrossRef]
35. Abdallah, M.M.S.; El Sebai, T.N.; Ramadan, A.A.E.M. El-Bassiouny, H.M.S. Physiological and biochemical role of proline, trehalose, and compost on enhancing salinity tolerance of quinoa plant. *Bull. Natl. Res. Cent.* **2020**, *44*, 275–282. [CrossRef]
36. Singh, R.; Jha, A.B.; Misra, A.N.; Sharma, P. Entrapment of enzyme in the presence of proline: Effective approach to enhance activity and stability of horseradish peroxidase. *3 Biotech* **2020**, *10*, 155. [CrossRef]
37. Funck, D.; Baumgarten, L.; Stift, M.; Wirén, N.; Schönemann, L. Differential contribution of P5CS isoforms to stress tolerance in arabidopsis. *Front. Plant Sci.* **2020**, *11*, 565134. [CrossRef] [PubMed]
38. Fabro, G.; Cislaghi, A.P.; Condat, F.; Borau, G.D.; Alvarez, M.E. The N-terminal domain of arabidopsis proline dehydrogenase affects enzymatic activity and protein oligomerization. *Plant Physiol. Biochem.* **2020**, *154*, 268–276. [CrossRef]
39. Koenigshofer, H.; Loeppert, H.G. The up-regulation of proline synthesis in the meristematic tissues of wheat seedlings upon short-term exposure to osmotic stress. *J. Plant Physiol.* **2019**, *237*, 21–29. [CrossRef] [PubMed]
40. Liu, H.; Pei, H.H.; Jiao, J.Q.; Jin, M.J.; Li, H.; Zhu, Q.G.; Ma, Y.P.; Rao, J.P. 1-methylcyclopropene treatment followed with ethylene treatment alleviates postharvest chilling injury of 'Xuxiang' kiwifruit during low-temperature storage. *Food Control* **2021**, *130*, 108340. [CrossRef]
41. Putnam, C.D.; Arvai, A.S.; Bourne, Y.; Tainer, J.A. Active and inhibited human catalase structures: Ligand and NADPH binding and catalytic mechanism. *J. Mol. Biol.* **1999**, *296*, 295–309. [CrossRef] [PubMed]
42. Yao, M.M.; Ge, W.Y.; Zhou, Q.; Zhou, X.; Luo, M.L.; Zhao, Y.B.; Wei, B.D.; Ji, S.J. Exogenous glutathione alleviates chilling injury in postharvest bell pepper by modulating the ascorbate-glutathione (AsA-GSH) cycle. *Food Chem.* **2021**, *352*, 129458. [CrossRef] [PubMed]
43. Kumari, P.; Kumar, V.; Kumar, R.; Pahuja, S.K. Sorghum polyphenols: Plant stress, human health benefits, and industrial applications. *Planta* **2021**, *254*, 47. [CrossRef] [PubMed]
44. Kim, D.O.; Lee, C.Y. Comprehensive study on vitamin C equivalent antioxidant capacity (VCEAC) of various polyphenolics in scavenging a free radical and its structural relationship. *Crit. Rev. Food Sci.* **2004**, *44*, 253–273. [CrossRef] [PubMed]

horticulturae

MDPI

Article

Non-Destructive Quality Estimation Using a Machine Learning-Based Spectroscopic Approach in Kiwifruits

Georgios Tziotzios [1], Xanthoula Eirini Pantazi [1], Charalambos Paraskevas [1], Christos Tsitsopoulos [1], Dimitrios Valasiadis [2], Elpida Nasiopoulou [2], Michail Michailidis [2,*] and Athanassios Molassiotis [2]

1 Laboratory of Agricultural Engineering, Faculty of Agriculture, Aristotle University of Thessaloniki, 54124 Thessaloniki, Greece; gtziotzio@agro.auth.gr (G.T.); xpantazi@agro.auth.gr (X.E.P.); paraskevasb@agro.auth.gr (C.P.); ctsitsop@agro.auth.gr (C.T.)

2 Laboratory of Pomology, Department of Horticulture, Aristotle University of Thessaloniki, 57001 Thessaloniki, Greece; dimitrisvala@gmail.com (D.V.); elpinasi@agro.auth.gr (E.N.); amolasio@agro.auth.gr (A.M.)

* Correspondence: msmichai@agro.auth.gr

Abstract: The current study investigates the use of a non-destructive hyperspectral imaging approach for the evaluation of kiwifruit cv. "Hayward" internal quality, focusing on physiological traits such as soluble solid concentration (SSC), dry matter (DM), firmness, and tannins, widely used as quality attributes. Regression models, including partial least squares regression (PLSR), bagged trees (BTs), and three-layered neural network (TLNN), were employed for the estimation of the above-mentioned quality attributes. Experimental procedures involving the Specim IQ hyperspectral camera utilization and software were followed for data acquisition and analysis. The effectiveness of PLSR, bagged trees, and TLNN in predicting the firmness, SSC, DM, and tannins of kiwifruit was assessed via statistical metrics, including R squared (R^2) values and the root mean square error (RMSE). The obtained results indicate varying degrees of efficiency for each model in predicting kiwifruit quality parameters. The study concludes that machine learning algorithms, especially neural networks, offer substantial accuracy, surpassing traditional methods for evaluating kiwifruit quality traits. Overall, the current study highlights the potential of such non destructive techniques in revolutionizing quality assessment during postharvest by yielding rapid and reliable predictions regarding the critical quality attributes of fruits.

Keywords: hyperspectral imaging; artificial intelligence; regression; machine learning; soluble solids; firmness; VisNIR; spectroscopy

Citation: Tziotzios, G.; Pantazi, X.E.; Paraskevas, C.; Tsitsopoulos, C.; Valasiadis, D.; Nasiopoulou, E.; Michailidis, M.; Molassiotis, A. Non-Destructive Quality Estimation Using a Machine Learning-Based Spectroscopic Approach in Kiwifruits. *Horticulturae* **2024**, *10*, 251. https://doi.org/10.3390/horticulturae10030251

Academic Editors: Alberto Pardossi and Adriana F. Sestras

Received: 20 December 2023
Revised: 30 January 2024
Accepted: 4 March 2024
Published: 6 March 2024

1. Introduction

Kiwifruit is harvested when it reaches a physiologically mature stage, but is not fully ripe, with a soluble solid content (SSC) higher than 6.25%. Consequently, it should be preserved to undergo the process of ripening before it is suitable for consumption [1]. One of the consistent quality traits of kiwifruit concerns its dry matter (DM), which remains constant during ripening and only sustains minor losses owing to transpiration or/and respiration [2]. Typically, DM is determined by employing ovens to remove the moisture content, and calculating the ratio (%) between the dry and fresh weights [3]. The "Hayward" kiwifruit places great importance on pericarp firmness as a key quality attribute after harvest. A minimum value of 20 N has been established for the transportation and wholesale of fruit, while for retail sale and direct consumption, the desired value of firmness is considered to be 10 N [4]. The consumption of high-firmness kiwifruit has been associated with greater astringency [5], which is mainly due to the elevated tannin content, and thus affects the aftertaste intensity of kiwifruit [6].

Assessing the qualitative characteristics of kiwifruit before it reaches the stage of ripeness for consumption is highly valuable [7]. An accurate estimation of internal properties, including dry matter (DM) and soluble solid concentration (SSC), under real-time conditions belongs to crucial factors often related to quality and customer choice [8,9]. The traditional measurement analysis of these properties can reflect the quality of the fruit. However, this type of analysis is labor-intensive and requires destruction of the fruit [10]. Spectral and hyperspectral imaging belong to advanced non-destructive technologies that have drawn a lot of interest in the last few decades for their potential to measure fruit quality attributes [10]. These methods are more efficient, user-friendly, and reliable in postharvest applications compared to conventional approaches [11]. Particularly, visible/near-infrared (Vis/NIR) spectroscopy is a nondestructive analytical technique that appears promising and does not require pre-sample preparation for quality assessment. The Vis/NIR spectroscopy technique has been utilized to evaluate the qualitative attributes of various fruits, such as apples, citrus fruits, and kiwifruits [12]. Non-destructive techniques like NIR, along with hyperspectral imaging, have been extensively employed to assess the qualitative characteristics of kiwifruit, such as firmness, pH, soluble solid content (SSC), and dry matter (DM) [2,13–22]. The integration of Vis/NIR into machine learning (ML) algorithms enhances the effectiveness of learning, estimating, predicting, and classifying crucial quality parameters [23]. Their effectiveness lies in their capability of extracting targeted information from the investigated dataset, securing fruit sustainability and resource consumption that comply with the fundamental principles of precision agriculture.

Moreover, artificial intelligence (AI) takes advantage of its computational potential, enabling the successful modeling of quality properties in fruits and vegetables [24], providing constant qualitative and quantitative tools for the assessment of several fruit profiles. The integration of both machine learning and visible/near-infrared spectroscopy has been utilized in the analysis of various fruits, particularly those with elevated levels of glucose, fructose, vitamins, and other vital nutrients [25]. Partial least square regression (PLSR) was used to predict the internal quality of *Actinidia chinensis* var. *deliciosa* (A. Chev.) cv. "Hayward" using Vis/NIR spectroscopy [26]. On the other hand, PLSR is known for its exceptional reliability and durability when it comes to creating models for SSC and other internal quality traits [17].

The current study aims to predict the quality traits of kiwifruit of individual fruit cv. "Hayward", including firmness, tannins, SSC, and DM, with the help of hyperspectral data obtained from the kiwifruit surfaces. Three different regression analysis methods were used for the assessment of the four investigated quality traits. The employed regression models aimed to extract possible correlations between the hyperspectral data and each of the four investigated quality parameters to enable predictions regarding the tested parameters. The overall results documented that the PLSR, bagged trees, and TLNN models may reflect the prediction results for firmness, SSC, DM, and tannins in individual kiwifruit during postharvest ripening.

2. Materials and Methods

2.1. Plant Material and Sampling Process

Kiwifruit (*Actinidia chinensis* var. *deliciosa* (A. Chev.) A. Chev. "Hayward") was harvested in early November (1–3) 2020 from 20 commercial orchards (from the area of Imathia, Northern Greece), and was immediately transferred to the Pomology Laboratory of the Aristotle University of Thessaloniki (AUTh). Regarding the applied sampling procedure, kiwifruit samples that demonstrated uniform appearance (with no defects) and weight were selected from each orchard to form a representative sample dataset. The samples were placed into 16-slot wooden shipping crates (the fruit and crates were marked) and stored at 0 °C (RH 95%) for 40 days, accomplishing a high variation in the tested kiwifruit quality traits. Afterward, kiwifruit was delivered to the Agricultural Engineering Laboratory of AUTh and the acquisition of the hyperspectral data was carried out on the same day at 20 °C. Each spectral shot was taken by placing a crate under halogen lamps Each spectral shot was acquired by manually placing the crate stationary under the halogen lamp at a distance of 60 cm from the camera. Each shot required 3 mins to be acquired due to lower artificial light intensity compared to natural sunlight. Following the camera shot, firmness and dry matter (DM) were destructively determined in individual fruit, and then each marked fruit was sampled with liquid nitrogen and stored at −80 °C for further analysis.

2.2. Hyperspectral Data Acquisition

For data acquisition, a Specim IQ (Specim Ltd., Oulu, Finland) hyperspectral camera was used. The camera was portable and easy to carry since it only weighed 1.3 kg. There was also the possibility mounting it on a tripod to take hyperspectral images of objects at a distance that exceeds 150 mm. In the current study, the camera was placed on a tripod at a static position, following the Specim IQ standards for spectral data acquisition and following the Specim's calibration procedure. It also provided the ability to capture data in the visible and near-infrared (Vis-NIR) spectra for 204 spectral wavelengths between 400 and 1000 nm. The device had a touch-screen and could be connected to a computer via a USB cable or a wireless Wi-Fi network; therefore, it was possible to immediately evaluate the quality. Principal component analysis (PCA) was performed on the attained "Hayward" kiwifruit Vis/NIR spectral data so as to extract the most relevant features from the acquired dataset. The extracted features denoted linear combinations of the original spectral bands that were captured through Specim IQ hyperspectral camera, conveying the most important characteristics of the attained hyperspectral data. This data reduction technique, implemented under the general framework of the data preprocessing procedure, was a crucial step for forming a more manageable dataset to be fed as input to the three employed regression models for the prediction of four investigated qualities in kiwifruit.

Hyperspectral captures were acquired using the default recording mode (DRM) option, in which raw data and reflection data are stored without being subjected to further processing by the device software. The device captures the absolute black (dark reference) with the shutter closed, and then the data are downloaded. The white reference confirmation was performed using simultaneous shooting by placing the special white reference panel provided by the manufacturer with the camera [27].

Image extraction was carried out with the help of specialized software, Specim IQ Studio version 2019.05.29.2, provided by the manufacturer of the hyperspectral camera. The software provides the ability to display the relevant downloads and reflectance data, create data libraries, and create groups in classification models. It also provides the ability to transfer the images from the camera environment to the computer environment.

2.3. Kiwifruit Quality Trait Determination

The kiwifruit firmness of pericarp was determined in each marked fruit using the Texture Analyzer TA XT2i (Stable Microsystems, Godalming, Surrey, UK), as previously described [28]. Initially, the peel was removed on two opposite sides at a depth of 1 mm; then, a steel cylinder with a 7.9 mm diameter (flat probe end) that was fitted in the machine branch was inserted into the pericarp (1 cm) at a speed of 20 mm/s. The results were expressed in Newton (N). After drying each marked fruit to a constant weight at 67 °C, the dry matter (DM) content (%) of a 5 mm-thick equatorial slice was found by dividing the fresh weights by the dried weights. The soluble solid concentration (SSC, %) was assessed using the juice of each marked fruit employing a digital refractometer (Atago PR-1, Atago Co., Ltd., Tokyo, Japan). The polyphenolic substances were extracted using 70% acetone and 0.5% acetic acid. Kiwifruit pericarp was ground up and added to the extraction solution at a ratio of 1:10 [29]. Tannins were determined using the Folin–Ciocalteu method, with some modifications [30]. Absorbance was measured at 760 nm on a microplate reader (Tecan infinite M200 PRO). The polyphenolic extract was incubated with polyvinyl polypyrrolidone (PVPP) to create a tannin–PVPP complex at the following ratio of extraction: PVPP, 30:1, v/w, and an absorbance at 760 nm of the residual phenolic solution was subtracted from that of total phenols. The results were expressed in equivalents of mg gallic acid kg^{-1} fresh weight (FW) [31].

2.4. Hyperspectral Visualization Software

The spectral data were entered into the Scyven program, which is a high-spatial (HS) image processing tool. The program has the capability to import and modify HS-type files (.hsz, .hdr, .tif) such as the Specim IQ downloads [32]. The Recover Reflectance option was utilized in the program to provide the reflectance data. Subsequently, the image label tool (specifically, the polygon tool) was employed to identify regions that matched the surface of each kiwifruit (Figure 1a,b).

Figure 1. (**a**) Acquisition of RGB images. (**b**) Polygon selection attained from the surface of the 8 + kiwifruit.

2.5. Hyperspectral Data Preprocessing

The attained hyperspectral data were exported to a Comma Separated Values, CSV-type file and then imported into the Excel environment where they were consolidated into an xlsx type file. A total of 104,394 spectral signatures were exploited for 110 different examined kiwifruits. During data import into the Matlab environment, data with zero values were removed, while missing reflectance values were replaced with the nearest acceptable value.

To avoid any potential issues and abnormalities that may occur during image acquisition that are often responsible for difficulties in analysis and misleading results, it was essential to ensure that data availability was of high quality to be subjected to further analysis. For this reason, one of the very well-known smoothing methods, called the Saviztky–Golay [33], was applied to the attained hyperspectral data (Figure 2).

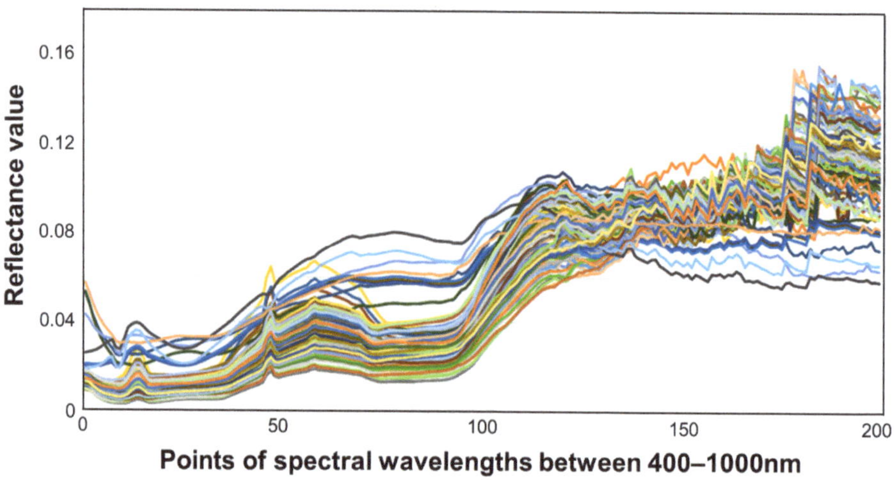

Figure 2. Raw reflectance values per kiwifruit at each spectral point between 400–1000 nm.

This technique belongs to a common technique that is primarily based on the selection of sub-windows around a specific point, and the subsequent estimation of the points of the sub-window projection onto a polynomial fitting. Missing reflectance values were replaced with the nearest acceptable value. Moreover, outliers were removed with the help of the outlier function which removes values that are more than three standard deviations from the data dimension to secure and enhance the reliability and integrity of the investigated dataset. A 10-fold cross-validation was performed by splitting the dataset into 10 subsets, with each subset being further divided into training–validation–testing sets at 70%, 20%, and 10% respectively.

2.6. Partial Least Square Regression (PLSR)

PLSR is performed with the plsregress function in Matlab in the general form given as follows:

[Xloadings,Yloadings,Xscores,Yscores,beta,pctVar,mse,stats] = plsregress(X,y,n)

where **X** represents the dependent variables matrix, **y** is the independent variable matrix, **n** denotes the number of components, **Xloadings** denotes the matrix with coefficients that define linear combinations of the components and simulate the initial data **X**, **Yloadings** is the table with coefficients that define linear combinations of the components and simulate the initial data **y**, **Xscores** denotes the table where each row corresponds to an observation of the **X** table and each column to a component, **Yscores** represents the table where each row corresponds to an observation of the y table and each column to a component, **beta** is the table with the PLS coefficients, **pctVar** denotes the table containing one row with the percentage of variation (variance) of **X** and one row with the percentage of variation of y, **mse** is the table with the mean squared error (mean squared error) for the values of **X**, **y**, and **stats** represents the statistics of the model with values of the weights (weights) and the R^2 error for the values of **Xscores**.

For carrying out the analysis, a high number of components ($n > 40$) were initially selected to find the percentage of the model that is successfully predicted according to the number of components used by employing the **pctVar**.

2.7. Bagged Trees Regression

The bagged trees algorithm [34], also known as bootstrap aggregating or bagging, is a powerful machine learning technique designed to improve the predictive performance and robustness of decision trees. Its function is characterized by the construction of multiple decision tree models, each being trained on a subset of training data derived by repeated random sampling from the original training data, the samples being randomly selected with replacement. By aggregating the predictions of these individual trees, typically through "majority voting" for classification or averaging for regression problems, bagged trees reduce the risk of overtraining, improve model generalization, and increase overall accuracy. Additionally, this ensemble method leverages the diversity of individual trees to capture various patterns and noise in the data, resulting in a more robust and stable predictive model that is less sensitive to variations in the training data.

2.8. Three-Layered Neural Network (TLNN)

A three-layer neural network is a machine learning model that consists of an input layer, an intermediate layer, and an output layer [35]. Each layer contains neurons that are connected to each other. At the input layer, the input data access the network, then the information is transferred to the intermediate layer, where many parallel processes are performed, such as calculating weights and applying activation functions. The output layer extracts the final prediction or output of the model. Training the three-layer neural network consists of optimizing connection weights based on the training data, allowing it to learn complex relationships and patterns in the data. The structure of this three-layer neural network allows for an efficient representation and extraction of features from data, making it ideal for many applications such as pattern recognition, classification, and prediction. Layers with 10 parallel neurons per layer and ReLu activation function [36] were used in the model.

3. Results

Kiwifruit quality attributes, such as firmness (determined in kilograms), soluble solid concentration (SSC, determined as a percentage), dry matter (DM, determined as a percentage), and tannins (determined in milligrams per gram), were determined for 10 marked fruits from 20 different orchards after 40 days of cold storage at 0 °C (RH 95%). The kiwifruit quality feature measurements are demonstrated in Table 1, while the kiwifruit quality measurements of each orchard are provided in Supplementary Table S1. Concisely, the mean kiwifruit firmness was 37 N, with the lowest and highest values being 4 and 66 N, respectively. The mean SSC was 12.7%, ranging from a minimum of 9.7% to a maximum of 15.7%. The average DM value was 17.6%, the lowest value was 14.1%, and the highest value was 23.5%. The mean tannin content was determined to be 198 mg kg^{-1}, ranging from a minimum of 0 to a maximum of 562 mg kg^{-1}.

Table 1. Kiwifruit quality extracted features.

	Firmness (N)	SSC (%)	DM (%)	Tannins (mg/kg)
Mean	36.9	12.68	17.59	198
Median	37.7	12.60	17.29	192
Min. Value	4.4	9.70	14.11	0
Max. Value	65.7	15.70	23.50	562
Range	61.3	6.00	9.39	562
SD	14	1.25	1.93	116

The regression analysis regarding kiwifruit firmness, SCC, DM, and tannins was performed by applying three different regression techniques, including partial least squares regression (PLSR), bagged trees, and TLNN (Table 2). The models were evaluated by using R^2 (coefficient of determination) and the root mean square error (RMSE). R^2 is generally used for determining the correlation of the variance in the dependent variable with its corresponding independent variables. It takes values between 0 and 1, where higher values indicate better performance. A value of 1 signifies the model's optimal prediction capability.

Table 2. Regression analysis results for firmness, SSC, DM, and tannins using the PLSR, bagged trees, and TLNN algorithm models.

		PLSR	Bagged Trees	TLNN
Firmness (n^* = 28)	R^2	0.90	0.93	0.97
	RMSE	0.3419	0.3486	0.2819
SSC (n^* = 30)	R^2	0.90	0.91	0.91
	RMSE	0.3479	0.3700	0.3651
DM (n^* = 19)	R^2	0.90	0.87	0.93
	RMSE	0.3685	0.7244	0.5181
Tannins (n^* = 32)	R^2	0.91	0.88	0.94
	RMSE	0.3418	0.4141	0.2819

n^* indicates the number of PLS components.

On the other hand, the RMSE denotes the square root of the mean squared error (MSE), which represents the average squared deviation between the predicted and actual values of the data. An RMSE equal to 0 indicates a perfect prediction, so values closer to 0 indicate better performance. As seen in Table 2, the method showcased distinct performances based on the R squared (R^2) values and root mean square error (RMSE).

For each of the applied ML models, the relevant linear regression equation was extracted, demonstrating that the mathematical relationship between the predicted and observed values for the four investigated quality traits was demonstrated (Table 3). The linear regression equations hold the typical form, given as follows (Equation (1)):

$$Y = m \times x + b \quad (1)$$

where m denotes the slope, indicating the rate at which each of the predicted quality traits changes with respect to changes in the observed variable; while b denotes the intercept, representing the predicted value when the observed variable is zero.

Table 3. Linear regression equations between the predicted and observed values regarding firmness, SSC, dry matter, and tannin content prediction derived from the PLSR, bagged trees, and TLNN models.

	Firmness	SSC	Dry Matter	Tannin Content
PLSR	y = 0.9005x + 0.3667	y = 0.9034x + 1.225	y = 0.9047x + 1.676	y = 0.9071x + 0.01835
Bagged Trees	y = 0.8884x + 0.4081	y = 0.872x + 1.665	y = 0.8012x + 3.511	y = 0.7691x + 0.04888
TLNN	y = 0.9706x + 0.07884	y = 1.002x − 0.02039	y = 1.001x − 0.009929	y = 1.004x + 0.005643

Moreover, statistical tests on the intercept and slope were performed to evaluate the prediction adaptability and consistency of each of the employed MLs with the actual observed values (Table 4).

Table 4. Results of the comparison of each linear regression with the corresponding 1:1 line. The n.s. symbol denotes statistically non-significant differences, and the * and ** symbols represent statistically significant differences between the lines, at $p < 0.05$ and 0.01, respectively.

	Firmness		SSC		Dry Matter		Tannin Content	
	Comparison with a 1:1 Line		Comparison with a 1:1 Line		Comparison with a 1:1 Line		Comparison with a 1:1 Line	
	Slope	Intercept	Slope	Intercept	Slope	Intercept	Slope	Intercept
PLSR	ns	**	ns	**	*	**	ns	**
Bagged Trees	*	**	*	**	ns	**	ns	**
TLNN	ns	**	**	**	**	**	*	**

For each combination consisting of one of the investigated traits and one of the three employed ML models, the fitted regression line was compared with the 1:1 line in order to explore statistically significant differences with respect to the slope and intercept. At this point, it is worthy to note that a non-significant intercept or slope may suggest the model's capacity to capture an essential aspect of the relationship between the observed and predicted variables. Among all three applied ML regression models, the PLSR model demonstrated statistically significant relationships ($p < 0.05$) between firmness and all predictors, while the bagged trees model showed significant relationships for firmness, dry matter, and tannin content, and marginally significant for soluble solid content. The TLNN indicated significant associations only between firmness and tannin content, with all other predicted values being statistically non-significant (Table 4).

3.1. Kiwifruit Firmness Analysis

By analyzing the percentage of variance regarding firmness prediction, it is evident that the initial 10 components did not provide enough information to the model. These components only account for up to 50% of the data, as demonstrated in Figure 3a. Conversely, it was noted that 28 components were successfully predicted with a considerably high accuracy that exceeds 90% for all three utilized models, namely PLSR, bagged trees, and TLNN (Table 2). It is also worthy noting that the model's performance tends to stabilize for a larger number of components, while there is a slight increase in the prediction rate. This behavior can be attributed to the employed model's ability to capture some additional variations in the hyperspectral data that have not been fully accounted for with fewer components (Figure 3a).

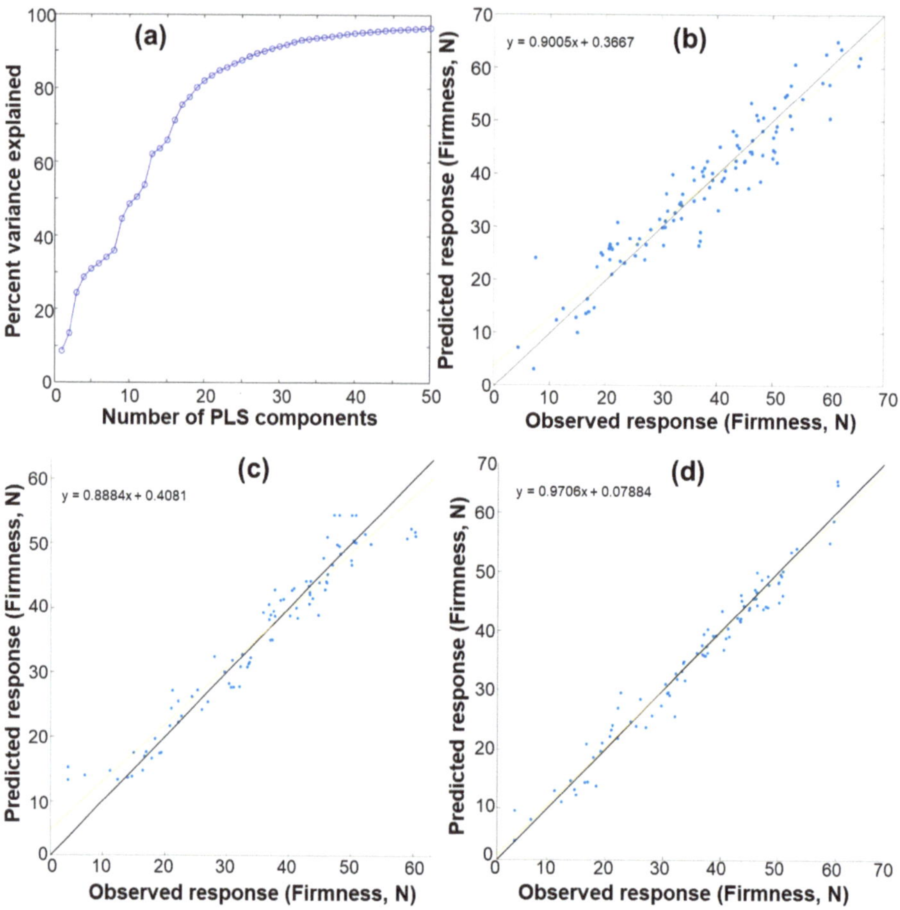

Figure 3. Firmness prediction results regarding variance. (**a**) Regression scatter plots comparing the predicted vs. observed values for firmness derived from the PLSR. (**b**) Bagged trees (**c**) and TNN (**d**) models. The black line in the graphs (**b–d**) denotes the theoretical 1:1 line, while the yellow line denotes the fitted regression line.

The PLSR model exhibited a strong performance with an R^2 of 0.90, signifying that approximately 90% of the variance in the dependent variable could be explained by the model (Figure 3b). Nevertheless, the root mean square error (RMSE) of 0.3419 signifies a substantial degree of residual error, implying a deviation between the expected outcomes and the actual observed values.

On the other hand, bagged trees demonstrated a higher R^2 of 0.93, indicating a better fit to the hyperspectral data than PLSR. Despite this, its RMSE equal to 0.3486 implies slightly higher error residuals compared to the PLSR, indicating a slightly less accurate predictive performance on average (Figure 3c and Table 2). However, the most notable performance was observed in the case of TLNN, which demonstrated an R^2 of 0.97. This high value indicates an extremely strong correlation between the predicted values and the target value, capturing 97% of the variance (Table 2). Moreover, the remarkably low RMSE of 0.246 signifies highly accurate predictions, indicating minimal deviation between the predicted and observed values. The aforementioned results emphasize the enhanced performance of the TLNN model in accurately representing the intricate connections within the kiwifruit data (Figure 3d). The PLSR and bagged trees models demonstrated satisfactory performances, but overall, the ANN model stands out among the other models for its remarkable predictive capacity and minimum error. This makes it a highly suitable and robust choice for predicting and modeling kiwifruit-related properties, as illustrated in Figure 3 and Table 2.

3.2. Soluble Solid Concentration (SSC) Results

Based on the variance percentage explained versus the number of components (n) regarding the SSC, it was noted that the first 12 components were proven insufficient for accurately predicting the model as they explain up to 50% of the data. When "n" is more than 30, the accuracy of the prediction increases over 90%, as demonstrated in Figure 4a. It was also observed that the curve tends to stabilize for a higher number of components, which slightly increases the prediction performance.

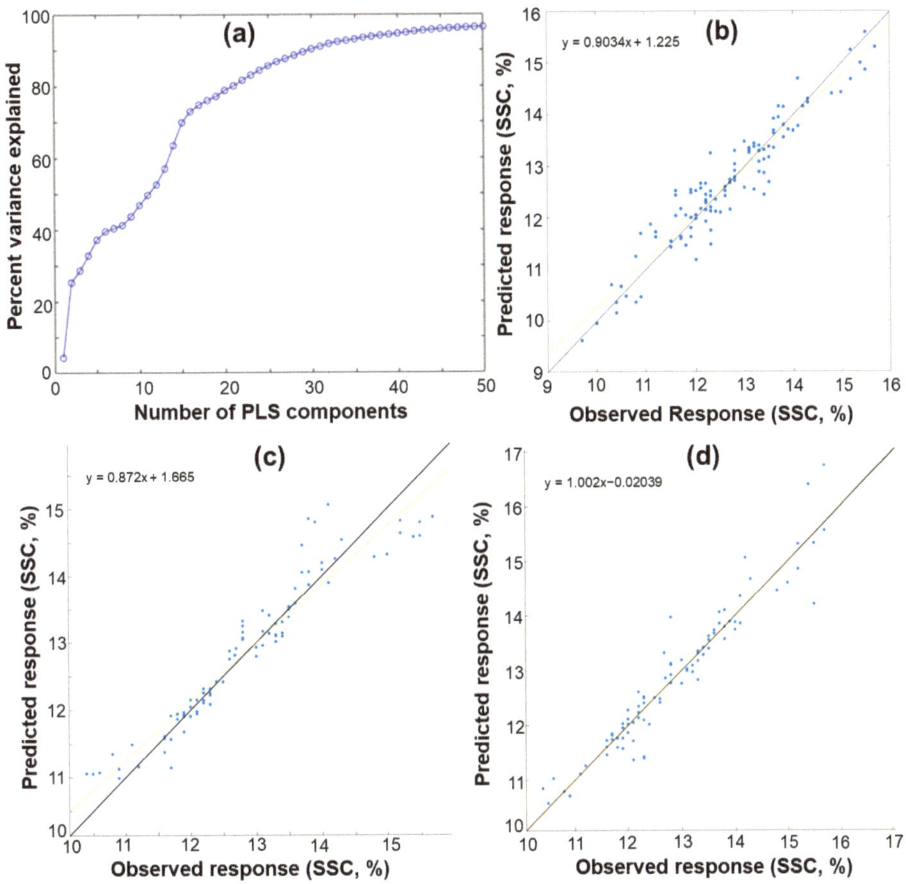

Figure 4. Soluble solid concentration (SSC) prediction results regarding variance. (**a**) Regression scatter plots comparing the predicted vs. observed values for SCC derived from the PLSR; (**b**) bagged trees algorithm; and (**c**) TLNN algorithm (**d**) models. The black line in the graphs (**b–d**) denotes the theoretical 1:1 line, while the yellow line denotes the fitted regression line.

The analysis of kiwifruit's soluble solid concentration using different regression techniques revealed interesting findings, as demonstrated in Table 2. Each employed method—PLSR, bagged trees, and TLNN—showed varying performances based on their R squared (R^2) values and RMSE. The PLSR displayed a strong performance with an R^2 of 0.90, indicating that approximately 90% of the variation in the dependent variable was accounted for by the model. However, it also showed a moderate RMSE of 0.3479, signifying some deviation between the predicted and actual values (Figure 4b and Table 2).

Both bagged trees and ANN models exhibited an R^2 value of 0.91, indicating a strong correlation between the predictors and the target variable. This implies that these models are able to explain 91% of the variance in the data. However, bagged trees had an RMSE of 0.3700, slightly higher than the PLSR, while TLNN showed an RMSE of 0.3651, indicating a marginally less accurate prediction on average compared to the PLSR (Table 2). In predicting the soluble solid concentration of kiwifruit, all three models—PLSR, bagged trees, and TLNN—performed well, explaining a significant portion of the variation. The PLSR seemed to have a slight edge in predictive accuracy, closely followed by bagged trees and ANN (Figure 4c,d and Table 2).

3.3. Dry Matter (DM) Results

Based on the DM variance percentage depicted in Figure 5a, it was noticed that the first five components predict up to 70% of the model. Moreover, the curve shows a gradual increase and for $n > 19$, the prediction is greater than 90%.

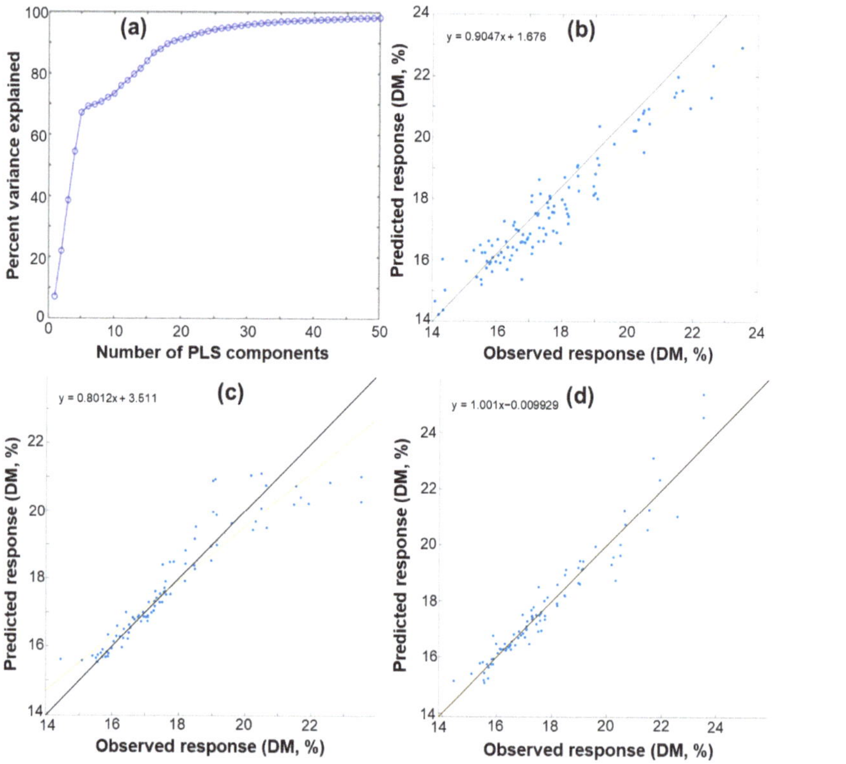

Figure 5. Dry matter (DM) content regression results regarding variance. (**a**) Regression scatter plots comparing the predicted vs. observed values for DM, derived from the PLSR; (**b**) bagged trees; (**c**) and TNN (**d**) models. The black line in the graphs (**b–d**) denotes the theoretical 1:1 line, while the yellow line denotes the fitted regression line.

The current assessment of kiwifruit dry matter using diverse regression methods revealed some interesting information. Each of the three employed approaches demonstrated varying performances, indicated by their relative R^2 and RMSE values. The PLSR model showcased respectable performance with an R^2 of 0.90, signifying that approximately 90% of the variance in the dependent variable was captured by the model. However, it displayed an RMSE of 0.3685, indicating a moderate level of deviation between the predicted and actual values (Figure 5b and Table 2). The bagged trees model exhibited an R^2 value of 0.87, indicating a strong fit to the data. However, the higher RMSE, equal to 0.7244, in comparison to the PLSR model indicates a less precise predictive performance (Figure 5c and Table 2). On the other hand, TLNN exhibited the highest R^2, equal to 0.93, suggesting a robust correlation between the predicted and the target value, capturing 93% of the variance. However, its RMSE of 0.5181, which is lower than that of bagged trees, still indicates a slightly higher level of error compared to the PLSR (Figure 5d and Table 2). For predicting the DM of kiwifruits, each method—PLSR, bagged trees, and TLNN—displayed distinctive performances. In general, the PLSR showcased a strong predictive capacity, followed by TLNN and bagged trees (Figure 5 and Table 2).

3.4. Tannin Results

Upon examination of Figure 6a, it is evident that the PLSR model prediction was observed for values of n less than 20 at a rate that is below 70%. When n is more than 32, the model's successful prediction rate exceeds 90%. The curve appears to reach a stable state as the number of components increases, resulting in a minor improvement in the prediction percentage.

Analysis of tannin levels in kiwifruit using different regression techniques yielded varying results. The effectiveness of each applied strategy was assessed based on their R squared (R^2) values and root mean square error (RMSE), as presented in Table 2.

The PLSR model demonstrated a solid performance with an R^2 of 0.91, capturing about 91% of the variation. Its RMSE was 0.3418, indicating some deviation between the predicted and actual values (Figure 6b and Table 2). Regarding the bagged trees algorithm, it displayed an R^2 equal to 0.88 and an RMSE 0.4141, indicating slightly lower accuracy compared to the PLSR (Figure 6c and Table 2). On the other hand, TLNN showcased the highest R^2 of 0.94, suggesting a strong correlation between the predicted and target variables, capturing 94% of the variance. Its RMSE of 0.2819 indicates a slight deviation between the predicted and observed values (Figure 6d and Table 2). Overall, in the evaluation of tannin levels in kiwifruits, artificial neural networks (ANN) have shown strong predictive capability, with partial least squares regression (PLSR), and bagged trees following closely behind.

Taking all the above into consideration, it is evident that all the utilized regression models demonstrate a significant ability to accurately predict the four investigated quality characteristics. The accuracy of these predictions ranges from 87% to 97%. The diverse regression models exhibited varying performances across the prediction of kiwifruit characteristics. Notably, notable predictions were achieved with the help of the three employed regression models, yielding high and constant predictive performances. Regarding the employed models' efficiency, the PLSR has consistently shown strong correlations, while, on the other hand, demonstrating moderate variations in terms of prediction accuracy. Both bagged trees and TLNN yielded competitive results for certain attributes; however, they presented different levels of accuracy and error rates, recommending that the selection of the appropriate model should consider the trade-offs between predictive power, accuracy, and computational complexity (Figures 3–6 and Table 2).

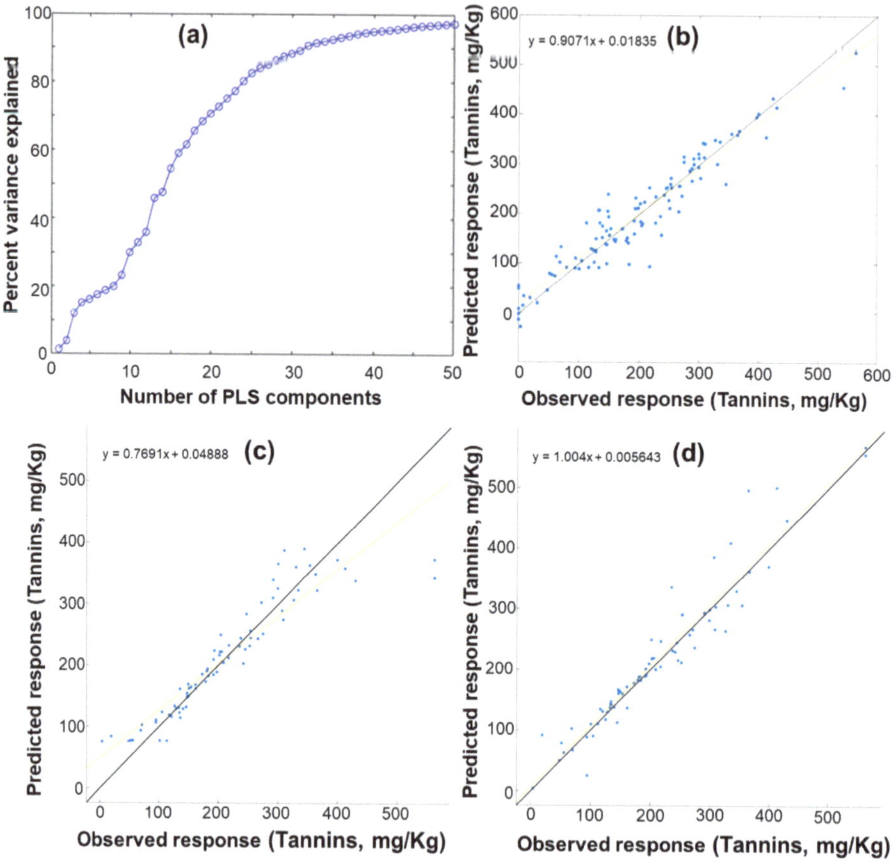

Figure 6. Tannin content regression results regarding variance. (**a**) Regression scatter plots regarding tannin prediction comparing the predicted vs. observed values derived from the (**b**) PLSR; (**c**); bagged trees algorithm; and (**d**) TLNN algorithm models. The black line in the graphs (**b–d**) denotes the theoretical 1:1 line, while the yellow line denotes the fitted regression line.

4. Discussion

After a short period of 40 days of cold storage, four "Hayward" kiwifruit quality traits were destructively tested, namely firmness, soluble solid concentration (SSC), dry matter (DM), and tannin content, from fruit harvested from different orchards using a Specim IQ hyperspectral camera. The firmness of kiwifruit is a critical qualitative characteristic that plays a key role in determining its storage life and commercial viability. This is because the firmness of the fruit affects how long it can be stored after being harvested [37]. Thus, accurately assessing variations in firmness over time may assist in developing effective storage and marketing approaches for kiwifruit [18]. The firmness of kiwifruit can be influenced by various factors, including the mineral composition, particularly the calcium level, which has a significant impact on the inherent quality of the fruit and its ripening process [38]. Furthermore, firmness is strongly influenced by both harvest and postharvest handlings; for example, a dramatic softening could be observed in wounded or even damaged (with no visual symptoms) kiwifruit during postharvest life [28,39]. In the current study, despite the above-mentioned severe difficulties in predicting the firmness of kiwifruits, the applied PLS models achieved optimal accuracies in predicting kiwifruit firmness, ranging from 90 to 97%. It is also worthy mentioning that the light scattering effect caused by the kiwifruit surface and the cell wall structure did not affect the performance of the applied models, since the PCA approach was performed for the attained reflectance data.

Two further significant indexes of kiwifruit quality are the content of soluble solids (SSC) and the proportion of dry matter (DM). Typically, kiwifruits have an SSC greater than 6.2% at the time of harvest [40], while a high dry matter content (over 16%) has been suggested as a predictor of quality in kiwifruit [3]. Furthermore, Hanker et al. [41] have suggested that SSC in ready-to-eat kiwifruits should fall between 10 and 14%. In a recent study, Titeli et al. [6] established an association between elevated levels of DM (dry matter), SSC (soluble solid content), and acidity, and a greater intensity of taste in kiwifruit. They emphasized the importance of these factors in determining the overall eating quality of kiwifruit. Non-destructive methodologies have been widely used to determine the quality features of kiwifruit, including firmness, SSC, and DM [2,13–19]. Therefore, these quality features significantly contribute to the overall quality and postharvest performance of kiwifruit. In the present study, the slightly lower performance in predicting the SSC can be attributed to the spectral range, as the data were acquired covering a spectral area from 400 to 1000 nm. It would be expected that more useful information for SSC prediction would be acquired once the data selection has been conducted within a wider range, covering up to 1450 nm. Since wavelengths at 1198 nm are associated to with C-H single bonds, while the wavelengths near 1450 nm are correlated with O-H bonds [22], this could possibly yield useful information, enhancing the employed models' performance regarding SSC prediction. On the other hand, the employed models' performances achieved for DM were slightly worse compared to the other investigated quality traits, which could be attributed to possible variability in ripeness during the 40 days of cold storage.

Kiwifruit and persimmon fruits have been identified as high sources of tannin [42], and in addition to enhancing the antioxidant capacity of fruit, non-polymerized tannins in both fruits are associated with astringency that affects taste [43,44]. It has been supported that the high aftertaste intensity of kiwifruit was linked to a lower taste intensity [6], indicating a negative quality trait for consumption. Moreover, tannin content in fruit wines has been correlated with bitterness and astringency [45]. Therefore, the level of tannins in kiwifruit may be associated with the strength of the aftertaste, which is an important aspect in determining the overall flavor quality of the fruit. Compared to the DM prediction, the performances of the employed models regarding tannin prediction were slightly better.

The current study offers a comprehensive insight into the application of non-destructive technologies, particularly spectral and hyperspectral imaging, for the assessment of the above-mentioned kiwifruit quality traits. Vis/NIR spectroscopy represents a significant advancement over traditional destructive techniques used to assess internal quality traits such as soluble solid concentration (SSC), dry matter (DM), firmness, and tannins. This transition to non-destructive techniques aligns with the growing need for efficient, reliable, and rapid quality assessment in the postharvest industry [13,14,46–50].

The obtained results underscore the potential of Vis/NIR spectroscopy in accurately predicting the internal quality parameters of kiwifruit, a finding consistent with earlier studies on other fruits [51]. The use of principal component analysis (PCA), partial least square regression (PLSR), and artificial neural networks (ANN) further enhanced the predictive accuracy, each demonstrating unique strengths and limitations [51]. The performance of PLSR, in particular, stood out, demonstrating a strong correlation to the quality parameters while maintaining a balance between predictive power and computational complexity [52]. The TLNN model revealed superior predictive ability, most notably in terms of the R^2 and RMSE values (Table 2), indicating its effectiveness in capturing complex patterns within the data. This highlights the increasing importance of machine learning (ML) techniques in agricultural research which offer a more comprehensive understanding of fruit quality assessment. Future studies should focus on optimizing the employed models for specific quality attributes, considering the trade-offs between accuracy, efficiency, and computational demands.

The bagged trees model has found wide application in various fields due to its ability to improve model performance and reduce the effects of data variability and noise [23]. A minimum number of eight "leafs" (leaf size) was used, i.e., the algorithm takes into account at least eight samples in each decision tree. The number of trees was set equal to 30 "learners" which feed the algorithm with their decisions individually (Table 2).

The partial least squares regression (PLSR) model is well recognized and scientifically validated as a methodology for effectively identifying as well as accurately estimating the internal quality characteristics of kiwifruit [15–17,53]. Moreover, this study serves as a compelling testament to the enhanced efficacy and acumen showcased by neural networks in delivering superior performance when juxtaposed against their traditional statistical counterparts. As such, it was concluded that machine learning algorithms provide sufficient accuracy in predicting kiwifruit characteristics.

The current study involved a thorough examination of internal quality parameters in kiwifruits with the help of three discrete regression models. During these analyses, the reflected data obtained from the surface of kiwifruits underwent extensive analysis (Figure 2). These reflections were then thoroughly compared to the obtained laboratory measurements of four distinct quality characteristics: fruit firmness, SSC, DM, and tannin content (Figures 3–6). The main goal of the regression model was to precisely find and describe the strong correlations between the spectral data in each of the quality attributes (Table 2).

This intricate and nuanced interconnection facilitated the subsequent predictive models for these attributes, thoroughly grounded in the established relationships forged. The incorporation of advanced regression models like PLSR, bagged trees, and TLNN was pivotal in this study. Each model exhibited varying degrees of effectiveness in predicting different quality parameters, with PLSR consistently showing strong correlations across all parameters (Table 2). The TLNN model exhibited outstanding prediction accuracy and minimum error, highlighting the promise of ML approaches in agricultural applications. This conclusion is particularly remarkable (Table 2). The focus on the above critical internal quality parameters has highlighted the capability of artificial intelligence approaches to provide rapid, reliable, and efficient quality assessment, surpassing traditional destructive methods, demonstrating performances ranging from 87% to 97% (Table 2). All this has been achieved by astutely employing a judicious and optimal array of components (max = 32) drawn from a vast gamut of 198 distinct wavelengths examined across the visible and near-infrared spectra. Moreover, the predictive model tailored for determining DW showcased an upsurge in performance, manifesting the remarkable ability to deliver comparable prediction percentages while mandating fewer components, hovering approximately at 2/3 of the previously required quantity (Table 2).

5. Conclusions

The results of the present study confirm that the effective combination of a non-destructive approach using Vis/NIR spectroscopy and machine learning algorithms offers a promising alternative to traditional methods, providing a comprehensive understanding regarding "Hayward" kiwifruit quality assessment, especially for the qualitative firmness, SSC, DM, and tannin content after postharvest cold storage. Among the three employed ML algorithms, namely PLSR, bagged trees, and TLNN, the latter demonstrated the highest prediction capability, demonstrating performances ranging from 87% to 97%. The high efficiency of the TNN model demonstrated the remarkable capabilities in transforming quality assessment processes by offering several advantages such as rapidity, reliability, and efficiency, often surpassing traditional destructive techniques. The ML techniques have not only been proven capable of improving the efficiency of quality evaluation, but also aid in more effective resource allocation and reduction in food waste through precise sorting and grading. The effective integration of spectral imaging and machine learning techniques can revolutionize quality assessment in the postharvest sector. The current approach has significant implications for the entire supply chain, from growers and processors to retailers and consumers, ensuring the provision of high-quality kiwifruits. Subsequent studies should prioritize the enhancement of these models, investigating their suitability for a broader variety of fruits. In addition, it is important to focus on integrating these nondestructive approaches into real-time, on-field quality assessment systems. This integration has the potential to significantly enhance the efficiency and sustainability of postharvest operations.

Supplementary Materials: The following supporting information can be downloaded at: https://www.mdpi.com/article/10.3390/horticulturae10030251/s1, Table S1. Kiwifruit quality traits of the 20 orchards evaluated.

Author Contributions: Conceptualization, X.E.P. and A.M.; methodology, X.E.P.; software, G.T. and X.E.P., validation, X.E.P. and A.M.; formal analysis, G.T., X.E.P. and C.P., investigation, C.P., G.T., M.M., D.V. and E.N.; resources, A.M.; data curation, G.T., X.E.P., C.P. and C.T.; writing—original draft preparation, X.E.P., M.M. and C.T.; writing—review and editing, X.E.P., C.P., M.M. and A.M.; visualization, G.T.; supervision, X.E.P. and A.M.; project administration, A.M.; funding acquisition, A.M. All authors have read and agreed to the published version of the manuscript.

Funding: This work was supported by E.P. Competitiveness, Entrepreneurship & Innovation (EPANEK), within the framework of the national action "RESEARCH-CREATE-INNOVATE" (ESPA 2014–2020) (project code: T2EΔK-03007; Premium Kiwi).

Data Availability Statement: The original contributions presented in the study are included in the article, further inquiries can be directed to the corresponding author.

Conflicts of Interest: The authors declare no conflicts of interest.

References

1. Minas, I.S.; Tanou, G.; Karagiannis, E.; Belghazi, M.; Molassiotis, A. Coupling of physiological and proteomic analysis to understand the ethylene- and chilling-Induced kiwifruit ripening syndrome. *Front. Plant Sci.* **2016**, *7*, 120. [CrossRef]
2. McGlone, V.A.; Kawano, S. Firmness, dry-matter and soluble-solids assessment of postharvest kiwifruit by NIR spectroscopy. *Postharvest Biol. Technol.* **1998**, *13*, 131–141. [CrossRef]
3. Crisosto, G.M.; Hasey, J.; Zegbe, J.A.; Crisosto, C.H. New quality index based on dry matter and acidity proposed for Hayward kiwifruit. *Calif. Agric.* **2012**, *66*, 70–75. [CrossRef]
4. Crisosto, C.H.; Garner, D.; Saez, K. Kiwifruit size influences softening rate during storage. *Calif. Agric.* **1999**, *53*, 29–31. [CrossRef]
5. Stec, M.G.H.; Hodgson, J.A.; Macrae, E.A.; Triggs, C.M. Role of fruit firmness in the sensory evaluation of kiwifruit (Actinidia deliciosa cv Hayward). *J. Sci. Food Agric.* **1989**, *47*, 417–433. [CrossRef]
6. Titeli, V.S.; Michailidis, M.; Tanou, G.; Molassiotis, A. Physiological and Metabolic Traits Linked to Kiwifruit Quality. *Horticulturae* **2023**, *9*, 915. [CrossRef]
7. Burdon, J.N. *Kiwifruit Biology: The Commercial Implications of Fruit Maturation*; Horticultural Reviews Volume 46; Wiley: Hoboken, NJ, USA, 2018; Chapter 8; pp. 385–421. ISBN 9781119521082. [CrossRef]
8. Burdon, J.; McLeod, D.; Lallu, N.; Gamble, J.; Petley, M.; Gunson, A. Consumer evaluation of "Hayward" kiwifruit of different at-harvest dry matter contents. *Postharvest Biol. Technol.* **2004**, *34*, 245–255. [CrossRef]
9. Nardozza, S.; Gamble, J.; Axten, L.G.; Wohlers, M.W.; Clearwater, M.J.; Feng, J.; Harker, F.R. Dry matter content and fruit size affect flavour and texture of novel Actinidia deliciosa genotypes. *J. Sci. Food Agric.* **2011**, *91*, 742–748. [CrossRef]
10. Chandrasekaran, I.; Panigrahi, S.S.; Ravikanth, L.; Singh, C.B. Potential of Near-Infrared (NIR) Spectroscopy and Hyperspectral Imaging for Quality and Safety Assessment of Fruits: An Overview. *Food Anal. Methods* **2019**, *12*, 2438–2458. [CrossRef]
11. Walsh, K.B.; McGlone, V.A.; Han, D.H. The uses of near infra-red spectroscopy in postharvest decision support: A review. *Postharvest Biol. Technol.* **2020**, *163*, 111139. [CrossRef]
12. Hussain, A.; Pu, H.; Sun, D.W. Innovative nondestructive imaging techniques for ripening and maturity of fruits—A review of recent applications. *Trends Food Sci. Technol.* **2018**, *72*, 144–152. [CrossRef]
13. Li, M.; Pullanagari, R.R.; Pranamornkith, T.; Yule, I.J.; East, A.R. Applying visible-near infrared (Vis-NIR) spectroscopy to classify "Hayward" kiwifruit firmness after storage. *Acta Hortic.* **2017**, *1154*, 1–7. [CrossRef]
14. Li, M.; Pullanagari, R.R.; Pranamornkith, T.; Yule, I.J.; East, A.R. Quantitative prediction of post storage 'Hayward' kiwifruit attributes using at harvest Vis-NIR spectroscopy. *J. Food Eng.* **2017**, *202*, 46–55. [CrossRef]
15. Yang, B.; Guo, W.; Li, W.; Li, Q.; Liu, D.; Zhu, X. Portable, visual, and nondestructive detector integrating Vis/NIR spectrometer for sugar content of kiwifruits. *J. Food Process Eng.* **2019**, *42*, e12982. [CrossRef]
16. Ciccoritti, R.; Paliotta, M.; Amoriello, T.; Carbone, K. FT-NIR spectroscopy and multivariate classification strategies for the postharvest quality of green-fleshed kiwifruit varieties. *Sci. Hortic.* **2019**, *257*, 108622. [CrossRef]

17. Tian, S.; Liu, W.; Xu, H. Improving the prediction performance of soluble solids content (SSC) in kiwifruit by means of near-infrared spectroscopy using slope/bias correction and calibration updating. *Food Res. Int.* **2023**, *170*, 112988. [CrossRef]
18. Mohammadi Torkashvand, A.; Ahmadi, A.; Gómez, P.A.; Maghoumi, M. Using artificial neural network in determining postharvest LIFE of kiwifruit. *J. Sci. Food Agric.* **2019**, *99*, 5918–5925. [CrossRef]
19. Valasiadis, D.; Georgios, M.; Michailidis, M.; Polychroniadou, C.; Tanou, G.; Bazakos, C.; Molassiotis, A. Postharvest Biology and Technology Wide-characterization of high and low dry matter kiwifruit through spatiotemporal multi-omic approach. *Postharvest Biol. Technol.* **2024**, *209*, 112727. [CrossRef]
20. Ma, T.; Zhao, J.; Inagaki, T.; Su, Y.; Tsuchikawa, S. Rapid and nondestructive prediction of firmness, soluble solids content, and pH in kiwifruit using Vis–NIR spatially resolved spectroscopy. *Postharvest Biol. Technol.* **2022**, *186*, 111841. [CrossRef]
21. Cevoli, C.; Iaccheri, E.; Fabbri, A.; Ragni, L. Data fusion of FT-NIR spectroscopy and Vis / NIR hyperspectral imaging to predict quality parameters of yellow flesh "Jintao" kiwifruit. *Biosyst. Eng. J.* **2024**, *237*, 157–169. [CrossRef]
22. Ma, T.; Xia, Y.; Inagaki, T.; Tsuchikawa, S. Non-destructive and fast method of mapping the distribution of the soluble solids content and pH in kiwifruit using object rotation near-infrared hyperspectral imaging approach. *Postharvest Biol. Technol.* **2021**, *174*, 111440. [CrossRef]
23. Zhang, M.; Zhang, B.; Li, H.; Shen, M.; Tian, S.; Zhang, H.; Ren, X.; Xing, L.; Zhao, J. Determination of bagged 'Fuji' apple maturity by visible and near-infrared spectroscopy combined with a machine learning algorithm. *Infrared Phys. Technol.* **2020**, *111*, 103529. [CrossRef]
24. Salehi, F. Quality, physicochemical, and textural properties of dairy products containing fruits and vegetables: A review. *Food Sci. Nutr.* **2021**, *9*, 4666–4686. [CrossRef]
25. Borba, K.R.; Spricigo, P.C.; Aykas, D.P.; Mitsuyuki, M.C.; Colnago, L.A.; Ferreira, M.D. Non-invasive quantification of vitamin C, citric acid, and sugar in 'Valência' oranges using infrared spectroscopies. *J. Food Sci. Technol.* **2021**, *58*, 731–738. [CrossRef]
26. Sarkar, S.; Basak, J.K.; Moon, B.E.; Kim, H.T. A comparative study of PLSR and SVM-R with various preprocessing techniques for the quantitative determination of soluble solids content of hardy kiwi fruit by a portable Vis/NIR spectrometer. *Foods* **2020**, *9*, 1078. [CrossRef]
27. Burger, J.; Gowen, A. Data handling in hyperspectral image analysis. *Chemom. Intell. Lab. Syst.* **2011**, *108*, 13–22. [CrossRef]
28. Polychroniadou, C.; Michailidis, M.; Adamakis, I.D.S.; Karagiannis, E.; Ganopoulos, I.; Tanou, G.; Bazakos, C.; Molassiotis, A. Mechanical stress elicits kiwifruit ripening changes in gene expression and metabolic status. *Postharvest Biol. Technol.* **2022**, *194*, 112102. [CrossRef]
29. Asami, D.K.; Hong, Y.J.; Barrett, D.M.; Mitchell, A.E. Comparison of the total phenolic and ascorbic acid content of freeze-dried and air-dried marionberry, strawberry, and corn grown using conventional, organic, and sustainable agricultural practices. *J. Agric. Food Chem.* **2003**, *51*, 1237–1241. [CrossRef]
30. Michailidis, M.; Karagiannis, E.; Polychroniadou, C.; Tanou, G.; Karamanoli, K.; Molassiotis, A. Metabolic features underlying the response of sweet cherry fruit to postharvest UV-C irradiation. *Plant Physiol. Biochem.* **2019**, *144*, 49–57. [CrossRef]
31. Makkar, H.P.S.; Blümmel, M.; Borowy, N.K.; Becker, K. Gravimetric determination of tannins and their correlations with chemical and protein precipitation methods. *J. Sci. Food Agric.* **1993**, *61*, 161–165. [CrossRef]
32. Habili, N.; Oorloff, J. ScyllarusTM: From research to commercial software. *ACM Int. Conf. Proceeding Ser.* **2015**, *28*, 119–122. [CrossRef]
33. Ruffin, C.; King, R.L.; Younan, N.H. A combined derivative spectroscopy and Savitzky-Golay filtering method for the analysis of hyperspectral data. *GIScience Remote Sens.* **2008**, *45*, 1–15. [CrossRef]
34. Rao, J.S.; Tibshirani, R.J. The out-of-bootstrap method for model averaging and selection. *Tech. Rep.* **1997**, 1–23.
35. Wang, C.J.; Wickert, M.A.; Wu, C.H. Three-layer neural networks for spectral estimation. In Proceedings of the ICASSP 90, 1990 International Conference on Acoustics, Speech and Signal Processing, Albuquerque, NM, USA, 3–6 April 1990; Volume 2, pp. 881–884. [CrossRef]
36. Banerjee, C.; Mukherjee, T.; Pasiliao, E. An empirical study on generalizations of the RelU activation function. In Proceedings of the 2019 ACM Southeast Conference, ACM SE '19, Kennesaw, GA, USA, 18–20 April 2019; pp. 164–167. [CrossRef]
37. Feng, J.; MacKay, B.R.; Maguire, K.M. Variation in firmness of packed hayward kiwifruit. *Acta Hortic.* **2003**, *610*, 211–217. [CrossRef]
38. Feng, J.; Maguire, K.M.; MacKay, B.R. Discriminating batches of "Hayward" kiwifruit for storage potential. *Postharvest Biol. Technol.* **2006**, *41*, 128–134. [CrossRef]
39. Polychroniadou, C.; Karagiannis, E.; Michailidis, M.; Adamakis, I.D.S.; Ganopoulos, I.; Tanou, G.; Bazakos, C.; Molassiotis, A. Identification of genes and metabolic pathways involved in wounding-induced kiwifruit ripening. *Plant Physiol. Biochem.* **2022**, *179*, 179–190. [CrossRef]
40. Wang, M.Y.; MacRae, E.; Wohlers, M.; Marsh, K. Changes in volatile production and sensory quality of kiwifruit during fruit maturation in Actinidia deliciosa "Hayward" and A. chinensis "Hort16A". *Postharvest Biol. Technol.* **2011**, *59*, 16–24. [CrossRef]
41. Harker, F.R.; Carr, B.T.; Lenjo, M.; MacRae, E.A.; Wismer, W.V.; Marsh, K.B.; Williams, M.; White, A.; Lund, C.M.; Walker, S.B.; et al. Consumer liking for kiwifruit flavour: A meta-analysis of five studies on fruit quality. *Food Qual. Prefer.* **2009**, *20*, 30–41. [CrossRef]
42. Pérez-Burillo, S.; Oliveras, M.J.; Quesada, J.; Rufián-Henares, J.A.; Pastoriza, S. Relationship between composition and bioactivity of persimmon and kiwifruit. *Food Res. Int.* **2018**, *105*, 461–472. [CrossRef]

43. Kim, J.G.; Beppu, K.; Kataoka, I. Varietal differences in phenolic content and astringency in skin and flesh of hardy kiwifruit resources in Japan. *Sci. Hortic.* **2009**, *120*, 551–554. [CrossRef]
44. Wu, W.; Zhu, Q.G.; Wang, W.Q.; Grierson, D.; Yin, X.R. Molecular basis of the formation and removal of fruit astringency. *Food Chem.* **2022**, *372*, 131234. [CrossRef]
45. Wen, J.; Wang, Y.; Cao, W.; He, Y.; Sun, Y.; Yuan, P.; Sun, B.; Yan, Y.; Qin, H.; Fan, S.; et al. Comprehensive Evaluation of Ten Actinidia arguta Wines Based on Color, Organic Acids, Volatile Compounds, and Quantitative Descriptive Analysis. *Foods* **2023**, *12*, 3345. [CrossRef]
46. Maniwara, P.; Nakano, K.; Boonyakiat, D.; Ohashi, S.; Hiroi, M.; Tohyama, T. The use of visible and near infrared spectroscopy for evaluating passion fruit postharvest quality. *J. Food Eng.* **2014**, *143*, 33–43. [CrossRef]
47. Wang, Z.; Künnemeyer, R.; McGlone, A.; Burdon, J. Potential of Vis-NIR spectroscopy for detection of chilling injury in kiwifruit. *Postharvest Biol. Technol.* **2020**, *164*, 111160. [CrossRef]
48. Xu, S.; Lu, H.; Ference, C.; Qiu, G.; Liang, X. Rapid Nondestructive Detection of Water Content and Granulation in Postharvest "Shatian" Pomelo Using Visible/Near-Infrared Spectroscopy. *Biosensors* **2020**, *10*, 41. [CrossRef]
49. Giovanelli, G.; Sinelli, N.; Beghi, R.; Guidetti, R.; Casiraghi, E. NIR spectroscopy for the optimization of postharvest apple management. *Postharvest Biol. Technol.* **2014**, *87*, 13–20. [CrossRef]
50. Cortés, V.; Blasco, J.; Aleixos, N.; Cubero, S.; Talens, P. Visible and Near-Infrared Diffuse Reflectance Spectroscopy for Fast Qualitative and Quantitative Assessment of Nectarine Quality. *Food Bioprocess Technol.* **2017**, *10*, 1755–1766. [CrossRef]
51. Cozzolino, D. Advantages and limitations of using near infrared spectroscopy in plant phenomics applications. *Comput. Electron. Agric.* **2023**, *212*, 108078. [CrossRef]
52. Pandiselvam, R.; Prithviraj, V.; Manikantan, M.R.; Kothakota, A.; Rusu, A.V.; Trif, M.; Mousavi Khaneghah, A. Recent advancements in NIR spectroscopy for assessing the quality and safety of horticultural products: A comprehensive review. *Front. Nutr.* **2022**, *9*, 973457. [CrossRef]
53. Afonso, A.M.; Antunes, M.D.; Cruz, S.; Cavaco, A.M.; Guerra, R. Non-destructive follow-up of 'Jintao' kiwifruit ripening through VIS-NIR spectroscopy—Individual vs. average calibration model's predictions. *Postharvest Biol. Technol.* **2022**, *188*, 111895. [CrossRef]

horticulturae

MDPI

Article

Study of the Effects of Spraying Non-Bagging Film Agent on the Contents of Mineral Elements and Flavonoid Metabolites in Apples

Fang Wang [1], Xiaomin Wu [2], Yuduan Ding [3], Xuan Liu [4], Xiaojing Wang [1], Yingyin Gao [1], Jianwen Tian [5,*] and Xiaolong Li [5,*]

1 Institute of Quality Standard and Testing Technology for Agroproducts of Ningxia, Yinchuan 750002, China; wangfang3159@126.com (F.W.); 13709501756@163.com (X.W.); gyy72525@outlook.com (Y.G.)
2 School of Agriculture, Ningxia University, Yinchuan 750021, China; 18648145650@163.com
3 College of Horticulture, Northwest A&F University, Xianyang 712100, China; dingyuduan@nwafu.edu.cn
4 Institute of Food Science and Technology, Chinese Academy of Agricultural Sciences (CAAS), Beijing 100000, China; liuxuancaas@126.com
5 Ningxia Academy of Agricultural and Forestry Sciences, Yinchuan 750002, China
* Correspondence: tjw6789@163.com (J.T.); xiaolong85115@163.com (X.L.)

Citation: Wang, F.; Wu, X.; Ding, Y.; Liu, X.; Wang, X.; Gao, Y.; Tian, J.; Li, X. Study of the Effects of Spraying Non-Bagging Film Agent on the Contents of Mineral Elements and Flavonoid Metabolites in Apples. *Horticulturae* **2024**, *10*, 198. https://doi.org/10.3390/horticulturae10030198

Academic Editor: Yuepeng Han

Received: 9 October 2023
Revised: 16 November 2023
Accepted: 15 December 2023
Published: 20 February 2024

Abstract: There has been growing interest in examining the potential of non-bagging patterns due to the decline of fruit inner quality and the increase in labor force cost and ecological pollution. Spraying a non-bagging film agent is an important method for non-bagging cultivation. This paper aims to study the effects of non-bagging film agents on the contents of mineral elements and flavonoid metabolites in apple fruits and determine the feasibility of this method. Fuji apples were used as the sample material and treated individually with two non-bagging film agents, namely, humic acid film (ABM) and Pirrio calcium film (CAM). Also, two control groups, namely, the clear water spraying without bagging group (CK) and the bagging group (TCK), were set in this study to measure the contents of mineral elements and flavonoid metabolites in these apples. Compared with those two control groups, the spraying treatment groups with two kinds of non-bagging film agents present a significant difference between their total contents of mineral elements, with the total content of mineral elements of apples in the ABM treatment group being 1.36 times the content of apples in the CK group. In terms of the flavonoid metabolites, only Astragalin, Tiliroside, Homoplantaginin, Phlorizin, Apigenin, Hesperidin, Oroxin A, and Kaempferol present significant differences in their proportions in apples, and there are no significant differences among the proportions of other compounds. Individual spraying of two kinds of non-bagging film agents can significantly increase the total contents of mineral elements in apples, with slight effects on the contents of flavonoid metabolites in these fruits. Therefore, both film agents can be used for cultivating Fuji apples.

Keywords: apples; mineral elements; flavonoid metabolites

1. Introduction

Apple (*Malus domestica*), one of the most widely cultivated tree fruits in many regions of the world, is the most important fruit and is liked by all classes of people due to its pleasant taste and established nutritional and economical value. It is a rich source of antioxidant compounds, carbohydrates, essential minerals, and dietary fiber [1–3]. China is the largest apple-producing and -consuming country in the world [4].

The commonly used fruit bagging material in the current market is generally the film-coated paper bag. With good effects on water retention, freshness preservation, and weight increase, this bagging material can improve the appearance of fruits and their market value at the same time. However, such a cultivation method has changed the growth and development environments of apples, thus leading to problems affecting fruit quality, such as reduced contents of phenolic compounds, diminished flavor quality, and some other problems [5–7]. In recent years, there has been growing interest in examining the potential of non-bagging patterns due to the decline of fruit inner quality and the increase in labor force cost and ecological pollution. It has been recognized that the bagging-free cultivation pattern is an inevitable trend in the apple industry development in China [8].

Spraying non-bagging film agents is an important method for practicing the non-bagging cultivation of fruits, with non-bagging film agents specifically used in the surface maintenance of fruits during development. With the spraying of film agent, a layer of "soft bi-layer biofilm" with a thickness of about 5–10 nanometers can be quickly formed, covering the surface layer of the fruit. Meanwhile, gaps between membrane molecules of the layer can allow the free passing-through of oxygen, carbon dioxide, and sunlight, thus providing all-around protection for preventing pesticides and pests from contaminating and harming plants.

At present, there are few studies on the influence of spraying non-bagging film agents on the fruit quality of apples. Under such a circumstance, this study has measured the contents of forty-two mineral elements and flavonoid metabolites in apple fruits with the assistance of an inductively coupled plasma emission spectrometer (ICP-MS). With Fuji apples taken as the research object, two control groups of clear water treatment and bagging treatment, as well as experimental treatment groups with individual spraying of two kinds of non-bagging film agents, were set in this study to analyze the contents of mineral elements and flavonoid metabolites in fruits under these four different treatment methods. Also, this study has explored the influences of spraying non-bagging film agents on the contents of mineral elements and flavonoid metabolites in Fuji apple fruits, with the study results providing basic data support for the further application of non-bagging cultivation.

2. Materials and Methods

2.1. Analytical Objects

A total of four test groups, namely, the clear water control group (CK), bagging treatment group (TCK), non-bagging humic acid film agent treatment group (ABM), and non-bagging Piriol calcium film agent treatment group (CAM), were designed in this study. Upon the removal of their fuzz, apples in the clear water control group, ABM group, and CAM group were sprayed with clear water and film agents correspondingly, with a spraying frequency of once every month for a consecutive period of three months. After the harvest of apple fruits, they were packed in cartons and delivered to the laboratory on the same day. Then, ten apples with similar shapes, sizes, and colors were picked from twenty apple samples under each treatment to investigate the differences among the physical and chemical indexes, mineral element contents, and flavonoid metabolite contents of apple fruits treated with different methods.

This study analyzed the physicochemical indicators such as sugar and acid and determined the contents of trace elements using a fresh sample; for mineral elements, a dried sample was used.

2.2. Instruments and Reagents

The instruments used for samples' preparation and analysis include: a microwave digestion system (MARS 6, CME, New York, NY, USA), an ICP-MS inductively coupled plasma mass spectrometer (ELAN DRC-e, Perkin Elmer, Waltham, MA, USA), an ultrapure water manufacturing system (Milli Plus 2150, MILLIPORE, Burlington, MA, USA),and a 6875D fully-automatic frozen grinder (Retsch, Haan, Germany). Flavonoids contents were detected by MetWare based on the AB Sciex QTRAP 6500 LC-MS/MS platform (Singapore, USA).

The reagents used for samples' preparation and analysis include: mixed standards of Al, As, B, Ba, Be, Bi, Cd, Ce, Co, Cr, Cs, Cu, Gd, Hg, Li, Mg, Mn, Mo, Na, Nb, Nd, Ni, Pr, Rb, Sb, Sc, Sm, Sn, Sr, Ti, Th, U, V, Fe, Zn, P, Ru, Au, Ga, Ca, Zr, and Pb (1000 mg·L^{-1}), provided by the National Institute of Metrology China, and prepared to the required concentration when used.

HPLC-grade acetonitrile (ACN) and methanol (MeOH) were purchased from Merck (Darmstadt, Germany). MilliQ water (Millipore, Molsheim, France) was used in all experiments. Formic acid was purchased from Sigma-Aldrich (St. Louis, MO, USA). All of the standards were purchased from MCE (MedChemExpress, Shanghai, China). The stock solutions of standards were prepared at the concentration of 10 mmol·L^{-1} in 70% MeOH. All stock solutions were stored at −20 °C. The stock solutions were diluted with 70% MeOH to working solutions before analysis.

HNO_3, $HClO_4$, and other chemicals used were of standard analytical grade. Deionized and distilled water was used throughout.

2.3. Measurement of Physical and Chemical Indexes

A 1% electronic balance was used to measure the single-fruit weight, a pointer-type fruit hardness tester was applied to measure the hardness of apples, and a Brix-acidity meter was used to measure the soluble solid contents and titratable acidity of apples.

2.4. Determination of Mineral Elements

Based on the method proposed by Liu et al. [9], a slightly modified analysis of inductively coupled plasma mass spectrometry was carried out in this study with the following specific operation steps:

(1) Dissolving the samples using the microwave digestion method, with the specific operation steps as follows: Weigh 0.5 g of Lycium barbarum sample (with a precision of 0.0001 g), and put it in a microwave digestion tube. After the addition of 10 mL of nitric acid, let the sample stand at room temperature for 3 h, and then put it in a microwave digestion instrument for sample digestion. Select temperature control to let the temperature rise to 120 °C in 5 min, and keep the temperature there for 10 min. Then, let the temperature rise to 150 °C in 5 min and keep it there for 20 min. After that, let the temperature further rise to 185 °C in 5 min and keep it there for 30 min. After the completion of digestion, the sample is cooled. Then, gently unscrew the lid and place the microwave digestion tube on an acid-driven processor to perform acid removal at 120 °C for 2 h. Then, cool the sample to room temperature and wash it with ultra-pure water in a test tube with scale of 25 mL. After that, dilute it with ultra-pure water to volume and shake the solution well. Meanwhile, prepare a reagent blank.

(2) Measure the contents of elements in apple fruits through the ICP-MS method under the following specific working conditions: Generator power: 1300 W; Atomizer flow rate: 0.95 L·min^{-1}; Plasma torch cooling gas flow rate: 17.0 L·min^{-1}; Auxiliary device flow rate: 1.20 L·min^{-1}; Detector analog stage voltage: −2350 V; Ion lens voltage: 6.00 V.

All samples were analyzed through the ICP-MS method, with each sample analyzed in duplicate. The working curve corresponding to each element was plotted, with its specific linear equation and correlation coefficient listed in Table 1. All correlation coefficients R^2 calculated are higher than 0.99, indicating good linear regressions of standard solution concentrations and absorption peak areas and the feasibility of using these curves to quantify the contents of mineral elements.

2.5. Metabolites Extraction

The sample was freeze-dried, ground into powder (30 Hz, 1.5 min), and stored at −80 °C until needed. In total, 20 mg powder was weighted and extracted with 0.5 mL 70% methanol. In total, 10 μL internal standard (4000 nmol·L^{-1}) was added into the extract as an internal standard (IS) for the quantification. The extract was sonicated for 30 min and centrifuged at 12,000× g under 4 °C for 5 min. The supernatant was filtered through a 0.22 μm membrane filter for further LC-MS/MS analysis.

Table 1. Influences of different treatment methods on the physical and chemical indexes of fruits.

Treatment	Single Fruit Weight (g)	Hardness ($kg·cm^{-2}$)	Soluble Solid Content (%)	Titratable Acid Content (%)
CK	231 ± 43.8 ^{b}	8.48 ± 0.82 ^{a}	19.4 ± 1.80 ^{a}	0.35 ± 0.10 ^{b}
TCK	247 ± 44.2 ^{b}	9.17 ± 0.87 ^{a}	13.7 ± 1.96 ^{a}	0.35 ± 0.13 ^{b}
ABM	250 ± 30.7 ^{a}	9.25 ± 1.16 ^{a}	16.6 ± 1.34 ^{a}	0.46 ± 0.19 ^{a}
CAM	208 ± 18.7 ^{b}	8.86 ± 1.03 ^{a}	17.3 ± 1.38 ^{a}	0.41 ± 0.05 ^{ab}

Values with different letters in the same column (a,b) are significantly different ($p < 0.05$) from each other. Values are given as the mean ± standard deviation.

2.6. UPLC-ESI-MS/MS Analysis

The sample extracts were analyzed using a UPLC-ESI-MS/MS system. The analytical conditions were as follows: UPLC column, Waters ACQUITY UPLC HSS T3 C18 (1.8 μm, 100 mm × 2.1 mm i.d.); solvent system, water with 0.05% formic acid (A), acetonitrile with 0.05% formic acid (B). The gradient elution program was set as follows: 0–1 min, 10–20% B; 1–9 min, 20–70% B; 9–12.5 min, 70–95% B; 12.5–13.5 min, 95% B; 13.5–13.6 min, 95–10% B; 13.6–15 min, 10% B. The flow rate was set at 0.35 $mL·min^{-1}$ and the temperature was set at 40 °C. The injection volume is 2 μL.

The ESI source operation parameters were as follows: ion source, ESI +/−; source temperature 550 °C; ion spray voltage (IS) 5500 V (positive), −4500 V (negative); curtain gas (CUR) was set at 35 psi, respectively. Flavonoids were analyzed using scheduled multiple reaction monitoring (MRM).

2.7. Data Processing and Statistical Analysis

The mean value (MEAN) and standard deviation (SD) of the traits were calculated using Microsoft Excel 2016 and were statistically analyzed using IBM SPSS Statistics 26.0 software. Independent-samples *t*-tests were used to determine the significance of differences among samples at a level of 0.05. Values were reported as means ± standard deviation (SD) from triplicate experiments. Mass spectrometric data conducted in MultiQuant 3.0.3.

3. Results

3.1. Effects of Non-Bagging Film Agent Treatment on the Physical and Chemical Indexes of Fruits

Single-fruit weight, hardness, soluble solids, and total acidity are important indicators for evaluating apple quality. We found that there is a significant difference between the ABM treatment group, which has the largest single-fruit weight, and the CAM treatment group, which has the smallest single-fruit weight. However, in terms of the single-fruit weight, the control groups are not significantly different from the CAM treatment group. Therefore, it can be inferred that the single-fruit weights of apples in the ABM treatment group have significantly increased (Table 1). Also, it was found that in terms of apple hardness, there was no significant difference among all treatment groups (Table 1). An analysis of the TSS contents shows that there was no significant difference among all treatment groups, which shows the insignificant influence on the TSS of apples. In terms of the influence on titratable acid, there were significant differences among different treatment groups, with the titratable acid contents in the non-bagging film agent treatment groups being significantly higher than those contents in the control groups ($p < 0.05$). In summary, the treatment of non-bagging film agents increases the single-fruit weights of apples and improves the accumulation of titratable acid.

3.2. Effects of Non-Bagging Film Agent Treatment on the Contents of Mineral Elements of Fruits

Mineral element content is one way to measure fruit quality traits, as mineral elements are closely related to fruit size, pulp hardness, and soluble solids, and play important roles in fruit disease resistance storage resistance, and maintaining good quality and flavor [10,11]. Mineral elements are also very important nutrients that are essential for healthy human growth and development [12].

Among all treatment groups, the contents of P, Mg, Ca, and Na are significantly higher than those contents of other trace elements, with their relative ranges between 0.000482 mg· kg^{-1} and 1789 mg·kg^{-1} (Table 2). Among all these elements, the content of element P, which ranges from 1185 mg·kg^{-1} to 1789 mg·kg^{-1}, is the highest. The contents of the elements Ca and Na are in the same order of magnitude as the content of the element Mg (109–417 mg·kg^{-1}). Also, the contents of elements Al, B, and V are in the same order of magnitude as the content of the element Fe (11.5–154 mg·kg^{-1}), and the contents of the elements As, Cr, Cu, Mn, Rb, Sr, Zn, and Ga are in the same order of magnitude (1.4–6.06 mg·kg^{-1}), with the contents of all of the other elements being lower than 1 mg·kg^{-1} (Table 2). An analysis of correlation variance shows that among those forty-one elements of apple fruits, there are no significant differences among the contents of fifteen elements (B, Cu, Mg, Mo, Pr, Sb, Sm, Sn, Ti, Th, Ru, Ga, Zr, and Ca) ($p > 0.05$), indicating that there are no significant differences among the apple fruits' absorption capacities of these elements.

Table 2. Effects of different treatment methods on the contents of mineral elements of fruits (mg·kg^{-1}).

Element	CK	TCK	ABM	CAM
Al	38.4 ± 5.7 ^{b}	29.7 ± 7.5 ^{b}	154 ± 51 ^{a}	43.4 ± 9.7 ^{b}
As	2.36 ± 0.15 ^{a}	2.35 ± 0.14 ^{a}	2.1 ± 0.20 ^{ab}	1.97 ± 0.23 ^{b}
B	69.6 ± 8.01 ^{a}	73.4 ± 9.08 ^{a}	77.2 ± 4.15 ^{a}	80.8 ± 5.86 ^{a}
Ba	0.91 ± 0.0884 ^{c}	2.48 ± 0.5922 ^{b}	1.24 ± 0.2521 ^{c}	4.86 ± 0.9407 ^{a}
Be	0.0015 ± 0.001 ^{b}	0.0065 ± 0.003 ^{ab}	0.0105 ± 0.007 ^{a}	0.0052 ± 0.001 ^{ab}
Bi	0.000652 ± 0.0002 ^{b}	0.000564 ± 0.0002 ^{b}	0.0013 ± 0.0004 ^{a}	0.000482 ± 0.0004 ^{b}
Cd	0.00116 ± 0.0001 ^{b}	0.00161 ± 0.0007 ^{b}	0.00151 ± 0.0006 ^{b}	0.00251 ± 0.0007 ^{a}
Ce	0.0166 ± 0.0038 ^{b}	0.021 ± 0.0074 ^{ab}	0.0274 ± 0.0047 ^{a}	0.014 ± 0.0073 ^{b}
Co	0.0228 ± 0.0067 ^{b}	0.0265 ± 0.0047 ^{ab}	0.0362 ± 0.0096 ^{a}	0.0344 ± 0.0047 ^{a}
Cr	1.86 ± 0.24 ^{b}	2.17 ± 0.37 ^{b}	2.61 ± 0.28 ^{a}	1.91 ± 0.28 ^{b}
Cs	0.0085 ± 0.0015 ^{b}	0.0180 ± 0.0042 ^{a}	0.0122 ± 0.0035 ^{b}	0.0085 ± 0.0014 ^{b}
Cu	1.69 ± 0.08 ^{a}	2.17 ± 0.31 ^{a}	2.15 ± 0.57 ^{a}	1.78 ± 0.28 ^{a}
Gd	0.000993 ± 0.0002 ^{b}	0.00125 ± 0.0003 ^{b}	0.00217 ± 0.0006 ^{a}	0.00142 ± 0.0003 ^{b}
Hg	0.0039 ± 0.0008 ^{b}	0.0024 ± 0.0003 ^{b}	0.0055 ± 0.0001 ^{b}	0.0708 ± 0.0138 ^{a}
Li	0.831 ± 0.065 ^{a}	0.5213 ± 0.077 ^{b}	0.6255 ± 0.17 ^{b}	0.3773 ± 0.11 ^{c}
Mg	363 ± 24.7 ^{a}	361 ± 39.3 ^{a}	417 ± 47 ^{a}	405 ± 37.9 ^{a}
Mn	1.66 ± 0.19 ^{d}	2.25 ± 0.31 ^{c}	3.44 ± 0.53 ^{a}	2.69 ± 0.23 ^{b}
Mo	0.177 ± 0.033 ^{a}	0.1562 ± 0.023 ^{a}	0.1197 ± 0.016 ^{a}	0.1772 ± 0.055 ^{a}
Na	109 ± 5.18 ^{b}	126 ± 16.7 ^{b}	257 ± 12.1 ^{a}	130 ± 18 ^{b}
Nb	0.0116 ± 0.0057 ^{a}	0.0058 ± 0.0026 ^{b}	0.0117 ± 0.0035 ^{a}	0.0047 ± 0.0011 ^{b}
Nd	0.0069 ± 0.001 ^{b}	0.0059 ± 0.0005 ^{bc}	0.0086 ± 0.0011 ^{a}	0.0053 ± 0.0014 ^{c}
Ni	0.3374 ± 0.028 ^{b}	0.3277 ± 0.107 ^{b}	0.3831 ± 0.035 ^{b}	0.5525 ± 0.131 ^{a}
Pr	0.0024 ± 0.0009 ^{a}	0.0025 ± 0.0009 ^{a}	0.0034 ± 0.0005 ^{a}	0.0033 ± 0.0014 ^{a}
Rb	2.81 ± 0.53 ^{b}	4.98 ± 0.87 ^{a}	4.4 ± 0.97 ^{a}	4.18 ± 0.24 ^{a}
Sb	0.0136 ± 0.0027 ^{a}	0.0087 ± 0.0026 ^{a}	0.0114 ± 0.0018 ^{a}	0.015 ± 0.0076 ^{a}
Sc	0.0927 ± 0.013 ^{a}	0.059 ± 0.0029 ^{b}	0.0815 ± 0.035 ^{ab}	0.0527 ± 0.014 ^{b}
Sm	0.0013 ± 0.0007 ^{a}	0.0016 ± 0.0006 ^{a}	0.0014 ± 0.0002 ^{a}	0.0016 ± 0.0006 ^{a}
Sn	0.0618 ± 0.012 ^{a}	0.0524 ± 0.0077 ^{a}	0.0542 ± 0.0065 ^{a}	0.0591 ± 0.014 ^{a}
Sr	1.74 ± 0.23 ^{c}	6.06 ± 1.36 ^{a}	4.27 ± 0.83 ^{b}	1.4 ± 0.25 ^{c}
Ti	1.2 ± 0.26 ^{a}	1.33 ± 0.38 ^{a}	0.76 ± 0.09 ^{a}	0.93 ± 0.54 ^{a}
Th	0.0022 ± 0.0013 ^{a}	0.0024 ± 0.0026 ^{a}	0.0049 ± 0.0018 ^{a}	0.0022 ± 0.0008 ^{a}
U	0.0061 ± 0.0006 ^{a}	0.0044 ± 0.0008 ^{ab}	0.0049 ± 0.0006 ^{b}	0.0036 ± 0.0007 ^{c}
V	15.6 ± 1.79 ^{a}	16.6 ± 1.88 ^{a}	16.8 ± 2.74 ^{a}	11.5 ± 0.66 ^{b}
Fe	38.4 ± 8.2 ^{ab}	32.1 ± 4.9 ^{b}	45.7 ± 7.8 ^{a}	40.7 ± 7.6 ^{ab}
Zn	4.38 ± 0.41 ^{ab}	3.65 ± 0.45 ^{b}	4.39 ± 0.62 ^{ab}	5.07 ± 1.11 ^{a}
P	1185 ± 134 ^{c}	1395 ± 215 b^{c}	1567 ± 175 ^{ab}	1789 ± 219 ^{a}
Ru	0.038 ± 0.0089 ^{a}	0.0352 ± 0.013 ^{a}	0.0427 ± 0.023 ^{a}	0.0288 ± 0.0077 ^{a}
Au	0.0333 ± 0.0068 ^{a}	0.0165 ± 0.0023 ^{b}	0.0138 ± 0.0018 ^{b}	0.0171 ± 0.0008 ^{b}
Ga	2.02 ± 0.44 ^{a}	2.3 ± 0.68 ^{a}	2.38 ± 0.78 ^{a}	2.87 ± 0.78 ^{a}
Zr	0.0219 ± 0.0039 ^{a}	0.0251 ± 0.013 ^{a}	0.0276 ± 0.0017 ^{a}	0.0227 ± 0.0049 ^{a}
Ca	242 ± 47 ^{a}	250 ± 28 ^{a}	262 ± 7.3 ^{a}	272 ± 32.9 ^{a}
Pb	0.027 ± 0.0054 ^{ab}	0.0167 ± 0.0055 ^{b}	0.0298 ± 0.0047 ^{ab}	0.0386 ± 0.018 ^{a}

Values with different letters in the same line (a–c) are significantly different ($p < 0.05$) from each other. Values are given as the mean ± standard deviation.

Among those forty-two elements mentioned above, thirteen elements (Al, As, Hg, Pb, Cd, Cr, Ni, Sn, Mo, V, Ti, Sb, and Ba) are harmful elements. The contents of the Pb and Cd elements of apples in each treatment group are all below the limits stipulated in the National Food Safety Standard—Maximum Levels for Contaminants in Food (GB 2762-2022) [13] on Pb (<0.1 mg·kg^{-1}) and Cd (<0.05 mg·kg^{-1}) contents in fruits. Among these thirteen elements, in this study, only high contents of Al elements were measured in apples. Also, the contents of the Al element varied significantly from group to group, with the Al element contents of apples in the ABM treatment group being 5.2 times those contents of apples in the TCK group. However, no relevant limit on the Al element is stipulated in the standard described above.

The results indicate that apples in the ABM treatment group present the highest total mineral element content, followed by the apples in the CAM treatment group, and the apples in the TCK and CK groups present the lowest total mineral element content (Figure 1). Also, the total mineral element content of apples in the ABM treatment group is 1.36 times the content of apples in the CK group. To sum up, the treatment of the non-bagging film agent will not change the composition of mineral elements in apples but will increase their total contents of mineral elements.

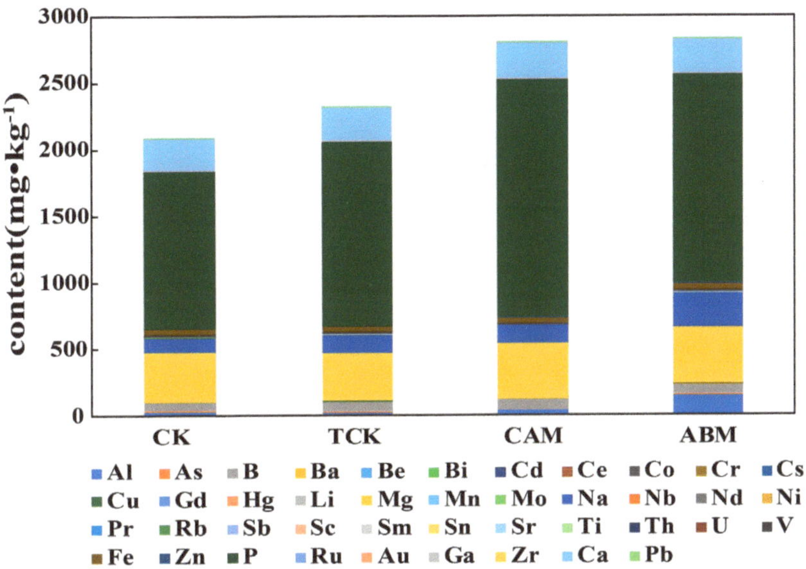

Figure 1. Influences of different treatment methods on the total mineral element content of apples.

3.3. *Effects of Non-Bagging Film Agent Treatment on the Contents of Flavonoid Metabolites in Apples*

In this study, a total of 204 types of metabolites were selected for testing. Some metabolite types had no content detected in apples, while some other metabolite types with relatively low contents detected in apples were not counted. Therefore, this study has performed a comprehensive analysis on a total of 39 metabolite types that were detected in apples with relatively high contents. From Table 3, it can be seen that flavonoid metabolites have a rich presence in apple fruits, with relatively great variations in contents ranging from 0.002 to 466 nmol·g^{-1}. Followed by Quercitrin, Astragalin, Rutin, Avicularin, Calycosin-7-O-β-D-glucoside, and Hyperoside, whose contents fall into a range between 1 nmol·g^{-1} and 100 nmol·g^{-1}, (-)-Epicatechin, (-)-Catechin, and Phlorizin have the highest contents, which are all above 100 nmol g^{-1}. Also, it shows that there are relatively small differences among the influences of different treatment methods on the contents of 39 metabolites. An analysis of correlation variance indicates that only Astragalin, Tiliroside, Homoplantaginin,

Phlorizin, Apigen, Hesperidin, Oroxin A, and Kaempferol present significant differences in their contents in apples, with no significant content differences among other metabolites, indicating insignificant influences of the non-bagging film agent treatment methods on the apple contents of flavonoid metabolites.

Table 3. Influences of different treatment methods on the contents of flavonoid metabolites in fruits ($nmol{\cdot}g^{-1}$).

Compounds	CK	TCK	ABM	CAM
Astilbin	0.165 ± 0.05 [a]	0.088 ± 0.01 [a]	0.128 ± 0.03 [a]	0.183 ± 0.04 [a]
Pinocembrin	0.302 ± 0.28 [a]	0.219 ± 0.18 [a]	0.260 ± 0.25 [a]	0.276 ± 0.22 [a]
Quercitrin	54.7 ± 4.9 [a]	44.9 ± 35.9 [a]	41.3 ± 12.8 [a]	50.1 ± 7.02 [a]
Narcissin	0.025 ± 0.008 [a]	0.022 ± 0.008 [a]	0.015 ± 0.004 [a]	0.015 ± 0.004 [a]
Astragalin	2.28 ± 0.73 [a]	0.372 ± 0.20 [b]	0.899 ± 0.05 [b]	1.48 ± 0.76 [ab]
Tiliroside	0.046 ± 0.006 [c]	0.062 ± 0.02 [bc]	0.084 ± 0.02 [ab]	0.108 ± 0.01 [a]
(-)-Epicatechin	358 ± 60 [a]	466 ± 59 [a]	461 ± 118 [a]	450 ± 46.4 [a]
(-)-Catechin gallate	0.107 ± 0.03 [a]	0.119 ± 0.02 [a]	0.146 ± 0.01 [a]	0.090 ± 0.02 [a]
Isosakuranetin	0.009 ± 0.005 [a]	0.005 ± 0.005 [a]	0.009 ± 0.008 [a]	0.012 ± 0.006 [a]
Apigenin 7-glucoside	0.033 ± 0.017 [a]	0.008 ± 0.002 [a]	0.008 ± 0.005 [a]	0.025 ± 0.01 [a]
Tectochrysin	0.013 ± 0.008 [a]	0.015 ± 0.01 [a]	0.021 ± 0.02 [a]	0.012 ± 0.008 [a]
Homoplantaginin	0.078 ± 0.01 [a]	0.045 ± 0.005 [b]	0.056 ± 0.008 [b]	0.051 ± 0.009 [b]
6-Hydroxyflavone	0.020 ± 0.003 [a]	0.016 ± 0.003 [a]	0.020 ± 0.003 [a]	0.018 ± 0.001 [a]
Naringenin-7-glucoside	0.526 ± 0.01 [a]	0.635 ± 0.15 [a]	0.950 ± 0.24 [a]	0.939 ± 0.15 [a]
Phloretin	0.058 ± 0.009 [a]	0.065 ± 0.004 [a]	0.070 ± 0.02 [a]	0.077 ± 0.017 [a]
Spiraeoside	0.039 ± 0.009 [a]	0.022 ± 0.019 [a]	0.024 ± 0.015 [a]	0.030 ± 0.008 [a]
Quercimeritrin	0.562 ± 0.16 [a]	0.525 ± 0.59 [a]	0.423 ± 0.11 [a]	0.404 ± 0.11 [a]
Isorhamnetin 3-O-glucoside	0.149 ± 0.02 [a]	0.085 ± 0.04 [a]	0.120 ± 0.02 [a]	0.111 ± 0.02 [a]
Neohesperidin dihydrochalcone	0.010 ± 0.001 [a]	0.010 ± 0.002 [a]	0.011 ± 0.002 [a]	0.011 ± 0.001 [a]
Quercetin	0.065 ± 0.009 [a]	0.045 ± 0.05 [a]	0.012 ± 0.01 [a]	0.035 ± 0.03 [a]
(-)-Catechin	105 ± 10.6 [a]	173 ± 21.4 [a]	164 ± 67.6 [a]	152 ± 37.8 [a]
Rutin	1.38 ± 0.67 [a]	1.19 ± 1.31 [a]	0.597 ± 0.12 [a]	0.640 ± 0.16 [a]
Chrysin	0.089 ± 0.06 [a]	0.083 ± 0.06 [a]	0.100 ± 0.02 [a]	0.139 ± 0.09 [a]
Phlorizin	139 ± 14.2 [ab]	111 ± 9.13 [b]	157 ± 30.2 [ab]	182 ± 29.2 [a]
Galangin	0.085 ± 0.06 [a]	0.071 ± 0.05 [a]	0.105 ± 0.02 [a]	0.127 ± 0.09 [a]
Apigenin	0.019 ± 0.003 [a]	0.008 ± 0.004 [b]	0.006 ± 0.001 [b]	0.014 ± 0.005 [ab]
Avicularin	14.9 ± 1.02 [a]	15.5 ± 15.4 [a]	9.63 ± 2.52 [a]	13.2 ± 1.02 [a]
Cynaroside	0.928 ± 0.19 [a]	0.561 ± 0.14 [a]	0.721 ± 1.45 [a]	0.905 ± 0.32 [a]
Calycosin-7-O-β-D-glucoside	12.8 ± 6.21 [a]	6.88 ± 4.07 [a]	8.22 ± 4.06 [a]	7.088 ± 0.13 [a]
Naringin Dihydrochalcone	0.025 ± 0.002 [a]	0.025 ± 0.002 [a]	0.023 ± 0.003 [a]	0.027 ± 0.006 [a]
Afzelechin	0.085 ± 0.034 [a]	0.085 ± 0.034 [a]	0.014 ± 0.001 [a]	0.031 ± 0.009 [a]
Hesperidin	0.024 ± 0.002 [a]	0.024 ± 0.002 [c]	0.002 ± 0.002 [b]	0.013 ± 0.004 [c]
Taxifolin	0.077 ± 0.007 [a]	0.077 ± 0.007 [a]	0.063 ± 0.012 [a]	0.057 ± 0.015 [a]
Engeletin	0.013 ± 0.003 [a]	0.013 ± 0.003 [a]	0.006 ± 0.001 [a]	0.012 ± 0.009 [a]
Oroxin A	0.04 ± 0.006 [a]	0.04 ± 0.006 [c]	0.009 ± 0.001 [bc]	0.019 ± 0.001 [ab]
Baicalin	0.012 ± 0.005 [a]	0.012 ± 0.005 [a]	0.012 ± 0.004 [a]	0.009 ± 0.004 [a]
Kaempferol	0.05 ± 0.015 [a]	0.05 ± 0.015 [b]	0.021 ± 0.001 [b]	0.023 ± 0.008 [ab]
Hyperoside	9.14 ± 1.42 [a]	9.14 ± 1.42 [a]	9.09 ± 0.01 [a]	6.49 ± 2.47 [a]
Isorhamnetin	0.014 ± 0.001 [a]	0.014 ± 0.001 [a]	0.014 ± 0.001 [a]	0.013 ± 0.001 [a]

Values with different letters in the same line (a–c) are significantly different ($p < 0.05$) from each other. Values are given as the mean ± standard deviation.

From Figure 2A, it can be seen that the major identified metabolites in apple fruits treated with different methods include chalcones, flavanones, flavanonols, flavanones, flavones, flavonols, flavanols, and isoflavanones. Among these metabolites, followed by flavones (28%), chalcones (10%), flavanols (10%), flavanones (8%), flavanonols (8%), flavanones (3%), and isoflavanones (3%), flavonols (31%) present the largest proportion. It can be seen that flavonols and flavones are two major metabolites in apples. This finding is consistent with the study results of Chen et al. [14].

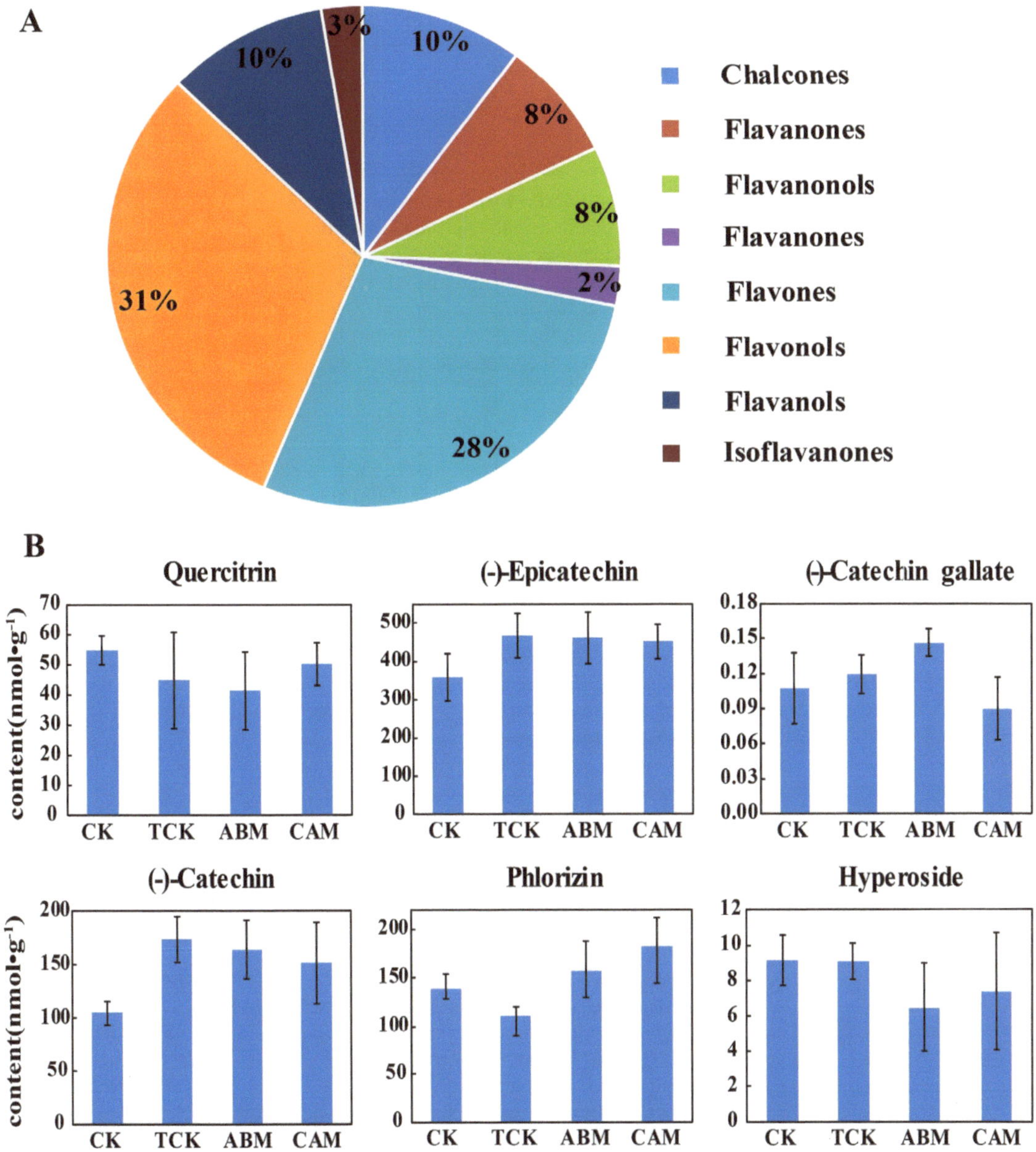

Figure 2. Classification results of flavonoid metabolites (**A**); Analysis results of important flavonoid metabolites in fruits (**B**).

Figure 2B presents the analysis results of important flavonoid metabolites in apple fruits. The results indicate that there are not great variations in the contents of Quercitrin, (-)-Epicatechin, (-)-Catechin gallate, (-)-Catechin, and Hyperoside between treatment groups and control groups, with no uniform interaction pattern among these groups. Apples in the treatment groups present more significant influences on the content of Phlorizin than those apples in the control groups, with less significant effects on the content of Hyperoside presented by apples in the treatment groups than those effects presented by apples in the control groups. Among those six important metabolites, only

Phlorizin presents lower contents in apples in the bagging treatment group than those contents in apples in the non-bagging groups, indicating that the syntheses of most of these metabolites do not rely on sunlight and are not affected by the spraying of non-bagging film agents.

4. Discussion

At present, there are few research reports on the effects of spraying non-bagging film agents on the quality and mineral elements of fruits. Under such a circumstance, this paper has systematically investigated the influences of spraying non-bagging film agents on the physical and chemical indexes and the contents of mineral elements and flavonoid metabolites of apple fruits. The study results show that spraying a non-bagging film agent is conducive to the accumulation of mineral elements in fruits and has a relatively small effect on the contents of their flavonoid metabolites, thus being feasible to be used in the non-bagging cultivation of apples, and these results are consistent with the study results of Cheng [15] et al.

In this study, compared with apples in the control groups, apples in the non-bagging film agent spraying treatment groups all present significantly increased contents of titratable acid, with no significant differences among their hardness and contents of soluble solids. This result is similar to the previous study results of other scholars, with some minor differences. One possible reason is that the effects of those bio-protective films applied in this study have coincided with the development and metabolic time of fruits, and another possible reason lies in the fact that these effects are related to the major components and concentrations of those bio-protective films applied.

Apples contain a great deal of elements such as magnesium (Mg), calcium (Ca), iron (Fe), and others that are beneficial to the human body [4]. Mineral element content plays an important role in judging fruit quality and in maintaining the normal physiological activities of the human body [16,17]. It is well known that phosphorus (P), potassium (K), calcium (Ca), and trace element iron (Fe) are essential for fruit quality [18]. In recent years, many studies have explored the effects of mineral nutrients on plant growth, fruit development, quality, and preservation of fruit, and their molecular mechanisms, especially the effects of P and K on fruit quality. The results of this research indicate that the contents of most elements of apples in the non-bagging film agent treatment groups are all higher than those contents of apples in the control groups, with increased total amounts of mineral elements. Therefore, this study has provided a new idea for the investigation of maximizing the nutritional value of apples and a reference for the further popularization and application of non-bagging cultivation. At the same time, this study has provided basic data support for the in-depth exploration of the molecular construction mechanisms of mineral elements in plant cell structures and the interactions and functions of these mineral elements.

Vitamin C, organic acid, sugar, flavonoids, and phenols are plant metabolites whose contents are regulated by the plant genetic and environmental factors. Some research has shown that humic acid substances, as an environmental factor, can affect plant metabolism by regulating the plant protease activity or inducing protease synthesis [19]. Polyphenols are a large group of bioactive plant compounds (over 8000) and have two main classes called non-flavonoids and flavonoids [20]. The non-flavonoids compounds include phenolic acids, stilbenes, and lignans, and the flavonoids compounds include flavonols, anthocyanidins, anthocyanins, isoflavones, flavones, flavanols (or cathechins), flavanones, and flavanonols [21]. The main polyphenols found in apples are anthocyanins, dihydrochalcones, flavanols, flavonols, hydroxybenzoic acid, and hydroxycinnamic acid that play an important role in the healthful properties of apples [22]. The major flavonoid metabolites contained in apples include dihydrochalcones, flavanols, and flavonols [13,23], which is consistent with the results obtained in this study. The influences of different treatment methods on the contents of 39 metabolite types in apples were analyzed in this study, with the results showing that only Astragalin, Tiliroside, Homoplantaginin, Phlorizin, Apigenin, Hesperidin, Oroxin A, and Kaempferol present significant differences in their contents, with insignificant differences among the contents of all other metabolites, indicating slight influences of non-bagging film agents on the contents of flavonoid metabolites in apples. Some literature has reported that the composition and antioxidant capacities of phenolic compounds will be affected by the bagging treatment of fruits [24], with significantly reduced contents of flavonoids in fruit pulp [25], which is similar to, but slightly different from, the results obtained in this study.

As a new material, non-bagging film agent has the prospect of replacing fruit bagging in the application and popularization of non-bagging cultivation of apples. However, because the non-bagging cultivation of apples has high requirements on tree vigor, orchard management level, and complementary measures in non-bagging cultivation, it is still necessary to carry out further in-depth research to clarify the complementary measures for this cultivation technique.

There are some limitations in this study. With the bagging of apples, the contact between fruits and pesticides is reduced, thus effectively reducing pesticide residues on apple fruits. Therefore, it is necessary to deeply explore the components of non-bagging film agents and monitor the pesticide residues of fruits sprayed with non-bagging film agents to evaluate their safety.

5. Conclusions

Most apples in China have been bagged during cultivating to prevent pests and diseases in earlier stages and promote coloration and mitigate fruit russeting in later stages. In this study, a comparative analysis has been performed on the bagging and non-bagging treatment methods of apples, with the conclusion that the total contents of mineral elements in apples can be significantly increased with the spraying of two kinds of non-bagging film agents, with relatively small effects on the contents of flavonoid metabolites in apples. Therefore, it is suitable to use these two non-bagging film agents in the non-bagging cultivation of Fuji apples.

Author Contributions: F.W. and Y.D. conceived the idea, compiled the information, and drafted the manuscript; X.L. (Xiaolong Li) and X.W. (Xiaomin Wu) performed mineral elements analyses; X.L. (Xuan Liu) and X.W. (Xiaojing Wang) performed the statistical analysis; F.W., Y.G. and J.T. reviewed and edited the manuscript. All authors have read and agreed to the published version of the manuscript.

Funding: This work was financially supported by the Key R&D project of the Autonomous Region (2021BBF02014, 2022BBF02035), the Independent Innovation Special Project of the Academy of Agricultural Sciences (NGSB-2021-1, NKYG-23-05), and the earmarked fund for the China Agriculture Research System (CARS-27).

Data Availability Statement: Data are contained within the article.

Conflicts of Interest: The authors declare that they have no known competing financial interests or personal relationships that could have appeared to influence the work reported in this paper.

References

1. Abid, M.; Jabbar, S.; Wu, T.; Hashim, M.M.; Hu, B.; Lei, S.; Zeng, X. Sonication Enhances Polyphenolic Compounds, Sugars, Carotenoids and Mineral Elements of Apple Juice. *Ultrason. Sonochem.* **2014**, *21*, 93–97. [CrossRef] [PubMed]
2. Preti, R.; Tarola, A.M. Study of Polyphenols, Antioxidant Capacity and Minerals for the Valorisation of Ancient Apple Cultivars from Northeast Italy. *Eur. Food Res. Technol.* **2021**, *247*, 273–283. [CrossRef]
3. Karaaslan, S.; Ekinci, K. Effect of Pretreatments on Solar Dehydration of Different Varieties of Apple (*Malus domestica*). *Czech J. Food Sci.* **2022**, *40*, 93–101. [CrossRef]
4. Liu, X.; Zhao, Y.; Mu, J.; Zhang, J.; Zhang, A. Determination of Geographical Origin of Concentrated Apple Juice through Analysis of Stable Isotopic and Mineral Elemental Fingerprints: Preliminary Results. *J. Sci. Food Agric.* **2021**, *101*, 3795–3803. [CrossRef] [PubMed]
5. Gou, X.; Zhang, W.; Zhang, J.; Zhang, J.; Zhang, J. Study on Quality Characteristics and Feasibility Analysis of Hail-Proof Plastic Bagging of 5000 Mu in Gansu. *IOP Conf. Ser. Earth Environ. Sci.* **2020**, *440*, 022048. [CrossRef]
6. Zhi, C.; Ali, M.M.; Zhang, J.; Shi, M.; Ma, S.; Chen, F. Effect of Paper and Aluminum Bagging on Fruit Quality of Loquat (*Eriobotrya japonica* Lindl.). *Plants* **2021**, *10*, 2704. [CrossRef] [PubMed]

7. Gao, C.; Zhang, Y.; Li, H.; Gao, Q.; Cheng, Y.; Ogunyemi, S.O.; Guan, J. Fruit Bagging Reduces the Postharvest Decay and Alters the Diversity of Fruit Surface Fungal Community in 'Yali' Pear. *BMC Microbiol.* **2022**, *22*, 239. [CrossRef]
8. Wang, D.; Wang, G.; Wang, J.; Zhai, H.; Xue, X. Inhibitory Effect and Underlying Mechanism of Cinnamon and Clove Essential Oils on Botryosphaeria Dothidea and Colletotrichum Gloeosporioides Causing Rots in Postharvest Bagging-Free Apple Fruits. *Front. Microbiol.* **2023**, *14*, 1109028. [CrossRef]
9. Liu, C.; Li, H.; Ren, A.; Chen, G.; Ye, W.; Wu, Y.; Ma, P.; Yu, W.; He, T. A Comparison of the Mineral Element Content of 70 Different Varieties of Pear Fruit (*Pyrus ussuriensis*) in China. *PeerJ* **2023**, *11*, e15328. [CrossRef]
10. Mészáros, M.; Hnátková, H.; Čonka, P.; Náměstek, J. Linking Mineral Nutrition and Fruit Quality to Growth Intensity and Crop Load in Apple. *Agronomy* **2021**, *11*, 506. [CrossRef]
11. Maity, A.; Marathe, R.A.; Sarkar, A.; Basak, B.B. Phosphorus and Potassium Supplementing Bio-Mineral Fertilizer Augments Soil Fertility and Improves Fruit Yield and Quality of Pomegranate. *Sci. Hortic.* **2022**, *303*, 111234. [CrossRef]
12. Gorący, I.; Rębacz-Maron, E.; Korbecki, J.; Gorący, J. Concentrations of Mg, Ca, Fe, Cu, Zn, P and Anthropometric and Biochemical Parameters in Adults with Chronic Heart Failure. *PeerJ* **2021**, *9*, e12207. [CrossRef] [PubMed]
13. *GB 2762-2022*; Limits of Contaminants in Food. National Food Safety Standards. National Health Commission: Beijing, China, 2022.
14. Chen, C.-S.; Zhang, D.; Wang, Y.-Q.; Li, P.-M.; Ma, F.-W. Effects of Fruit Bagging on the Contents of Phenolic Compounds in the Peel and Flesh of 'Golden Delicious', 'Red Delicious', and 'Royal Gala' Apples. *Sci. Hortic.* **2012**, *142*, 68–73. [CrossRef]
15. Cheng, S.; Lu, Y.; Sun, L.; Wang, C.; Li, Y.; Gao, H. Effect of different biological protective films on fruit quality of Ruixue apple. *J. Gansu Agric. Univ.* **2022**, *57*, 96–102. [CrossRef]
16. Bai, Q.; Shen, Y.; Huang, Y. Advances in Mineral Nutrition Transport and Signal Transduction in Rosaceae Fruit Quality and Postharvest Storage. *Front. Plant Sci.* **2021**, *12*, 620018. [CrossRef] [PubMed]
17. Huang, X.; Chen, T.; Zhou, P.; Huang, X.; Liu, D.; Jin, W.; Zhang, H.; Zhou, J.; Wang, Z.; Gao, Z. Prediction and Optimization of Fruit Quality of Peach Based on Artificial Neural Network. *J. Food Compos. Anal.* **2022**, *111*, 104604. [CrossRef]
18. Ban, Q.; Liu, T.; Ning, K.; Fan, J.; Cui, Q.; Guo, Y.; Zai, X. Effect of Calcium Treatment on the Browning of Harvested Eggplant Fruits and Its Relation to the Metabolisms of Reactive Oxygen Species (ROS) and Phenolics. *Food Sci. Nutr.* **2021**, *9*, 5567–5574. [CrossRef]
19. He, X.; Zhang, H.; Li, J.; Yang, F.; Dai, W.; Xiang, C.; Zhang, M. The Positive Effects of Humic/Fulvic Acid Fertilizers on the Quality of Lemon Fruits. *Agronomy* **2022**, *12*, 1919. [CrossRef]
20. Zizkova, P.; Stefek, M.; Rackova, L.; Prnova, M.; Horakova, L. Novel Quercetin Derivatives: From Redox Properties to Promising Treatment of Oxidative Stress Related Diseases. *Chem.-Biol. Interact.* **2017**, *265*, 36–46. [CrossRef]
21. Gouveia, H.J.C.B.; Urquiza-Martínez, M.V.; Manhães-de-Castro, R.; Costa-de-Santana, B.J.R.; Villarreal, J.P.; Mercado-Camargo, R.; Torner, L.; De Souza Aquino, J.; Toscano, A.E.; Guzmán-Quevedo, O. Effects of the Treatment with Flavonoids on Metabolic Syndrome Components in Humans: A Systematic Review Focusing on Mechanisms of Action. *Int. J. Mol. Sci.* **2022**, *23*, 8344. [CrossRef]
22. Sut, S.; Zengin, G.; Maggi, F.; Malagoli, M.; Dall'Acqua, S. Triterpene Acid and Phenolics from Ancient Apples of Friuli Venezia Giulia as Nutraceutical Ingredients: LC-MS Study and In Vitro Activities. *Molecules* **2019**, *24*, 1109. [CrossRef] [PubMed]
23. Wang, X.; Wei, Z.; Ma, F. The Effects of Fruit Bagging on Levels of Phenolic Compounds and Expression by Anthocyanin Biosynthetic and Regulatory Genes in Red-Fleshed Apples. *Process Biochem.* **2015**, *50*, 1774–1782. [CrossRef]
24. Yuri, J.A.; Neira, A.; Fuentes, M.; Razmilic, I.; Lepe, V.; González, M.F. Bagging Cv. Fuji, Raku Raku Apple Fruit Affects Their Phenolic Profile and Antioxidant Capacity. *Erwerbs-Obstbau* **2020**, *62*, 221–229. [CrossRef]
25. Feng, F.; Li, M.; Ma, F.; Cheng, L. The Effects of Bagging and Debagging on External Fruit Quality, Metabolites, and the Expression of Anthocyanin Biosynthetic Genes in 'Jonagold' Apple (*Malus domestica* Borkh.). *Sci. Hortic.* **2014**, *165*, 123–131. [CrossRef]

horticulturae

Article

Exploitation of Post-Ripening Treatment for Improving Cold Tolerance and Storage Period of Jin Huang Mango

Ying-Che Lee [1], Meng-Chieh Yu [1], Chi-Yun Yen [1], Jyh-Shyan Tsay [2], Chih-Yao Hou [3], Po-Hsien Li [4], Ping-Hsiu Huang [5,*,†] and Yu-Shen Liang [1,*,†]

1 Department of Plant Industry, National Pingtung University of Science and Technology, Pingtung 91201, Taiwan; leeyingche1954@gmail.com (Y.-C.L.); jamie880819@gmail.com (M.-C.Y.); aa0939599516@gmail.com (C.-Y.Y.)

2 Department of Horticulture and Landscape Architecture, National Taitung Junior College, Taitung 950, Taiwan; jishyan@ntc.edu.tw

3 Department of Seafood Science, College of Hydrosphere, National Kaohsiung, University of Science and Technology, Kaohsiung 81157, Taiwan; chihyaohou@gmail.com

4 Department of Food and Nutrition, Providence University, 200, Sec. 7, Taiwan Boulevard, Shalu Dist., Taichung City 43301, Taiwan; pohsien0105@pu.edu.tw

5 School of Food, Jiangsu Food and Pharmaceutical Science College, No. 4, Meicheng Road, Higher Education Park, Huaian 223003, China

* Correspondence: hugh0530@gmail.com (P.-H.H.); justinliang@mail.npust.edu.tw (Y.-S.L.); Tel.: +86-1301622982 (P.-H.H.); +886-8-7703202 (ext. 6247) (Y.-S.L.)

† These authors contributed equally to this work.

Abstract: The limited cold tolerance of the Jin Huang mango represents a significant impediment to its potential for international trade. Therefore, this study evaluated the quality index changes of green maturity Jin Huang mangoes with different post-ripening treatments and then when stored at different storage periods (7, 14, 21, and 28) at 4 °C followed by 6 days at 20 °C. This study showed that the mangoes treated with 500 ppm ethylene were slow to ripen during 4 °C storage, which could be sustainable even under 20 °C storage. In addition, the control (CK) group failed to mature or ripen unevenly after storage at 4 °C. Moreover, the T3 group (ethylene ripening for 1 day and post-ripening at 20 °C for 1 day) minimized the occurrence of CI during storage compared to the CK group while contributing to a 30% decrease in anthracnose incidence and a decrease in firmness and titratable acid (TA), while total soluble solids (TSS) notably increased, yet the ascorbic acid content in this group was lower. Hence, the treatment conditions of Jin Huang mango using T3 helped extend its shelf-life at 20 °C, stocking and minimizing CI and anthracnose, thereby maintaining a certain quality.

Keywords: mango; post-harvest; ripening; chilling injury; shipping; shelf-life

Citation: Lee, Y.-C.; Yu, M.-C.; Yen, C.-Y.; Tsay, J.-S.; Hou, C.-Y.; Li, P.-H.; Huang, P.-H.; Liang, Y.-S. Exploitation of Post-Ripening Treatment for Improving Cold Tolerance and Storage Period of Jin Huang Mango. *Horticulturae* **2024**, *10*, 103. https://doi.org/10.3390/horticulturae10010103

Academic Editor: Elazar Fallik

Received: 14 December 2023
Revised: 15 January 2024
Accepted: 19 January 2024
Published: 20 January 2024

1. Introduction

Mangifera indica Linn. was successfully hybridized with ♂keitt × ♀white and has long been circulating on the market with the breeder's name (trade name) Jin Huang mango [1]. It has a large size, high sugar content, low acidity, and high fiber content, with flat seeds and a high pulp ratio, and it is appreciated by consumers both for its bright yellow appearance and sweet pulp [2]. However, it is a typical perennial fruit with a short storage period (shelf-life), which will ripen rapidly post-harvest while undergoing drastic physiological changes during the ripening process, with respiration and ethylene peaks occurring. Moreover, anthracnose is a widespread factor that reduces the storage life of mangoes and is characterized by dark brown or black spots, pits, and decay on the mango peel, which results in severe economic losses [3].

Currently, it is common practice to extend the shelf-life of ripe mangoes using low-temperature storage. Yet, Jin Huang mango has poor cold tolerance; more specifically, it experiences chilling injury (CI), the inability to change peel color period, browning, depression, ripening disorders, poor taste, and sensitivity to spoilage, which are the adverse effects of low-temperature (5 °C–8 °C) storage [4–6]. Additionally, the market receives unripe fruits that are unevenly ripened, necessitating re-ripening treatment and repackaging, while these extra processes result in increased mechanical damage risks and labor costs [7], which are not economically feasible. Consequently, this contributes to the high prices of fruits at the final market end. Unfortunately, the mango faces significant challenges in international distribution due to its inherent difficulty in post-harvest and ripening. This poses a considerable obstacle for the mango industry and limits the potential commercial viability. Therefore, the appropriate mitigation and control of anthracnose, including suitable harvesting techniques, post-harvest handling, and storage practices, are essential. These would effectively prolong the shelf life of mangoes, minimize economic losses, and ensure a steady supply of quality mangoes.

Typically, mango fruits self-ripen on the tree, and most mango varieties can be considered fully ripe on the tree as the standard for harvesting, where the fruits have the best flavor for immediate consumption. In commercial marketing practices, it is generally advised against harvesting fruits when they are fully ripe; however, such fruits are more vulnerable to injury and have a shorter shelf life, which can present significant challenges during transportation and marketing. Since the Jin Huang mango tends to deteriorate in the flesh during ripening, large-scale cultivations commonly harvest at a definite point, such as 100–120 days after anthesis. Yet, the ripening of each fruit differs at the time of harvest, which causes uneven ripening speeds and, hence, variable quality in different fruits. Therefore, the artificial ripening technique involves an exogenous ripening agent, such as exogenous ethylene, which accelerates the ripening time to ensure more uniform ripening of the same batch of fruits and provides consistent quality after ripening [8,9]. Simultaneously, it offers superior quality fruit (appearance, color, smell, and flavor) to satisfy consumers' demands. Moreover, the cold sensitivity of ripened fruits decreases. Zhao et al. [10] reported that the CI index of unripened mangoes was higher than that of light yellow and yellow ones for 12 days of storage at 2 °C and rewarming to 25 °C. In addition, treating tomatoes with ethylene before storage or shipping can be more effective than post-storage treatments, which results in faster and more consistent ripening, leading to a more extended storage period and minimizing the risk of CI during low-temperature storage [7].

In this study, the aim of ripening was combined with refrigerated storage at low temperatures while simulating the transportation conditions of rewarming storage to investigate the different post-ripening treatments (no ripening, ethylene-ripening for 1 day, ethylene-ripening for 1 day + post-ripening for 1 day, and ethylene-ripening for 1 day + post-ripening for 2 days) for green maturity Jin Huang mangoes before low-temperature storage in the refrigerated storage at 4 °C. Simultaneously, we investigated the changes in the anthracnose incidence, CI index, ripening index, respiration rate, ethylene production, color analysis, firmness, total soluble solids (TSS), titratable acid (TA), and ascorbic acid contents of mango during the 28-day storage period and after 20 °C rewarming for 6 days, which could facilitate the evaluation of the feasibility of mango in the actual commercial operation, namely, the long-term low-temperature marine shipping.

2. Materials and Methods

2.1. Materials

The green maturity Jin Huang mangoes used in this study (three batches were harvested from May to July 2023, within 110–120 days after anthesis [11], and the detailed screening color parameters are shown in Figure A1) were purchased from a local fruit farmer (Pingtung, Taiwan). For this study, fruits of commensurate size and free from any signs of disease were selected, while any fruit that displayed signs of bruising was excluded. All mangoes were placed in PVC carriers and transported via a vehicle back to the laboratory (within 1 h of travel time). Afterward, the abraded fruits were excluded, and similar-sized fruits were selected for subsequent trials. All chemicals were purchased from Sigma-Aldrich® (Merck KGaA, Darmstadt, Germany) and used without any treatment unless otherwise stated.

2.2. Grouping and Pre-Processing

All the experimental mangoes were randomly arranged in sponge-lined baskets and divided into 4 treatment groups ($n = 90$) as follows: the control group (CK) was without any treatment; the T1 group was ripened with 500 ppm ethylene at 20 °C for 1 day; the T2 group was ripened with 500 ppm ethylene at room temperature for 1 day and ripened at 20 °C for 1 day; the T3 group was ripened with 500 ppm ethylene at 20 °C for 1 day and ripened at 20 °C for 2 days. Then, the mangoes with different treatments were refrigerated at 4 °C, and during the storage period, the chilling injury (CI) index, anthracnose incidence, ripening index, respiration rate, ethylene production color change, and physiological and quality changes were investigated every 7 days, and some of the mangoes were shifted to the 20 °C refrigerators for rewarming investigation on the 14th, 21st, and 28th days. Moreover, the CI index, anthracnose incidence, and color change were investigated daily during the warming period. In addition, peel color analysis was conducted on the 2nd, 4th, and 6th days, while fruit quality was analyzed on the 3rd and 6th days of the warming period.

Group	Treatment method
Control (CK)	Without any treatment
T1	Ripened with 500 ppm ethylene at 20 °C for 1 day
T2	Ripened with 500 ppm ethylene at 20 °C for 1 day and ripened at 20 °C for 1 day
T3	Ripened with 500 ppm ethylene at 20 °C for 1 day and ripened at 20 °C for 2 days

2.3. Chilling Injury (CI) Index

The determination of the CI index of mango in this study was performed as described by Kong et al. [12] with slight modifications. The symptoms of CI were observed with visual inspection for the presence of brown spots, localized browning, and pitting. Specifically, the appearance of scald-like spots (gray) on the epidermis of mangoes also includes browning and discoloration of the pulp [13]. The CI index was categorized into five levels, and the indexes were determined as follows: level 0: Without CI, while the CI area of level 1 is less than 2%; level 2: 2–10%; level 3: 10–20%; and level 4: more than 20%.

2.4. Anthracnose Incidence

The visual observation of the occurrence and determination of anthracnose in mangoes was based on an approach described by Mokgalapa et al. [14] with minor modifications. The symptoms of anthracnose present as black, irregular, and sunken lesions. Affected fruits, especially those of larger size, can undergo a prolonged incubation period, delaying the onset of disease until the final stages of ripening [15]. Prompt diagnosis and treatment are essential to prevent the spread of the disease and minimize losses in yield. Mangoes found to be naturally infected (quiescent) with anthrax at the observation time points during the storage period of this study were immediately removed, and the incidence of anthrax in each group was calculated using the following equation.

$$\text{Anthracnose incidence}(\%) = \frac{\text{Number of anthracnose} - \text{affected fruits}}{\text{total number of fruit}} \times 100 \quad (1)$$

2.5. Ripening Index

The appearance of color change in the mango was observed by the naked eye and determined according to the description in Raghavendra et al. [16], with slight modifications. The color change levels were classified into 5 levels, defined as follows: Level 0: bright green peel color. Level 1: greenish to yellowish peel color area. Level 2: the peel color area is more yellow than green. Level 3: bright yellow peel color. Level 4: dark yellow peel color.

2.6. Respiration Rate and Ethylene Production

In this study, mango respiration rate and ethylene content were determined according to those described by Cheng et al. [17] and Lawson et al. [18], respectively, with some modifications. Briefly, mangoes were kept in a 6 L breathing tank and then sealed, followed by sampling every 2 days using a 1 mL syringe at the outlet of the breathing tank for analysis. Subsequently, sample analysis was performed using a gas chromatograph (GC-8A, Shimadzu Co., Kyoto, Japan) equipped with a thermal conductivity detector and column of Porapak P (80/100, 2 m, 2 mm ID, GL Sciences Inc., Torrance, CA, USA) under the following conditions: 100 °C at the injection, 90 °C at the column, and respiration rate expressed in mg CO_2 $kg^{-1}h^{-1}$. In addition, the ethylene production measurement was changed to a flame ionization detector with a Porapak Q (100/120, 2 m, 2 mm ID, GL Sciences Inc.) column under the following analytical conditions: 100 °C at the injection, 80 °C at the column, and the ethylene production expressed as μL kg^{-1} h^{-1}.

2.7. Color Analysis

The color appearance of the mango sample was determined following an approach described by Lin et al. [19] and Nkhata, S. G. [20], with minor modifications. The *L*, *a*, and *b* values of the samples were determined by a colorimeter (SD 5000, Nippon Denshoku Ins., Co., Ltd., Tokyo, Japan). The *L* value (brightness) ranges from 0 to 100, with higher values representing brighter colors; *a* value (red, with positive values representing red and negative values representing green); and *b* value (yellow, with positive values representing yellow and negative values representing blue). In addition, the measurements were performed in the middle section on one side of the fruit (2 points were measured randomly and averaged), which was covered with a black cloth to avoid affecting the data results. Then, the total color change (ΔE) was calculated by the following equation.

$$\text{Total color difference}(\Delta E) = \sqrt{\left(\Delta L - CK\right)^2 + (\Delta a - CK)^2 + \left(\Delta b - CK\right)^2} \quad (2)$$

where the ΔL, Δa, and Δb are the measured values of the sample (changes in lightness, redness, and yellowness) minus the difference from the CK (control).

Moreover, the hue angle (θ value) and color concentration (C value) determinations and calculations were performed according to the description of Lawson et al. [18], with minor modifications. The θ value was calculated as the absolute values of b/a multiplied by the arctangent function, which was used to indicate the color change of the fruit (0° represents reddish purple, 90° represents yellow, 180° represents blue-green, and 270° represents blue). The C value was calculated by (a value^2 + b value^2)$^{1/2}$, and the higher value indicated that the C value of the mango peel was more intense.

2.8. Firmness

The determination of the firmness of the mango sample was based on the method described by Cheng et al. [17]. The firmness of the mango was measured by using a texture analyzer (EZ-Test 500N, Shimadzu Co., Japan, Tokyo) with the following conditions: use of a No. 5 probe (0.5 cm in diameter and 10 mm deep into the sample). Each mango was measured at a single point at the equator, both the front and back sides were measured, the values were averaged, and the firmness (N) was determined.

2.9. Total Soluble Solid (TSS)

TSS determination in mango was based on the method described by Wu et al. [21]. TSS was determined by taking the juice from the mango pulp and measuring it using a handheld brix meter (N-1E, Atago Co., Ltd., Tokyo, Japan), expressed in terms of °Brix.

2.10. Titratable Acid (TA)

TA determination in mango was based on the method described by Wu et al. [21], with some modifications. A total of 10 g of mango pulp was added to 100 mL of reverse osmosis water and then used a homogenizer (Hsiangtai Machinery Industry Co., Ltd., New Taipei, Taiwan) for homogenization and then filtration. Afterward, 25 mL of the clarified solution was titrated with 0.1 N NaOH to pH 8.1 titration endpoint using an automatic titrator (T5, Mettler Toledo, Columbus, OH, USA). The titration result was determined by the equivalence of malic acid and NaOH to determine the TA content.

2.11. Ascorbic Acid

The determination of the ascorbic acid of the mango sample was based on the method described by Cheng et al. [17]. Mango pulp (5 g) was added to 50 mL of 3% metaphosphoric acid (HPO_3), mixed evenly, and filtered. Then, 5 mL of the filtrate was added with 5 mL of HPO_3, which was titrated with indophenol until the color of the solution turned pink. The above process was repeated for the standard of 1 mM ascorbic acid. Finally, it was calculated by the following equation:

$$\text{Ascorbic acid content(mg/100 g)} = \frac{\text{Volume of indophenol used for sample titration (mL)}}{\text{Volume of indophenol used for titration of ascorbic acid standard (mL)}} \times \frac{50\text{ mL}}{5\text{ mL}} \times \frac{1\text{ g}}{5\text{ g}} \quad (3)$$

2.12. Statistical Analysis

All data in this study were analyzed with Statistical Analysis System (V9.0, SAS Institute, Cary, NC, USA) for one-way analysis of variance (ANOVA), followed by Fisher's Least Significant Difference (LSD) to analyze the differences at the significance level of $p < 0.05$. Each treatment was observed for thirty fruits, with every ten fruits as one repetition ($n = 1$), while all assays used six fruits per group and performed triplicate trials ($n = 3$). All the data in the figures were expressed as mean ± standard deviation (SD), and the figures were plotted with SigmaPlot (V10.0, Systat Software, Inc., Chicago, IL, USA).

3. Results

3.1. Changes in Different Treatments on the Appearance of Mango Fruits

This study showed the peel of CK and T1 remained green and yellowish during storage, while the peel of T2 and T3 turned yellow (Figure 1A). However, CI occurred in CK and T1 during storage and was most severe in CK, whereas it occurred to a lesser extent in T2 and T3. This was attributed to the increased post-ripening period before storage, which assisted in minimizing the occurrence of CI in the mango fruit. Afterward, all mango fruits were stored at 4 °C for 14, 21, and 28 days, respectively. A portion of the mangoes was removed and kept at 20 °C for 6 days. The results indicated that both CK and T1 peels showed visible signs of CI and a change in the color of the peels (Figure 1B). Moreover, the peels on T2 showed a bright yellow color and some anthracnose symptoms, while the peels of CK and T3 showed more severe anthracnose symptoms.

3.2. Effect of the Post-Ripening Period on the Incidence of Anthracnose in Post-Harvest Mangoes

This study revealed that the incidence of anthracnose was significantly higher in T3 compared to others during storage; in particular, there were 23% and 47% incidences of anthracnose at 14 and 21 days of storage, respectively, whereas there was significant anthracnose incidence in CK at 28 days of storage, while T1 exhibited no anthracnose incidence during the storage period (Figure 2A). Next, all fruits after 14, 21, and 28 days of storage were rewarmed (20 °C) and stored for 6 days. In this study, the anthracnose incidence of all fruits was positively correlated with the subsequent storage time, while T3 was the highest and T1 was the lowest. (Figure 2B–D). Notably, CK showed a rapid expression of anthracnose symptoms after 21 and 28 days of storage and rewarming.

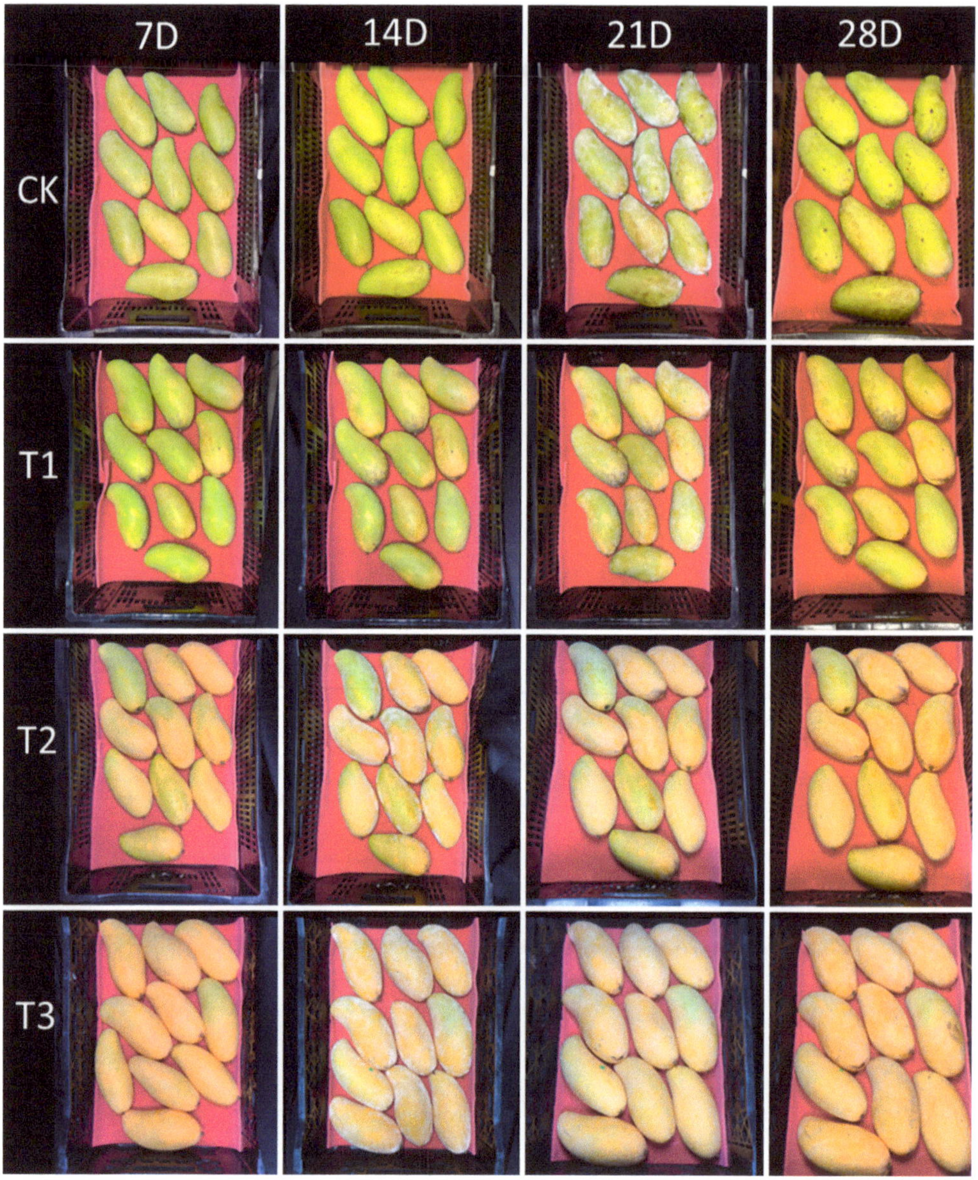

(A)

Figure 1. *Cont.*

(B)

Figure 1. Effects of the post-ripening period on the appearance of Jin Huang mango (**A**) stored at 4 °C for 7, 14, 21, and 28 days; (**B**) followed by those stored at 20 °C for 6 days. The control group (CK) was without any treatment; the T1 group was ripened with 500 ppm ethylene at 20 °C for 1 day; the T2 group was ripened with 500 ppm ethylene at 20 °C for 1 day and ripened at 20 °C for 1 day; the T3 group was ripened with 500 ppm ethylene at 20 °C for 1 day and ripened at 20 °C for 2 days.

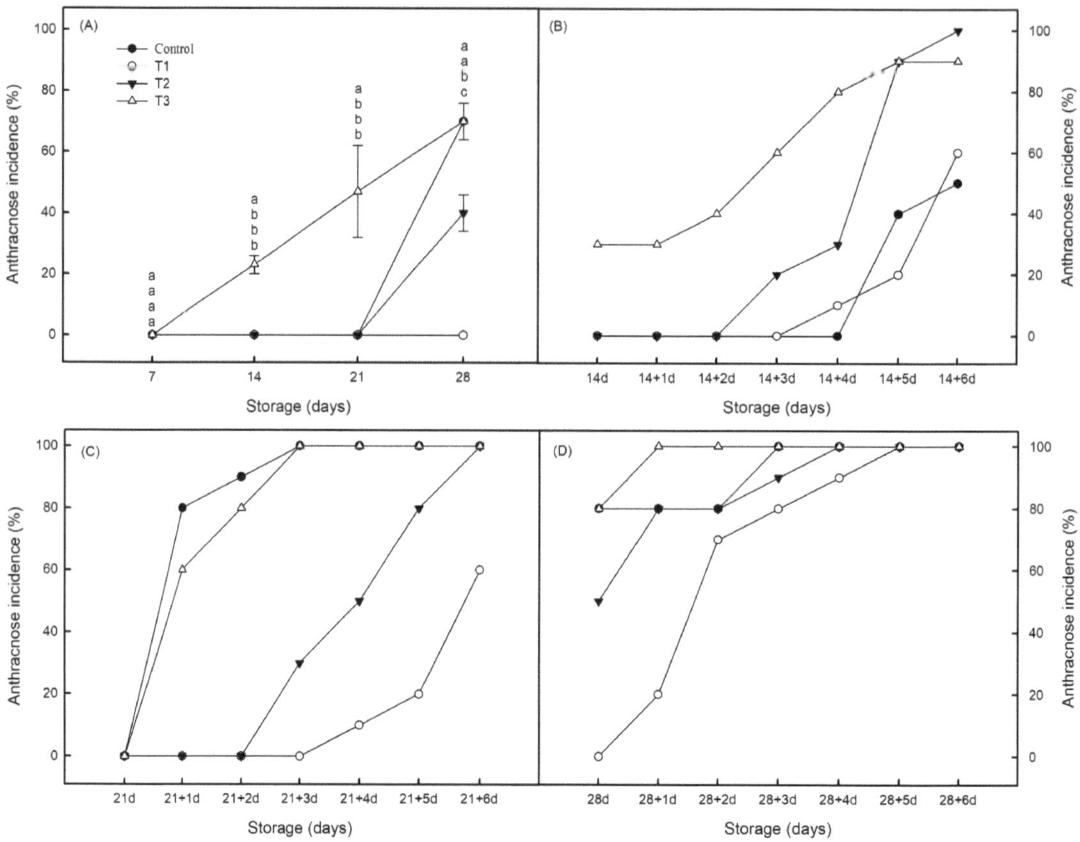

Figure 2. Effects of post-ripening time on the anthracnose incidence of post-harvest Jin Huang mango. (**A**) Storage at 4 °C; (**B**) storage at 4 °C for 14 days with rewarming (20 °C) for 6 days; (**C**) storage at 4 °C for 21 days with rewarming (20 °C) for 6 days; (**D**) storage at 4 °C for 28 days with rewarming (20 °C) for 6 days. The control group (CK; •) was without any treatment; the T1 group (○) was ripened with 500 ppm ethylene at 20 °C for 1 day; the T2 group (▼) was ripened with 500 ppm ethylene at 20 °C for 1 day and ripened at 20 °C for 1 day; the T3 group (△) was ripened with 500 ppm ethylene at 20 °C for 1 day and ripened at 20 °C for 2 days. The vertical bar represents the standard errors of the mean. Values with different letters indicate significant differences ($p < 0.05$).

3.3. Effects of the Post-Ripening Period on Chilling Injury (CI) Index of Post-Harvest Mangoes

This study showed that the CI index increased rapidly in both the CK and T1 groups during storage, while CK (3.5) was higher than the others during the 28-day storage period, and T3 was the lowest ($p < 0.05$) (Figure 3A). In addition, the CI index of CK and T1 were significantly higher ($p < 0.05$) for 14 and 21 days of storage at 4 °C with rewarming. This also indicated that the T2 and T3 treatments significantly minimized the CI index of mango, where the T3 treatment was the most effective, practically without any CI (Figure 3B,C). However, the CI index at 4 °C for 28 days of storage and rewarming in the CK and T3 groups was significantly higher than others ($p < 0.05$), whereas the T2 treatment was effective in minimizing ones (Figure 3D).

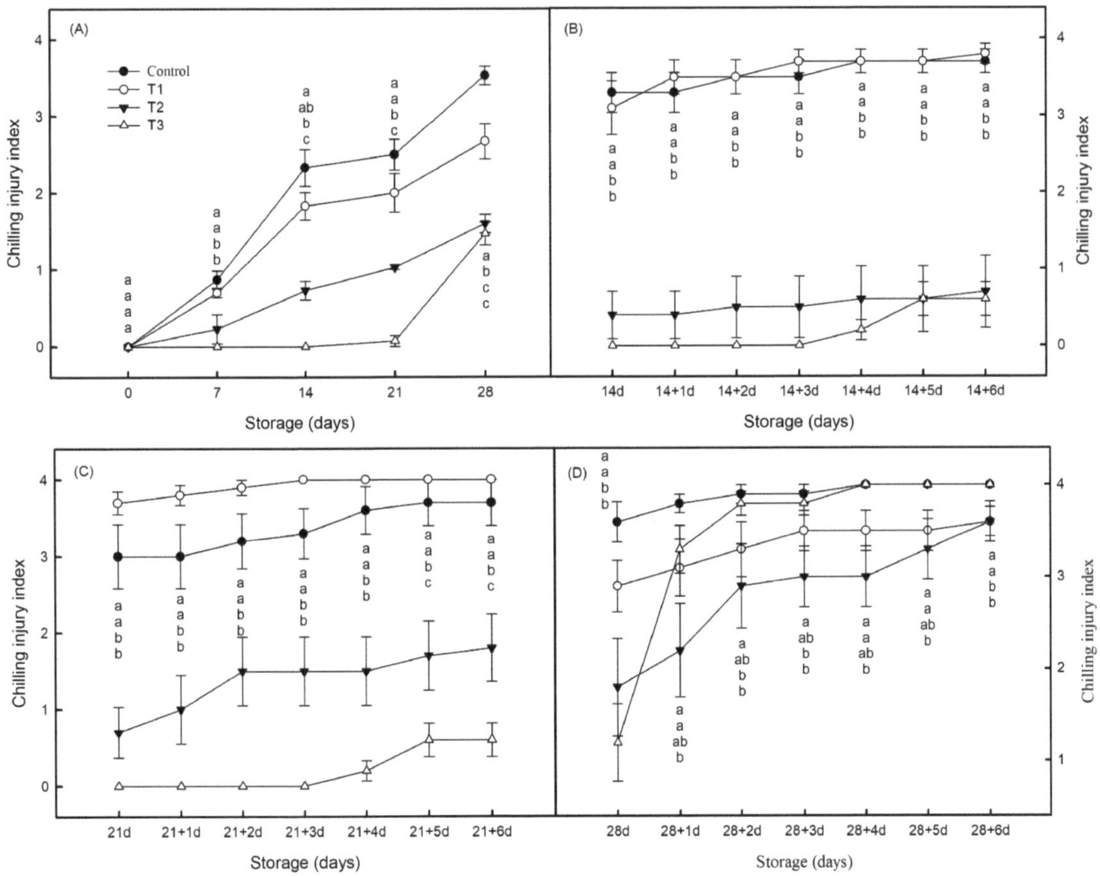

Figure 3. Effects of post-ripening time on the chilling injury index of post-harvest Jin Huang mango. (**A**) Storage at 4 °C; (**B**) storage at 4 °C for 14 days with rewarming (20 °C) for 6 days; (**C**) storage at 4 °C for 21 days with rewarming (20 °C) for 6 days; (**D**) storage at 4 °C for 28 days with rewarming (20 °C) for 6 days. The control group (CK; •) was without any treatment; the T1 group (○) was ripened with 500 ppm ethylene at 20 °C for 1 day; the T2 group (▼) was ripened with 500 ppm ethylene at 20 °C for 1 day and ripened at 20 °C for 1 day; the T3 group (△) was ripened with 500 ppm ethylene at 20 °C for 1 day and ripened at 20 °C for 2 days. The vertical bar represents the standard errors of the mean. Values with different letters indicate significant differences ($p < 0.05$).

3.4. Effects of the Post-Ripening Period on Ripening Index of Post-Harvest Mangoes

This study revealed the same trend of peel color change in mango during storage at 4 °C and rewarming (20 °C) at storage. However, CK without ripening treatment showed almost no discoloration, while the ripening index was lower than the ripened mango during the entire storage period (Figure A2A–D). Notably, the CK showed an abnormal post-ripening phenomenon after rewarming at storage. Moreover, the ripening index of fruits was higher and quicker as the duration of ripening treatment increased, where T3 exhibited the highest ripening index during the storage period, followed by T2, while T1 was slower ($p < 0.05$).

3.5. Effects of the Post-Ripening Period on Respiration Rate and Ethylene Production of Post-Harvest Mangoes

This study showed no noticeable change in the respiration rate of all groups during storage. In contrast, the respiration rate of the treatments with different ripening times was significantly higher than that of CK on the 7th day of storage ($p < 0.05$). In contrast, the T2 and T3 groups exhibited higher respiration rates of 12.6 and 13.9 mg CO_2 kg^{-1} h^{-1}, respectively (Figure 4A). In addition, the respiratory peak was reached in the T3 group at 14 days of storage (24.5 mg CO_2 kg^{-1} h^{-1}), whereas in the T1 group at 21 days of storage (16.6 mg CO_2 kg^{-1} h^{-1}), which was higher than others. Regarding ethylene production, CK showed a higher ethylene production of 1.1 μL kg^{-1} h^{-1} at 14 days of storage, whereas there was no significant difference between the groups, which were less than 1.0 μL kg^{-1} h^{-1} (Figure 4B).

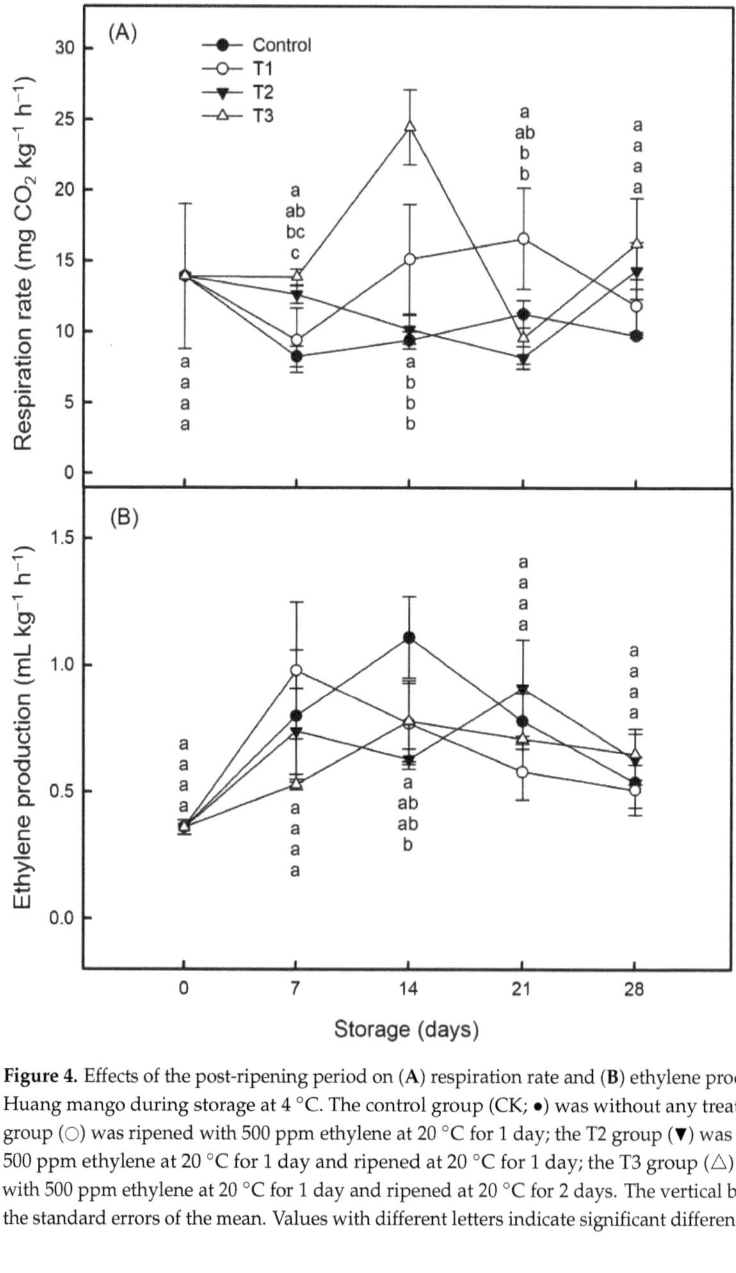

Figure 4. Effects of the post-ripening period on (**A**) respiration rate and (**B**) ethylene production of Jin Huang mango during storage at 4 °C. The control group (CK; •) was without any treatment; the T1 group (○) was ripened with 500 ppm ethylene at 20 °C for 1 day; the T2 group (▼) was ripened with 500 ppm ethylene at 20 °C for 1 day and ripened at 20 °C for 1 day; the T3 group (△) was ripened with 500 ppm ethylene at 20 °C for 1 day and ripened at 20 °C for 2 days. The vertical bar represents the standard errors of the mean. Values with different letters indicate significant differences ($p < 0.05$).

3.6. Effects of the Post-Ripening Period on the Post-Harvest Quality of Jin Huang Mangoes

3.6.1. Change in ΔE of the Peel

During the initial stage of treatment with varying post-ripening periods, the color parameters changed in all groups (Figure A3), which were significantly different from each other ($p < 0.05$). These implied that values increased over storage time, where the appearance of color in the T2 and T3 groups significantly differed from that of the CK group. In addition, the L^* and a^* values of all fruits under each treatment decreased compared to the initial values of each, which were reflected in the presentation of ΔE, meaning that the fruits tended to over-ripen; these results were also validated in the subsequent analysis (Section 3.7). However, during storage at 4 °C, the peel L^* value of the CK group was observed to have undergone a considerable decline, which was significantly more pronounced than that of the other groups, while this was also reflected in the value of ΔE. Regarding the storage in the warmed (20 °C) condition, the lowest peel L^* value was determined in the CK group, while the T1, T2, and T3 groups were significantly higher, which also leads to differences in the ΔE (Figure A3B–D). This was attributed to the lack of post-ripening treatment in the CK group, thereby contributing to CI during storage. In contrast, the other groups showed relatively slight CI due to the post-ripening treatment, which enhanced the tolerance to low temperatures. It was recommended that further investigations be conducted to elucidate the underlying mechanisms responsible for these observations.

3.6.2. Changes in the Hue Angle (θ Value) of the Peel

This study showed that during storage at 4 °C, the CK group had the highest peel θ value, which ranged from 99.6 ± 0.9 to 103.9 ± 0.6 ($p < 0.05$), as indicated in Figure A4A. However, the three ripening treatment groups showed a significantly lower peel θ value, while the T3 group exhibited the lowest value, ranging from 85.9 ± 0.8 to 86.7 ± 1.0 during storage. In contrast, the T2 group consistently declined the peel θ value throughout storage, starting from day 14 and persisting until day 28. This trend suggested a gradual deterioration in the mango quality of those that were stored. Moreover, a similar trend occurred for the 4 °C stored fruits, which were changed to the rewarming (20 °C) storage, and the peel θ value gradually decreased in all groups with increasing rewarming time (Figure A4B–D).

3.6.3. Changes in the Color Concentration (C Value) of the Peel

This study showed that the peel C value of mango with different ripening treatments during storage at 4 °C ranged from high to low in T3, T2, T1, and CK groups, respectively. This indicated that the peel C values of mango were significantly increased ($p < 0.05$) by ripening treatments (Figure A5). During storage at rewarming (20 °C), the C values of the ripening treatment groups were also significantly higher than that of the CK group (Figure A5B–D), which suggested that there were no significant differences in the effects of the two storage conditions on the C value of the mango peel.

3.6.4. Changes in the Firmness of the Mango

This study showed that the firmness of the ripened mango groups was significantly lower than that of the CK group during storage at 4 °C ($p < 0.05$) (Figure 5A), which was negatively correlated with the storage time. However, the firmness of mango fruits of each group in descending order was CK, T1, T2, and T3. Similar trends were also observed after rewarming, where increased post-ripening contributed to a decrease in firmness during storage, which also decreased with increased storage time (Figure 5B–D).

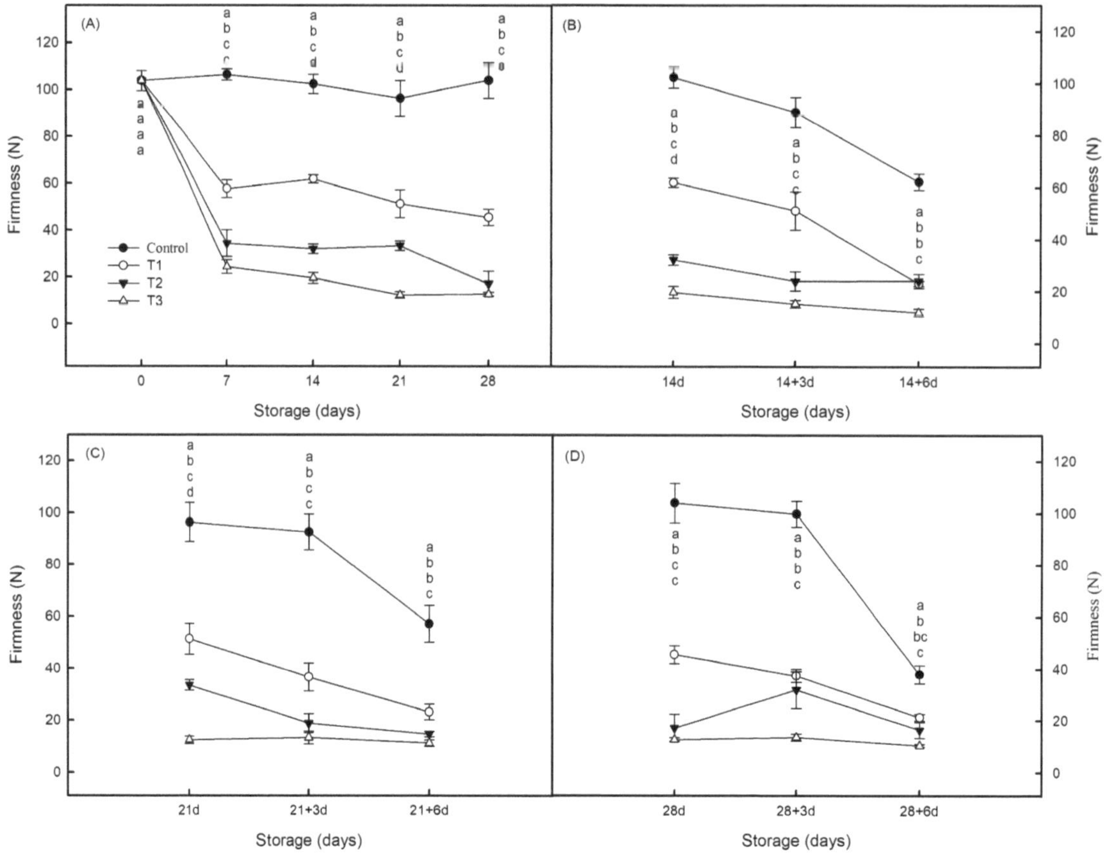

Figure 5. Effects of post-ripening time on the firmness of post-harvest Jin Huang mango. (**A**) Storage at 4 °C; (**B**) storage at 4 °C for 14 days with rewarming (20 °C) for 6 days; (**C**) storage at 4 °C for 21 days with rewarming (20 °C) for 6 days; (**D**) storage at 4 °C for 28 days with rewarming (20 °C) for 6 days. The control group (CK; •) was without any treatment; the T1 group (○) was ripened with 500 ppm ethylene at 20 °C for 1 day; the T2 group (▼) was ripened with 500 ppm ethylene at 20 °C for 1 day and ripened at 20 °C for 1 day; the T3 group (△) was ripened with 500 ppm ethylene at 20 °C for 1 day and ripened at 20 °C for 2 days. The vertical bar represents the standard errors of the mean. Values with different letters indicate significant differences ($p < 0.05$).

3.6.5. Changes in the Total Soluble Solids (TSS) Content of the Mango

This study showed that the CK group exhibited the lowest TSS content (ranging from 5.8 ± 0.7. Brix to 7.6 ± 1.4. Brix) during storage at 4 °C (Table 1). However, the TSS contents increased with post-ripening treatment time, particularly in the T2 and T3 groups, which exhibited higher TSS contents. In contrast, it was discovered that the TSS content of the post-ripening treatment groups was significantly greater than that of the CK group during the rewarming storage period (20 °C) ($p < 0.05$). However, no significant differences were observed among the treatment groups. Notably, the TSS content of the CK group tended to increase as the rewarming period of storage increased (14–28 days).

The control group (CK) was without any treatment; the T1 group was ripened with 500 ppm ethylene at 20 °C for 1 day; the T2 group was ripened with 500 ppm ethylene at 20 °C for 1 day and ripened at 20 °C for 1 day; the T3 group was ripened with 500 ppm ethylene at 20 °C for 1 day and ripened at 20 °C for 2 days.

Table 1. Effects of post-ripening time on the total soluble solids (TSS), titratable acid (TA), and ascorbic acid contents of post-harvest Jin Huang mango.

Treatment	Group/Measurement Item											
	CK	T1	T2	T3	CK	T1	T2	T3	CK	T1	T2	T3
	Total Soluble Solids (°Brix)				Titratable Acid (%)				Ascorbic Acid (mg/100 g)			
0 Day (D)	6.20 ± 0.32a				0.56 ± 0.03a				24.10 ± 1.14a			
4 °C, 7D	5.90 ± 0.32b	8.47 ± 1.26b	12.80 ± 0.65a	14.30 ± 0.76a	0.46 ± 0.00b	0.48 ± 0.05b	0.52 ± 0.02ab	0.59 ± 0.03a	20.91 ± 0.07b	23.40 ± 0.07a	21.08 ± 0.33b	18.82 ± 1.19c
4 °C,14D	7.57 ± 1.39c	10.73 ± 0.52b	12.87 ± 0.64ab	14.63 ± 0.37a	0.68 ± 0.01b	0.82 ± 0.03a	0.52 ± 0.07c	0.77 ± 0.03ab	20.69 ± 0.33c	25.8 ± 0.79a	22.42 ± 0.46b	21.88 ± 0.09bc
4 °C, 21D	6.50 ± 0.67c	11.7 ± 0.61b	12.8 ± 0.25ab	14.23 ± 0.12a	0.52 ± 0.08b	0.63 ± 0.09ab	0.87 ± 0.05a	0.68 ± 0.14ab	20.68 ± 0.82a	18.44 ± 0.62ab	19.80 ± 0.39ab	16.83 ± 1.74b
4 °C, 28D	5.77 ± 0.69c	13.50 ± 0.51b	15.77 ± 0.61a	15.3 ± 0.71ab	0.68 ± 0.03a	0.68 ± 0.11a	0.67 ± 0.04a	0.63 ± 0.01a	15.36 ± 0.30c	18.84 ± 0.60bc	25.19 ± 1.78a	20.12 ± 1.37b
4 °C, 7D	5.90 ± 0.32b	8.47 ± 1.26b	12.80 ± 0.65a	14.30 ± 0.76a	0.46 ± 0.00b	0.48 ± 0.05b	0.52 ± 0.02ab	0.56 ± 0.27a	20.91 ± 0.07b	23.40 ± 0.07a	21.08 ± 0.33b	18.82 ± 1.19c
4 °C, 7D + 20 °C, 3D	9.37 ± 0.26c	14.53 ± 0.67b	16.80 ± 1.01ab	17.73 ± 1.27a	0.54 ± 0.03c	0.67 ± 0.03b	0.78 ± 0.02a	0.63 ± 0.02b	17.61 ± 0.26a	17.97 ± 0.57a	17.13 ± 0.79a	16.93 ± 1.30a
4 °C, 7D + 20 °C, 6D	16.67 ± 0.86a	15.93 ± 1.34a	16.10 ± 0.61a	15.43 ± 1.23a	0.88 ± 0.12a	0.57 ± 0.02b	0.51 ± 0.02b	0.38 ± 0.02b	30.96 ± 3.39a	10.53 ± 0.96c	16.50 ± 0.52b	16.83 ± 0.74b
4 °C,14D	7.57 ± 1.39c	10.73 ± 0.52b	12.87 ± 0.64ab	14.63 ± 0.37a	0.69 ± 0.01b	0.82 ± 0.03a	0.52 ± 0.07c	0.77 ± 0.03ab	20.69 ± 0.33c	25.8 ± 0.79a	22.42 ± 0.46b	21.88 ± 0.09bc
4 °C,14D + 20 °C, 3D	10.70 ± 1.06c	13.70 ± 0.29ab	12.30 ± 0.23bc	14.93 ± 0.22a	0.83 ± 0.05a	0.72 ± 0.09a	0.87 ± 0.10a	0.79 ± 0.12a	21.27 ± 0.40a	23.52 ± 0.46a	24.73 ± 2.83a	20.41 ± 0.61a
4 °C, 14D + 20 °C, 6D	12.23 ± 0.67c	16.76 ± 0.35a	15.37 ± 0.26b	14.50 ± 0.26b	0.83 ± 0.06a	0.64 ± 0.00b	0.86 ± 0.04a	0.35 ± 0.02c	21.08 ± 0.95a	9.96 ± 0.80c	15.36 ± 0.54b	16.46 ± 0.68b
4 °C, 21D	6.50 ± 0.67c	11.7 ± 0.61b	12.8 ± 0.25a	14.23 ± 0.12ab	0.52 ± 0.08b	0.63 ± 0.09ab	0.87 ± 0.05a	0.68 ± 0.14ab	20.68 ± 0.82a	18.44 ± 0.62ab	19.80 ± 0.39ab	16.83 ± 1.74b
4 °C, 21D + 20 °C, 3D	8.53 ± 2.11b	16.00 ± 0.35a	16.97 ± 0.89a	16.27 ± 1.50a	0.76 ± 0.05a	0.74 ± 0.01ab	0.65 ± 0.04b	0.69 ± 0.01ab	22.91 ± 0.78a	13.65 ± 0.62c	22.24 ± 1.76ab	18.74 ± 0.83b
4 °C, 21D + 20 °C, 6D	15.70 ± 0.52a	16.57 ± 0.33a	16.80 ± 0.26a	15.7 ± 0.23a	0.71 ± 0.06a	0.53 ± 0.03bc	0.65 ± 0.05ab	0.42 ± 0.02c	18.56 ± 0.80a	17.36 ± 0.98a	18.31 ± 0.46a	19.98 ± 1.01a
4 °C, 28D	5.77 ± 0.69c	13.50 ± 0.51b	15.77 ± 0.61a	15.3 ± 0.71ab	0.68 ± 0.03a	0.69 ± 0.11a	0.67 ± 0.04a	0.63 ± 0.01a	15.36 ± 0.3c	18.84 ± 0.6bc	25.19 ± 1.78a	20.12 ± 1.37b
4 °C, 28D + 20 °C, 3D	7.67 ± 0.67b	15.60 ± 0.38a	14.60 ± 0.51a	16.67 ± 0.95a	0.87 ± 0.02a	0.61 ± 0.02b	0.63 ± 0.04b	0.44 ± 0.01c	26.80 ± 1.97a	18.33 ± 0.46c	23.34 ± 0.52ab	21.06 ± 1.98bc
4 °C, 28D + 20 °C, 6D	13.17 ± 0.64b	16.47 ± 0.65a	15.60 ± 0.89a	16.03 ± 0.33a	0.58 ± 0.06ab	0.61 ± 0.04a	0.43 ± 0.05b	0.47 ± 0.06ab	23.30 ± 1.45a	10.80 ± 0.95c	10.76 ± 0.33c	16.25 ± 0.76b

Different letters in lowercase in the same column represent significant differences ($p < 0.05$).

3.6.6. Changes in the Titratable Acid (TA) Content of the Mango

This study showed that the TA content of each group during 4 °C storage initially increased with the storage time, then decreased slightly until stabilized (Table 1). These results indicate that the storage time influenced the TA content of the mangoes. However, after 14 and 21 days of storage at 4 °C then storage at a rewarming temperature (20 °C) for 6 days, the TA contents of the T1 and T3 groups were significantly lower, whereas the CK and T2 groups exhibited higher TA contents. In addition, the lower TA contents of the T2 and T3 groups were detected after 28 days at 4 °C and then 6 days of storage at 20 °C. However, the TA contents of mangoes in all treatment groups tended to decrease significantly during the rewarming storage period.

3.6.7. Changes in the Ascorbic Acid Content of the Mango

In this study, except for the T2 group, which showed a higher ascorbic acid content after storage at 4 °C, other groups showed a gradual decrease in ascorbic acid contents as storage time increased; from highest to lowest, they are ranked as T2, T3, T1, and CK, which were significantly different from each other ($p < 0.05$) (Table 1). However, after 14 and 28 days of storage at 4 °C and then 6 days of rewarming (20 °C), the ascorbic acid contents tended to decrease in all groups. Interestingly, after 21 days of storage at 4 °C followed by rewarming, there was a decreasing trend in CK and T2, whereas there was an increasing trend in T1 and T3. Yet, there were no significant differences for all groups.

3.7. Changes in the Appearance of Mango Pulp

This study showed that the pulps of CK and T groups were light yellow, and there were no significant changes during storage at 4 °C, whereas the pulps of T2 and T3 groups were markedly yellow and turned more yellow as storage time increased (Figure 6A). Subsequently, in the 6 days of storage at a rewarming temperature (20 °C), the pulp of the CK group showed uneven ripening and was waterlogged, as opposed to the fully ripened and dark yellow pulps of the ripened groups (Figure 6B).

(A)

Figure 6. *Cont.*

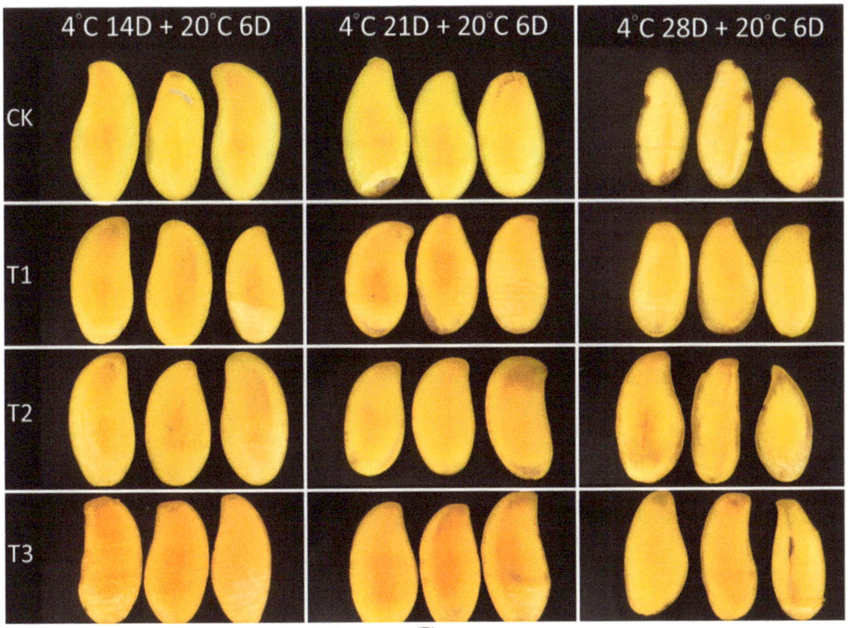

(B)

Figure 6. The effects of the post-ripening period on pulps appearance of post-harvest Jin Huang mangoes (**A**) stored at 4 °C for 14, 21, and 28 days; (**B**) followed by stored at 20 °C for 6 days. The control group (CK) was without any treatment; the T1 group was ripened with 500 ppm ethylene at 20 °C for 1 day; the T2 group was ripened with 500 ppm ethylene at 20 °C for 1 day and ripened at 20 °C for 1 day; the T3 group was ripened with 500 ppm ethylene at 20 °C for 1 day and ripened at 20 °C for 2 days.

4. Discussion

Anthracnose pathogens are known to form wounds on the fruit surface, whereas ethylene improves the permeability of the wounds and releases nutrients to facilitate the bio-viability of the attached pathogens, thus enhancing the probability of the fruit being infected by the pathogens [3,14,22]. Moreover, anthracnose is no longer a minor issue but a major concern for exporters, given the crucial demand for top-notch produce in international markets [23]. It was reported that the most common fungal isolates currently associated with mangoes involve 13 species, mainly *C. asianum* and *C. siamense*, which accounted for 60% of the germination [24]. It is worth mentioning that anthracnose has asexual spores (conidia) that can be transmitted, while spores contacting the susceptible tissues of the host begin to infect the host. The pathogen reproduces, culminating in developing the disease in the fruit [25]. More specifically, *Colletotrichum* spp. utilizes a semi-biotrophic mode of infection to penetrate, colonize, and spread within susceptible host plants; i.e., black adherent cysts are formed that penetrate the host and form harmless primary mycelium during the biotrophic stage of infection. Secondary hyphae are then formed during the necrotic stage, leading to spore colonization on the surface of infected tissues of the host, resulting in continuous repetitive pathogen transmission and infection [26]. However, this study showed that storage at 4 °C for 28 days in the T1 group and storage at 4 °C for 21 days in the T3 group effectively minimized anthracnose incidence development (Figure 2A). It is worth mentioning that the CK and T1 groups storage at 4 °C for 14–21 days before switching to rewarming storage for 2–4 days also effectively decreases the incidence of anthracnose (Figure 2B,C). As fruits ripened, anthracnose incidence increased [27–29]. Furthermore, it was suggested that fruit stored at 20 °C ripens faster and is more susceptible to disease than fruit stored at 10 °C and 15 °C, with a faster increase in disease severity [30].

The invasion of fruit pathogens will not show symptoms immediately and only show red pinpoints on the fruit surfaces, indicating latent infection. Rain or dew will also assist in spreading the infection, namely, the appearance of reddish-colored tear spots on the surface of the fruit [31]. The needle-like spots of latent infection expand after ripening, and black, sunken, irregularly shaped spots appear. However, the spots expand rapidly when the fruits are ripe, eventually becoming water-soaked and rotted, making it a severe storage disease [22,31].

Previously, this study confirmed that the T1 group treatment (ripened with 500 ppm ethylene at 20 °C for 1 day) and the modification of storage temperature contributed to the prolonged shelf-life and incidence of anthracnose during the post-ripening mango storage.

Moreover, CI is the primary limiting factor in mango quality maintenance during long-term refrigeration and ambient temperature shelf-life, where the typical symptom of CI was incomplete post-ripening [32]. In particular, exogenous ethylene supplementation during the storage of tomatoes [33] and mangoes [34–36] provided excellent relief of fruit CI symptoms. This was attributed to ethylene being a gaseous hormone that regulates fruit post-ripening, aging, and response to adversity [8,36], and the blockade of ethylene biosynthesis under low-temperature adversity was shown to be a key factor contributing to the CI of some fruits. Therefore, the high-maturity fruits were less susceptible to CI than poorly matured ones [33]. Simultaneously, ethylene treatment promotes mango peel color change during storage, associated with pigment metabolism in the peel involved in the endogenous enzyme activities, anthocyanin, and carotenoid accumulation [35].

However, this study confirmed that T2 and T3 treatments were significantly associated with decreased CI index (Figure 3A–C), where T3 treatment was the most effective (Figure 3B). Unfortunately, the CI symptoms were evident after rewarming the fruits from low-temperature storage to high, as observed in this study for 28 days of storage at 4 °C and subsequently storage at 20 °C of mangoes (Figure 3D). Similarly, Jiang et al. [37] reported that CI symptoms of 'Guifei' mango during storage at 5 °C were minor and exhibited a gradual darkening of the peel color with the appearance of small black spots while chilling symptoms of mango turned poorer rapidly after rewarming to 20 °C as severe browning, enlargement of the black spots, depression, and unripeness were observed. In terms of color appearance, this study found that in the CK group, which was severely chilled after 14 days of storage at 4 °C and then at 20 °C for another 6 days, there were still some green coloration, which might result from the chloroplast damage of CI [6,38–40].

Ethylene production and respiration rate serve as important indicators of mango physiological metabolism, whereas peaks of ethylene and respiration were characterized in the post-ripening process of mango [41]. Therefore, all groups in this study peaked in respiration rate and ethylene production during storage at 4 °C for 14–21 days (Figure 5A,B), despite a study by Jiang et al. [37], which reported that ethylene production in mango fruit was inhibited by storage at a low temperature of 5 °C. It was hypothesized that the possible explanation was caused by cultivar differences and differences associated with the ripening treatments before storage. Fruit softening is one of the annual ripening characteristics, significantly affecting shelf life and market value [42]. The softening process of fruits is caused by the degradation of polysaccharides such as starch, cellulose, and pectin by amylase, polygalacturonase, and pectin esterase enzymes, which leads to the structural degradation of the cell wall, thus contributing to the reduction of fruit firmness [21,43–45]. Therefore, the loss of cell expansion pressure and cell wall relaxation with fruit ripening and senescence resulted in a decrease in firmness [46], which agreed with the results of this study. Moreover, ethylene plays a crucial role in inducing gene expression, associated with starch conversion to sugar, with changes in the cell wall structure contributing to decreased fruit firmness [47]. However, in this study, treatment with ethylene prior to storage enhanced fruit ripening, leading to the gradual softening of fruits, presumably due to pectin hydrolysis [48,49].

Regarding TA and TSS, both influence the flavor of fruits and are indicators of their quality. In mango fruits, TSS content tends to increase with storage time due to the decomposition of large molecules such as starch and pectin into smaller ones, namely, the accumulation of sugar during respiration and the decrease in water content, thus leading to an increase in TSS content [43,50]. Therefore, this study confirmed that the TSS contents of mangoes treated with ripening agents were considerably higher than those of the CK group. The TSS contents kept increasing with ripening during storage. A similar study reported that the post-harvest application of exogenous ethylene increased TSS levels in Keitt mangoes [51]. Unfortunately, the TA content of mangoes is observed to decrease with prolonged storage time. This phenomenon is supported by previous research conducted by Montalvo et al. [36], who reported that the application of exogenous ethylene can cause a decrease in the TA of fruits. Moreover, Wang et al. [52] documented a rapid decline in the citric acid content of mangoes subjected to ethylene treatment. These findings highlight the significance of storage duration and ethylene treatment in preserving fruit quality.

Furthermore, ascorbic acid is one of the critical antioxidant components in the plant defense system, which can be minimized by producing free radicals during fruit ripening [53]. In this study, we observed that the ascorbic acid content of mango was significantly higher in the T1 and T2 groups during storage at 4 °C as the minimum CI, which was consistent with the findings of Taghipour et al. [54]. However, the occurrence of CI and lower ascorbic acid content in the CK group agreed with the findings of Ribeiro, B. S. and S. T. de Freitas [55], as CI is associated with the production of excessive reactive oxygen species (ROS) related to chloroplasts and mitochondria, such as hydrogen peroxide, superoxide radicals, and hydroxyl radicals, thereby leading to oxidative damage [12,27,37,38].

Taken together, the conditions of the T3 treatment in this study led to a positive effect on quality-related indicators, namely decreased TA content and increased TSS. This implies that mangoes improve overall quality (including flavor, color, texture, and nutrition) during post-harvest storage. Therefore, the results of this study may contribute to the improvement and control of the ripening of mangoes during storage and transportation operations with other potential practical applications.

5. Conclusions

This study confirmed that the pre-ripening treatment at 20 °C for 1 day and post-ripening for 1 day of Jin Huang mangoes with ethylene before storage was sustainable during storage (4 °C), while the accompanying change in temperature effectively mitigated the CI index of mangoes and provided the benefit of decreasing the anthracnose incidence. Moreover, maintaining fruit quality storage at 20 °C proved to be a practical approach to enhancing the storage and transportation of Jin Huang mangoes, which also contributed to the maintenance of the mango availability for consumption despite the decreased firmness, TA, and ascorbic acid content but increased TSS and favorable appearance. Additionally, this information could be highly beneficial for researchers and suppliers of Jin Huang mango, facilitating further development opportunities. However, it is necessary to conduct further exploration to comprehensively understand the underlying intricacies and develop a more effective combined approach for mitigating various effects on mango quality during storage.

Author Contributions: Conceptualization, Y.-C.L. and M.-C.Y.; methodology, Y.-C.L. and C.-Y.Y.; software, M.-C.Y., C.-Y.Y. and J.-S.T.; validation, J.-S.T., C.-Y.H. and P.-H.L.; formal analysis, Y.-C.L. and C.-Y.Y.; investigation, Y.-C.L., M.-C.Y., C.-Y.Y. and J.-S.T.; resources, Y.-S.L.; data curation, Y.-C.L., C.-Y.Y. and M.-C.Y.; writing—original draft preparation, Y.-C.L. and P.-H.H.; writing—review and editing, P.-H.H. and Y.-S.L.; visualization, Y.-C.L., M.-C.Y., C.-Y.Y. and J.-S.T.; supervision, C.-Y.H., P.-H.L. and Y.-S.L.; project administration, Y.-C.L. and Y.-S.L. All authors have read and agreed to the published version of the manuscript.

Funding: This research received no external funding.

Data Availability Statement: Data are contained within the article.

Acknowledgments: The authors would like to acknowledge all the individuals who volunteered for this study.

Conflicts of Interest: The authors declare that they have no known competing financial interests or personal relationships that could have appeared to influence the work reported in this paper.

Appendix A

Figure A1. The color of Jin Huang mango (harvested within 110–120 days after anthesis) on day 0 of each batch in this study. The specific conditions are as follows: (1) *L* value: 64.96 (average), 69.43 (maximum), 52.34 (minimum); (2) Hue angle (θ value): 103.89 (average), 107.91 (maximum), 98.99 (minimum); (3) Chroma (C value): 44.79 (average), 49.31 (maximum), 39.19 (minimum).

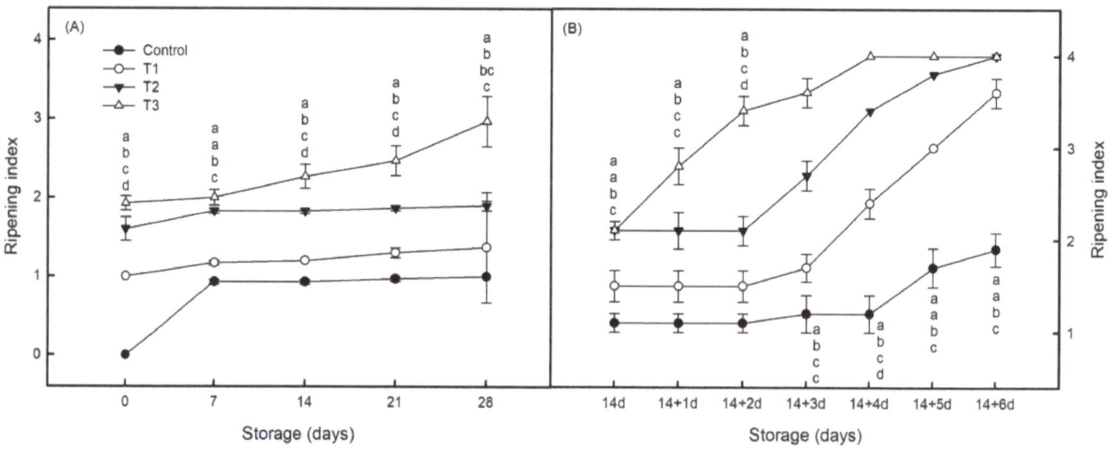

Figure A2. *Cont.*

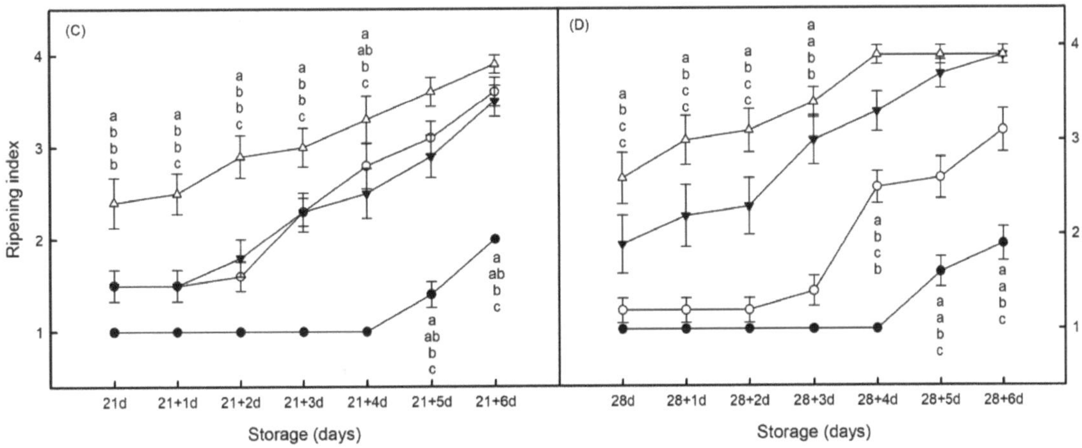

Figure A2. Effects of post-ripening time on the ripening index of post-harvest Jin Huang mango. (**A**) Storage at 4 °C; (**B**) storage at 4 °C for 14 days with rewarming (20 °C) for 6 days; (**C**) storage at 4 °C for 21 days with rewarming (20 °C) for 6 days; (**D**) storage at 4 °C for 28 days with rewarming (20 °C) for 6 days. The control group (CK; •) was without any treatment; the T1 group (○) was ripened with 500 ppm ethylene at 20 °C for 1 day; the T2 group (▼) was ripened with 500 ppm ethylene at 20 °C for 1 day and ripened at 20 °C for 1 day; the T3 group (△) was ripened with 500 ppm ethylene at 20 °C for 1 day and ripened at 20 °C for 2 days. The vertical bar represents the standard errors of the mean. Values with different letters indicate significant differences ($p < 0.05$).

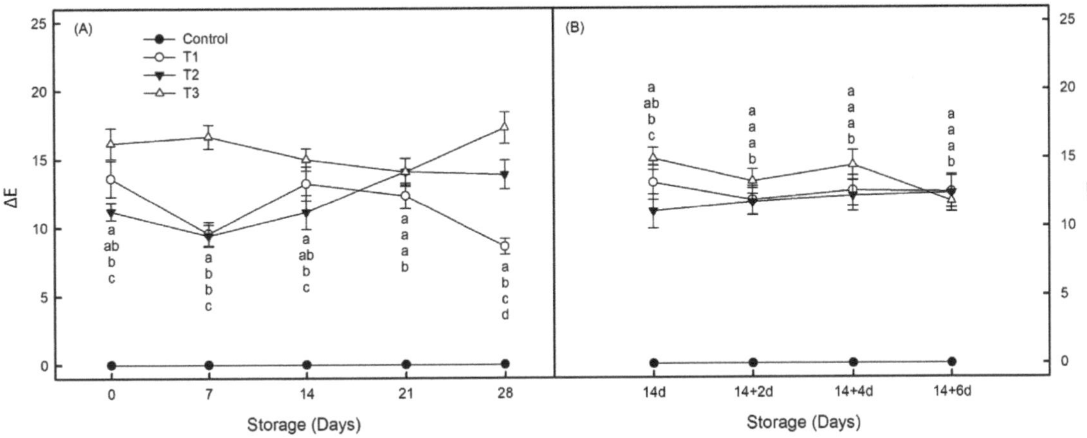

Figure A3. *Cont.*

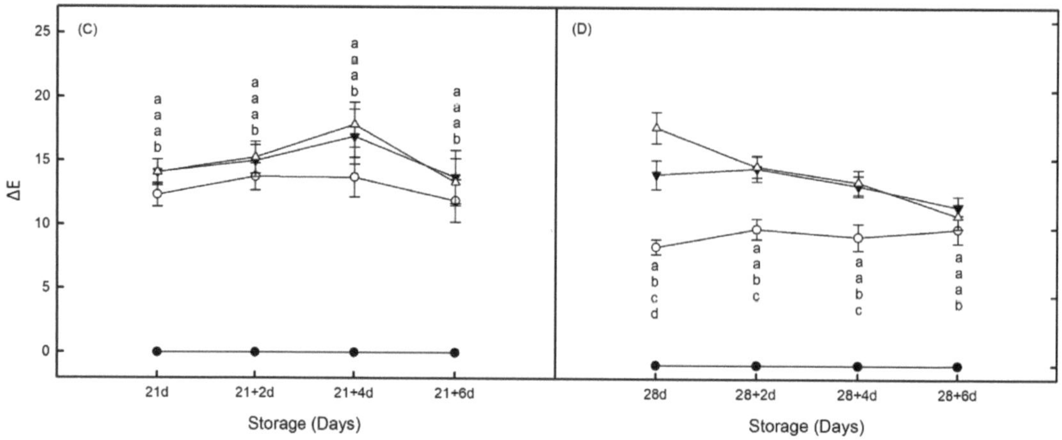

Figure A3. Effects of post-ripening time on the total color change (ΔE) of post-harvest Jin Huang mango. (**A**) Storage at 4 °C; (**B**) storage at 4 °C for 14 days with rewarming (20 °C) for 6 days; (**C**) storage at 4 °C for 21 days with rewarming (20 °C) for 6 days; (**D**) storage at 4 °C for 28 days with rewarming (20 °C) for 6 days. The control group (CK; •) was without any treatment; the T1 group (○) was ripened with 500 ppm ethylene at 20 °C for 1 day; the T2 group (▼) was ripened with 500 ppm ethylene at 20 °C for 1 day and ripened at 20 °C for 1 day; the T3 group (△) was ripened with 500 ppm ethylene at 20 °C for 1 day and ripened at 20 °C for 2 days. The vertical bar represents the standard errors of the mean. Values with different letters indicate significant differences ($p < 0.05$).

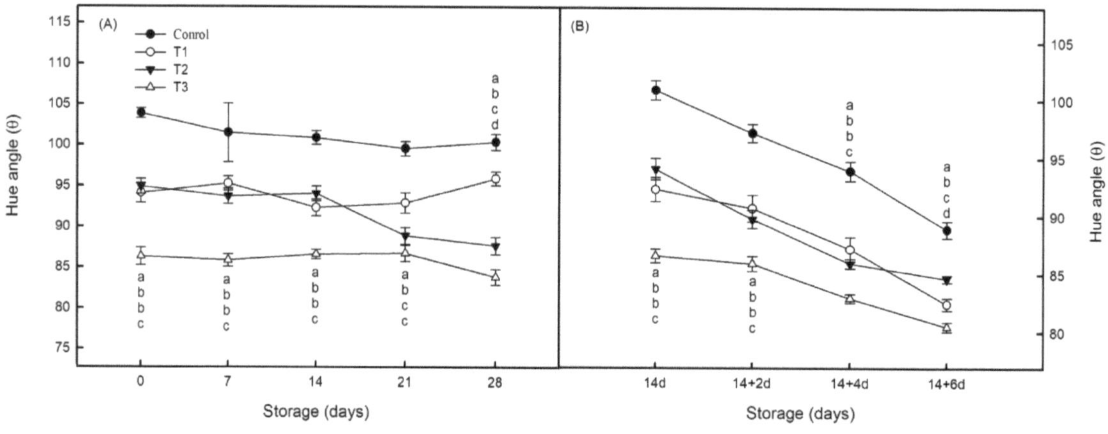

Figure A4. *Cont.*

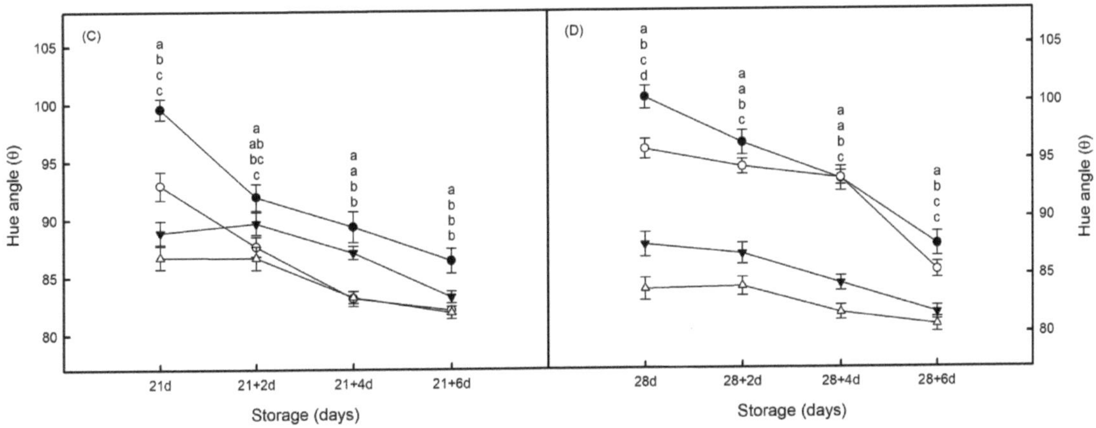

Figure A4. Effects of post-ripening time on the hue angle (θ value) of post-harvest Jin Huang mango. (**A**) Storage at 4 °C; (**B**) storage at 4 °C for 14 days with rewarming (20 °C) for 6 days; (**C**) storage at 4 °C for 21 days with rewarming (20 °C) for 6 days; (**D**) storage at 4 °C for 28 days with rewarming (20 °C) for 6 days. The control group (CK; •) was without any treatment; the T1 group (○) was ripened with 500 ppm ethylene at 20 °C for 1 day; the T2 group (▼) was ripened with 500 ppm ethylene at 20 °C for 1 day and ripened at 20 °C for 1 day; the T3 group (△) was ripened with 500 ppm ethylene at 20 °C for 1 day and ripened at 20 °C for 2 days. The vertical bar represents the standard errors of the mean. Values with different letters indicate significant differences ($p < 0.05$).

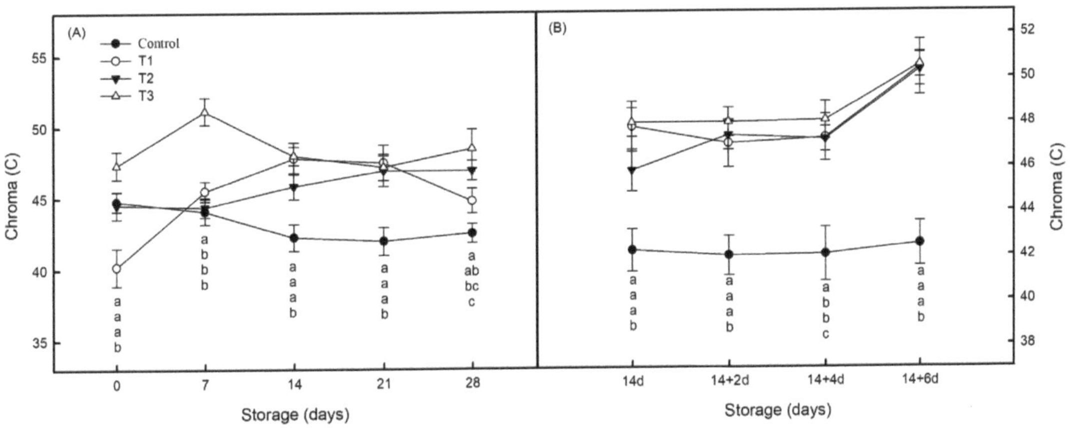

Figure A5. *Cont.*

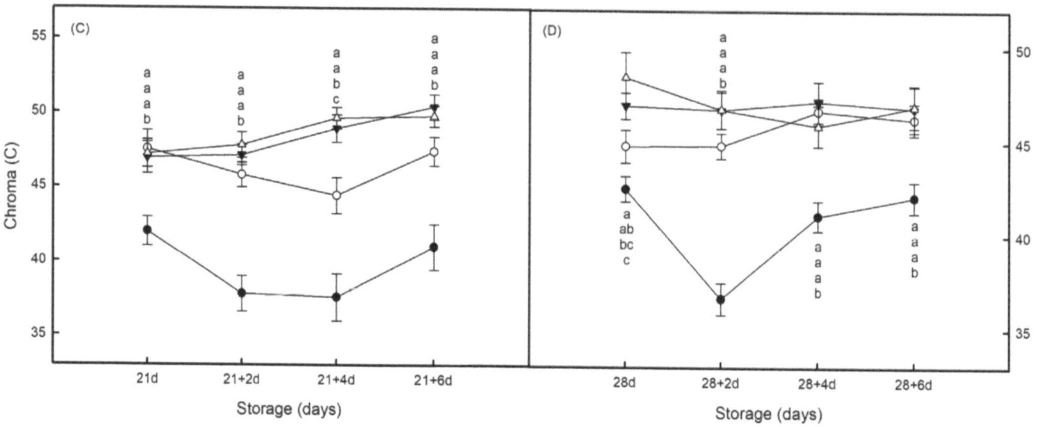

Figure A5. Effects of post-ripening time on the color concentration (C value) of post-harvest Jin Huang mango. (**A**) Storage at 4 °C; (**B**) storage at 4 °C for 14 days with rewarming (20 °C) for 6 days; (**C**) storage at 4 °C for 21 days with rewarming (20 °C) for 6 days; (**D**) storage at 4 °C for 28 days with rewarming (20 °C) for 6 days. The control group (CK; •) was without any treatment; the T1 group (○) was ripened with 500 ppm ethylene at 20 °C for 1 day; the T2 group (▼) was ripened with 500 ppm ethylene at 20 °C for 1 day and ripened at 20 °C for 1 day; the T3 group (△) was ripened with 500 ppm ethylene at 20 °C for 1 day and ripened at 20 °C for 2 days. The vertical bar represents the standard errors of the mean. Values with different letters indicate significant differences ($p < 0.05$).

References

1. David. Meet the Jing Huang Mango. 2019. Available online: https://youtu.be/2bwbFakMNsQ (accessed on 17 September 2023).
2. Lin, H.L.; Shiesh, C.C.; Chen, P.J. Physiological disorders in relation to compositional changes in mango (*Mangifera indica* L. 'Chiin hwang') fruit. *Acta Hortic.* **2013**, *984*, 357–363. [CrossRef]
3. Archana, T.J.; Gogoi, R.; Kaur, C.; Varghese, E.; Sharma, R.R.; Srivastav, M.; Tomar, M.; Kumar, M.; Kumar, A. Bacterial volatile mediated suppression of postharvest anthracnose and quality enhancement in mango. *Postharvest Biol. Technol.* **2021**, *177*, 111525. [CrossRef]
4. Phakawatmongkol, W.; Ketsa, S.; Doorn, W.G.v. Variation in fruit chilling injury among mango cultivars. *Postharvest Biol. Technol.* **2004**, *32*, 115–118. [CrossRef]
5. Sivankalyani, V.; Sela, N.; Feygenberg, O.; Zemach, H.; Maurer, D.; Alkan, N. Transcriptome dynamics in mango fruit peel reveals mechanisms of chilling stress. *Front. Plant Sci.* **2016**, *7*, 1579. [CrossRef] [PubMed]
6. Valenzuela, J.L.; Manzano, S.; Palma, F.; Carvajal, F.; Garrido, D.; Jamilena, M. Oxidative stress associated with chilling injury in immature fruit: Postharvest technological and biotechnological solutions. *Int. J. Mol. Sci.* **2017**, *18*, 1467. [CrossRef]
7. Chomchalow, S.; El Assi, N.M.; Sargent, S.A.; Brecht, J.K. Fruit maturity and timing of ethylene treatment affect storage performance of green tomatoes at chilling and nonchilling temperatures. *HortTechnology* **2002**, *12*, 104–114. [CrossRef]
8. Igiebor, F.A.; Odozi, E.B.; Ikhajiagbe, B. Chemical-based fruit ripening and the implications for ecosystem health and safety. In *One Health Implications of Agrochemicals and Their Sustainable Alternatives*; Ogwu, M.C., Chibueze Izah, S., Eds.; Springer Nature: Singapore, 2023; pp. 335–353.
9. Kou, X.; Feng, Y.; Yuan, S.; Zhao, X.; Wu, C.; Wang, C.; Xue, Z. Different regulatory mechanisms of plant hormones in the ripening of climacteric and non-climacteric fruits: A review. *Plant Mol. Biol.* **2021**, *107*, 477–497. [CrossRef]
10. Zhao, Z.; Cao, J.; Jiang, W.; Gu, Y.; Zhao, Y. Maturity-related chilling tolerance in mango fruit and the antioxidant capacity involved. *J. Sci. Food Agric.* **2009**, *89*, 304–309. [CrossRef]
11. Wongmetha, O.; Ke, L.-S.; Liang, Y.-S. The changes in physical, bio-chemical, physiological characteristics and enzyme activities of mango cv. Jinhwang during fruit growth and development. *NJAS—Wagening. J. Life Sci.* **2015**, *72–73*, 7–12. [CrossRef]
12. Kong, X.-m.; Ge, W.-y.; Wei, B.-d.; Zhou, Q.; Zhou, X.; Zhao, Y.-b.; Ji, S.-j. Melatonin ameliorates chilling injury in green bell peppers during storage by regulating membrane lipid metabolism and antioxidant capacity. *Postharvest Biol. Technol.* **2020**, *170*, 111315. [CrossRef]
13. Koolpluksee, M.; Ketsa, S.; Subhadrabandhu, S. Effect of modified atmospheres on quality and chilling injury of 'nam dok mai' mango fruits. *Agric. Nat. Resour.* **1993**, *27*, 115–124.
14. Mokgalapa, N.; Akinola, S.A.; Shoko, T.; Pillai, S.K.; Sivakumar, D. Chitosan molecular weights affect anthracnose incidence and elicitation of defence-related enzymes in avocado (*Persea americana*) cultivar 'fuerte'. *Int. J. Food Microbiol.* **2022**, *366*, 109561. [CrossRef]

15. Prusky, D.; Kobiler, I.; Miyara, I.; Alkan, N. *The Mango. Botany, Production and Uses*, 2nd ed.; CABI: London, UK, 2010; pp. 210–230.
16. Raghavendra, A.; Guru, D.S.; Rao, M.K.; Sumithra, R. Hierarchical approach for ripeness grading of mangoes. *Artif. Intell. Agric.* **2020**, *4*, 243–252. [CrossRef]
17. Cheng, Y.-T.; Huang, P.-H.; Chan, Y.-J.; Chen, S.-J.; Lu, W.-C.; Li, P.-H. A new strategy to design novel modified atmosphere packaging formulation maintains the qualities of postharvest strawberries (*Fragaria ananassa*) during low-temperature storage. *J. Food Saf.* **2023**, *43*, e13082. [CrossRef]
18. Lawson, T.; Lycett, G.W.; Ali, A.; Chin, C.F. Characterization of southeast asia mangoes (*Mangifera indica* L.) according to their physicochemical attributes. *Sci. Hortic.* **2019**, *243*, 189–196. [CrossRef]
19. Lin, Y.-W.; Tsai, C.-L.; Chen, C.-J.; Li, P.-L.; Huang, P.-H. Insights into the effects of multiple frequency ultrasound combined with acid treatments on the physicochemical and thermal properties of brown rice postcooking. *LWT* **2023**, *188*, 115423. [CrossRef]
20. Nkhata, S.G. Total color change (δe∗) is a poor estimator of total carotenoids lost during post-harvest storage of biofortified maize grains. *Heliyon* **2020**, *6*, e05173. [CrossRef]
21. Wu, M.-C.; Jiang, C.-M.; Huang, P.-H.; Wu, M.-Y.; Wang, Y.T. Separation and utilization of pectin lyase from commercial pectic enzyme via highly methoxylated cross-linked alcohol-insoluble solid chromatography for wine methanol reduction. *J. Agric. Food Chem.* **2007**, *55*, 1557–1562. [CrossRef]
22. Brown, G.E. The effect of ethylene on susceptibility of robinson tangerines to anthracnose. *Phytopathology* **1977**, *77*, 120. [CrossRef]
23. de Souza-Pollo, A.; de Goes, A. Mango pathology and diseases. In *Handbook of Mango Fruit*; Siddiq, M., Brecht, J.K., Sidhu, J.S., Eds.; John Wiley & Sons, Inc.: Hoboken, NJ, USA, 2017; pp. 91–104.
24. Li, Q.; Bu, J.; Shu, J.; Yu, Z.; Tang, L.; Huang, S.; Guo, T.; Mo, J.; Luo, S.; Solangi, G.S.; et al. Colletotrichum species associated with mango in southern china. *Sci. Rep.* **2019**, *9*, 18891. [CrossRef]
25. Paudel, A.; Yadav, P.; Poudel, P.; Yogi, M.; Bhattarai, A. Insights on the mango anthracnose and its management. *J. Plant Pathol. Res.* **2022**, *4*, 81–90. [CrossRef]
26. Marcianò, D.; Mizzotti, C.; Maddalena, G.; Toffolatti, S.L. The dark side of fungi: How they cause diseases in plants. *Front. Young Minds* **2021**, *9*, 560315. [CrossRef]
27. Perkins, M.L.; Joyce, D.C.; Coates, L.M. Possible contribution of impact injury at harvest to anthracnose expression in ripening avocado: A review. *Sci. Hortic.* **2019**, *246*, 785–790. [CrossRef]
28. Peralta-Ruiz, Y.; Rossi, C.; Grande-Tovar, C.D.; Chaves-López, C. Green management of postharvest anthracnose caused by colletotrichum gloeosporioides. *J. Fungi* **2023**, *9*, 623. [CrossRef]
29. Singh, J.; Daulta, B. Factors affecting the development of anthracnose of guava during storage. *Haryana J. Hortic. Sci.* **1983**, *12*, 41–43.
30. Reyes, M.U.; Paull, R.E. Effect of storage temperature and ethylene treatment on guava (*Psidium guajava* L.) fruit ripening. *Postharvest Biol. Technol.* **1995**, *6*, 357–365. [CrossRef]
31. De Silva, A.G.D.; Crous, P.; Ades, P.; Hyde, K.; Taylor, P. Life styles of colletotrichum species and implications for plant biosecurity. *Fungal Biol. Rev.* **2017**, *31*, 155–168. [CrossRef]
32. Singh, Z.; Singh, R.K.; Sane, V.A.; Nath, P. Mango—Postharvest biology and biotechnology. *Crit. Rev. Plant Sci.* **2013**, *32*, 217–236. [CrossRef]
33. Shu, P.; Li, Y.; Xiang, L.; Sheng, J.; Shen, L. Ethylene enhances tolerance to chilling stress in tomato fruit partially through the synergistic regulation between antioxidant enzymes and atp synthases. *Postharvest Biol. Technol.* **2022**, *193*, 112065. [CrossRef]
34. Nair, S.; Singh, Z. Pre-storage ethrel dip reduces chilling injury, enhances respiration rate, ethylene production and improves fruit quality of 'kensington' mango. *J. Food Agric. Environ.* **2003**, *1*, 93–97.
35. Chen, M.; Gu, H.; Wang, L.; Shao, Y.; Li, R.; Li, W. Exogenous ethylene promotes peel color transformation by regulating the degradation of chlorophyll and synthesis of anthocyanin in postharvest mango fruit. *Front. Nutr.* **2022**, *9*, 911542. [CrossRef] [PubMed]
36. Montalvo, E.; García, H.S.; Tovar, B.; Mata, M. Application of exogenous ethylene on postharvest ripening of refrigerated 'ataulfo' mangoes. *LWT—Food Sci. Technol.* **2007**, *40*, 1466–1472. [CrossRef]
37. Jiang, L.; Han, Z.; Liu, J.; Xiang, Y.; Xu, X.; Jiang, Y.; Jiang, G.; Zhang, Z. Intermittent stepwise cooling and warming ameliorate chilling injury and improve quality in postharvest 'guifei' mango fruit. *LWT* **2023**, *181*, 114740. [CrossRef]
38. Liu, X.; Zhou, Y.; Xiao, J.; Bao, F. Effects of chilling on the structure, function and development of chloroplasts. *Front. Plant Sci.* **2018**, *9*, 1715. [CrossRef] [PubMed]
39. El-Dakak, R.A.; Badr, R.H.; Zeineldein, M.H.; Swedan, E.A.; Batrawy, O.E.; Hassaballah, A.F.; Hassan, I.A. Effect of chilling and salinity stress on photosynthetic performance and ultrastructure of chloroplast in faba beans (*Vicia faba* L.) leaves. *Rend. Lincei. Sci. Fis. E Nat.* **2023**, *34*, 447–456. [CrossRef]
40. Marangoni, A.; Smith, A.; Yada, R.; Stanley, D. Ultrastructural changes associated with chilling injury in mature-green tomato fruit. *J. Am. Soc. Hort. Sci.* **1989**, *114*, 958–962. [CrossRef]
41. Sousa, F.F.; Pinsetta Junior, J.S.; Oliveira, K.T.E.F.; Rodrigues, E.C.N.; Andrade, J.P.; Mattiuz, B.-H. Conservation of 'palmer' mango with an edible coating of hydroxypropyl methylcellulose and beeswax. *Food Chem.* **2021**, *346*, 128925. [CrossRef]
42. Ge, Y.; Zhang, J.; Li, C.; Xue, W.; Zhang, S.; Lv, J. Trisodium phosphate delays softening of jujube fruit by inhibiting cell wall-degrading enzyme activities during ambient storage. *Sci. Hortic.* **2020**, *262*, 109059. [CrossRef]

43. Ali, S.; Ullah, M.A.; Nawaz, A.; Naz, S.; Shah, A.A.; Gohari, G.; Razavi, F.; Khaliq, G.; Razzaq, K. Carboxymethyl cellulose coating regulates cell wall polysaccharides disassembly and delays ripening of harvested banana fruit. *Postharvest Biol. Technol.* **2022**, *191*, 111978. [CrossRef]
44. Ali, Z.M.; Chin, L.-H.; Lazan, H. A comparative study on wall degrading enzymes, pectin modifications and softening during ripening of selected tropical fruits. *Plant Sci.* **2004**, *167*, 317–327. [CrossRef]
45. Kou, X.; Yang, S.; Chai, L.; Wu, C.; Zhou, J.; Liu, Y.; Xue, Z. Abscisic acid and fruit ripening: Multifaceted analysis of the effect of abscisic acid on fleshy fruit ripening. *Sci. Hortic.* **2021**, *281*, 109999. [CrossRef]
46. Zhao, Y.; Zhu, X.; Hou, Y.; Pan, Y.; Shi, L.; Li, X. Effects of harvest maturity stage on postharvest quality of winter jujube (*Zizyphus jujuba* mill. Cv. Dongzao) fruit during cold storage. *Sci. Hortic.* **2021**, *277*, 109778. [CrossRef]
47. Botton, A.; Tonutti, P.; Ruperti, B. Chapter 5—Biology and biochemistry of ethylene. In *Postharvest Physiology and Biochemistry of Fruits and Vegetables*; Yahia, E.M., Ed.; Woodhead Publishing: Sawston, UK, 2019; pp. 93–112.
48. Ebrahimi, A.; Khajavi, M.Z.; Ahmadi, S.; Mortazavian, A.M.; Abdolshahi, A.; Rafiee, S.; Farhoodi, M. Novel strategies to control ethylene in fruit and vegetables for extending their shelf life: A review. *Int. J. Environ. Sci. Technol.* **2022**, *19*, 4599–4610. [CrossRef]
49. Ahrens, M.J.; Huber, D.J. Physiology and firmness determination of ripening tomato fruit. *Physiol. Plant.* **1990**, *78*, 8–14. [CrossRef]
50. Abbas, M.e.F.; Fandi, B.S. Respiration rate, ethylene production and biochemical changes during fruit development and maturation of jujube (*Ziziphus mauritiana* lamk). *J. Sci. Food Agric.* **2002**, *82*, 1472–1476. [CrossRef]
51. Busatto, N.; Vittani, L.; Farneti, B.; Khomenko, I.; Caffini, M.; Faccini, S.; Boschetti, M.; Costa, F. Physiological and molecular characterization of the late ripening stages in *Mangifera indica* cv keitt. *Postharvest Biol. Technol.* **2022**, *183*, 111746. [CrossRef]
52. Wang, L.; Li, R.; Shi, X.; Wei, L.; Li, W.; Shao, Y. Ripening patterns (off-tree and on-tree) affect physiology, quality, and ascorbic acid metabolism of mango fruit (cv. Guifei). *Sci. Hortic.* **2023**, *315*, 111971. [CrossRef]
53. Kim, Y.; Brecht, J.K.; Talcott, S.T. Antioxidant phytochemical and fruit quality changes in mango (*Mangifera indica* L.) following hot water immersion and controlled atmosphere storage. *Food Chem.* **2007**, *105*, 1327–1334. [CrossRef]
54. Taghipour, L.; Rahemi, M.; Assar, P.; Ramezanian, A.; Mirdehghan, S.H. Alleviating chilling injury in stored pomegranate using a single intermittent warming cycle: Fatty acid and polyamine modifications. *Int. J. Food Sci.* **2021**, *2021*, 2931353. [CrossRef]
55. Ribeiro, B.S.; de Freitas, S.T. Maturity stage at harvest and storage temperature to maintain postharvest quality of acerola fruit. *Sci. Hortic.* **2020**, *260*, 108901. [CrossRef]

horticulturae

MDPI

Article

Effects of Different Storage Temperatures and Methyl Jasmonate on Grape Quality and Antioxidant Activity

Essam Elatafi [1,2], Abdelmonem Elshahat [3,4], Yu Xue [1], Li Shaonan [1], Lu Suwen [1], Dong Tianyu [1] and Jinggui Fang [1,*]

1 Key Laboratory of Genetics and Fruit Development, College of Horticulture, Nanjing Agricultural University, Nanjing 210095, China; eelatafi@mans.edu.eg (E.E.); 2021104009@stu.njau.edu.cn (Y.X.); 2023204003@stu.njau.edu.cn (L.S.); lusuwen@njau.edu.cn (L.S.); 2020204002@stu.njau.edu.cn (D.T.)
2 Department of Pomology, Faculty of Agriculture, Mansoura University, Mansoura 35516, Egypt
3 Department of Horticulture, Faculty of Agriculture, Kafrelsheikh University, Kafr El-Sheikh 33516, Egypt; a.elshahat@agr.kfs.edu.eg
4 Key Laboratory of Horticultural Plant Biology, College of Horticulture & Forestry Sciences, Huazhong Agricultural University, Wuhan 430070, China
* Correspondence: fanggg@njau.edu.cn

Abstract: The aim of this study was to find out how different postharvest temperatures and MeJA treatments affected the quality of table grapes, their antioxidant properties, and the amount of hydrogen peroxide and malondialdehyde they contained. For the investigation, postharvest Shine Muscat table grapes were treated with low and high temperatures and MeJA at concentrations of 10 and 100 μmol/L. The results indicated that treating grape berries with MeJA at concentrations of 10 and 100 μmol/L effectively reduced weight loss and mitigated the increase in soluble solid content while also mitigating the decrease in berry firmness and titratable acidity. Consequently, this treatment preserved the sensory and nutritional qualities of the berries and extended their shelf life. Meanwhile, the application of MeJA at a concentration of 10 μmol/L demonstrated superior effectiveness compared to the 100 μmol/L concentration and resulted in a significant enhancement of antioxidant activities by increasing levels of superoxide dismutase, catalase, ascorbate peroxidase, and polyphenol oxidase. Furthermore, the levels of hydrogen peroxide and malondialdehyde in the samples increased for all treatments throughout the storage period. Nevertheless, the levels of hydrogen peroxide and malondialdehyde generation following MeJA treatment remained much lower compared to samples treated at room temperature and low temperature. Therefore, the postharvest application of MeJA at a concentration of 10 μmol/L could play a critical role as a stimulator of fruit quality as well as enhance physicochemical parameters and antioxidant activities for extending the shelf life of grapes during storage.

Keywords: grape; postharvest; room temperature; low temperature; methyl jasmonate

Citation: Elatafi, E.; Elshahat, A.; Xue, Y.; Shaonan, L.; Suwen, L.; Tianyu, D.; Fang, J. Effects of Different Storage Temperatures and Methyl Jasmonate on Grape Quality and Antioxidant Activity. *Horticulturae* **2023**, *9*, 1282. https://doi.org/10.3390/horticulturae9121282

Academic Editor: Isabel Lara

Received: 26 October 2023
Revised: 18 November 2023
Accepted: 20 November 2023
Published: 29 November 2023

1. Introduction

Table grape cultivars are among the most widely consumed non-climacteric fruits globally. China has been the world's leading producer of table grapes since 2011, with over 582,728 hectares planted and an annual yield of 11,269,900 t in 2021 [1]. Over the past ten years, table grape agriculture in China has spread from the western and northern regions to the southwestern and southern provinces of Sichuan, Jiangsu, Guangxi, and Yunnan [2]. Table grapes have a low physiological activity rate, and traits like appearance, color, texture, flavor, and aroma determine their quality. The "veraison" stage marks the onset of ripening, during which the grapes undergo changes such as sugar accumulation, berry softening, anthocyanin synthesis, organic acid metabolism, and the accumulation of flavour compounds [3].

Soluble solid content (SSC) and sugar/acid ratios are key indicators of table grape quality, with specific minimum requirements set for each cultivar [4]. The synthesis of hundreds of different volatile compounds during ripening determines the flavour of table grapes, which is a complex and important aspect of their quality [5]. Table grapes are highly perishable after harvest and sensitive to water loss due to rachis and pedicel desiccation, resulting in browning, weight loss, and fruit softening. Additionally, the necrotrophic fungus *Botrytis cinerea* is primarily responsible for fungal degradation, which results in significant losses [6]. This fungus grows quickly and can spread through berries even at low temperatures (LT) around 0 °C. As a result, table grape preservation is difficult and depends on a variety of characteristics, with temperature and relative humidity being especially important.

Temperature is one of the most important factors affecting fruit storage. Fruit can have its shelf life extended after harvest by being stored at the right temperature, which also reduces quality deterioration and microbiological infection [7,8]. A growing body of research suggests that low-temperature storage plays a protective role in fruits by regulating antioxidant activity, thereby reducing the accumulation of reactive oxygen species (ROS). This mechanism helps to minimize nutrient consumption and preserve fruit quality [9–11]. According to Zhao et al. [12], preserving sweet cherry and nectarine at near-ice temperatures considerably increased the activities of superoxide dismutase (SOD), catalase (CAT), and ascorbate peroxidase (APX). Storage at 4 °C also kept the antioxidant enzyme activities in peach fruit active for a considerable amount of time [13]. The findings of the study demonstrated that low-temperature storage can effectively maintain high antioxidant enzyme activity and delay senescence in fruits. This preservation of antioxidant activity contributes to the preservation of stored fruit quality and extends the overall storage period.

Methyl jasmonate (MeJA) is a plant hormone that acts as a signal molecule involved in the growth and development of plants. It also plays a crucial role in the plant's response to various abiotic and biotic stresses [14–16]. Several previous studies and investigations show that the exogenous application of MeJA has a positive impact on the quality and shelf life of various harvested fruits, such as wine grape and bell pepper, which could be attributed to improved oxidation resistance. Moreover, MeJA applications have been shown to extend shelf life and be beneficial in mitigating symptoms of chilling injury in sweet orange and pomegranate [17,18]. It has been proposed that using MeJA lowers the activity of enzymes that hydrolyse glycosidic connections between cell wall components to promote cell wall softening in fruits, enhancing firmness and resistance to mechanical damage while indirectly lowering microbial attack [19].

MeJA has recently been used to improve the chilling resistance of fruits and vegetables such as pepper and pineapple [20,21]. It is possible that a stronger antioxidant defence system and a lower malondialdehyde (MDA) level are linked to MeJA's ability to protect fruit quality from chilling damage.

Treatment with MeJA raised the levels of ascorbic acid and carotenoids in cherry tomato fruits that had already been picked. It also caused bioactive metabolites to build up [22]. During storage, a particular quantity of exogenous MeJA treatment preserved the fruit quality of blood orange and retained greater total phenolic and anthocyanin content than in the control [23]. Additionally, postharvest MeJA treatment can preserve and even enhance the nutritional and medicinal quality of medlar, wine grape, mandarin fruit, and *Centella asiatica* [24–27].

It has been demonstrated in several studies that the application of MeJA can improve the innate disease resistance of plants against pathogen infection and ecological stresses such as cold [28]. Moreover, MeJA has been shown to have insecticidal properties in agricultural crops by enhancing the activities of chitinase and β-1,3 glucanases in plant leaves [29]. Furthermore, exogenous MeJA has been found to enhance chilling tolerance in horticultural crops by improving the expression of heat shock proteins (HSPs) and C-repeat binding factor (CBF) [30]. These findings suggest that treatment with MeJA could be a valuable approach to reducing postharvest diseases and enhancing fruit quality, ultimately resulting in an extended shelf life for horticultural produce. Therefore, the aim of our study is to investigate how temperature regulation and MeJA affect the antioxidant activities of the Shine Muscat grape from a physiochemical perspective during a period of storage.

2. Materials and Methods

2.1. Fruit Samples and Treatments

The investigation was conducted using white table grapes, namely the Shine Muscat varietal. The Shine Muscat grapes were manually collected from the experimental vineyard of Nanjing Agricultural University, situated in the Jiangsu Academy of Agricultural Sciences in China. The grape bunches were promptly transported to the laboratory in cardboard boxes within a span of one hour. Prior to the experiment, the bunches were organized to guarantee that they possessed identical dimensions and hues and were devoid of any imperfections or indications of illness. The experiment took place at the laboratory of the pomology department, situated at Nanjing Agricultural University in Jiangsu Province, China. The bunches were divided into four groups, as indicated in Table 1. There were 15 boxes in each group, and each group had three boxes that served as replications. Subsequently, the bunches were placed on filter paper and packed in plastic boxes weighing 500 g each prior to storage.

Table 1. Experimental Conditions for Storage of Shine Muscat Grape Bunches for a Duration of 30 Days.

Group	Treatment	Treatment Mode	Temperature	Relative Humidity
RT	Tap water	Soak for 5 min	20 (±5) °C	70 (±5) %
LT	Tap water	Soak for 5 min	4 °C	95%
M_1	MeJA 10 µmol/L	Soak for 5 min	4 °C	95%
M_2	MeJA 100 µmol/L	Soak for 5 min	4 °C	95%

Note: RT stands for room temperature, LT stands for low temperature, and MeJA stands for methyl jasmonate.

Analyses were conducted on days 0, 10, 20, and 30 following the treatments. For the RT treatment, no measurements were taken on the 30th day except for estimating the weight loss at the end of storage because it had reached the unacceptable limit for consumption and marketing. Following the physiological evaluation, five berries were taken as samples and tested three times for subsequent analyses. The samples were immediately frozen in liquid nitrogen and stored at −80 °C.

2.2. Measurement of Physicochemical Properties

The fruit's quality was assessed by measuring the weight loss ratio, firmness, soluble solid content (SSC), and titratable acidity (TA) using the methodology outlined in the AOAC [31] guidelines. Five berries were utilised in order to ascertain the weight reduction of the samples, and their initial weight was measured prior to storage. The weights of the remaining samples were measured at each sampling stage. Weight loss was determined by subtracting the weight at each sampling stage from the initial weight of each sample, dividing the difference by the initial weight, and then multiplying the quotient by 100.

A digital penetrometer (GY-4 digital fruit penetrometer, China) was used to measure the penetration force required for a 6 mm diameter probe to enter the berry at a rate of 5 mm/s to a depth of 5 mm. The experiment involved using five berries, which were tested three times, and the measurements were recorded in Newtons (N).

Five berries from each bunch were mixed and ground to make a homogeneous sample that was the same for SSC testing. The measurement was made in triplicate at 20 °C using a digital refractometer (3T, Atago Co., Ltd., Tokyo, Japan). A few drops of juice were put on the prism of the refractometer, and the results were expressed as a percentage.

Five mL of juice was taken and titrated with 0.1 N sodium hydroxide (NaOH) to a phenolphthalein endpoint using a colour indicator (clear to pink) to measure the percentage of titratable acidity. Total acidity was represented as a percentage of tartaric acid in the results.

2.3. Measurement of Antioxidant Enzyme Activities

The evaluation of antioxidant enzyme activity was conducted using the method described by Modesti et al. [25]. The frozen berry tissue powder was dissolved in an extraction buffer (2:5 *w*/*v*) containing 100 mM potassium phosphate buffer (pH 7.8), 100 mM sodium EDTA (pH 7), 1.25 mM polyethylene glycol, and 2 mM dithiothreitol to obtain the total soluble proteins. After being thoroughly mixed in the mortar with a 5% solution of polyvinylpolypyrrolidone, the sample was then transferred to a 2 mL Eppendorf tube. The samples were subjected to a 30 min 14,000× *g* centrifugation at 4 °C. The supernatant was used to conduct enzyme activity experiments.

A mix of 1.5 mL of 100 µL crude enzyme extract, 50 mM of potassium phosphate buffer (pH 7.8), 0.1 mM sodium EDTA, 13 mM of methionine, 75 µM NBT, and 2 µM riboflavin was made to measure SOD activity. Riboflavin was added to start the reaction, and the absorbance at 560 nm was measured after 15 min of incubation at room temperature with constant illumination. The amount of enzyme that, under the abovementioned assay conditions, inhibits the rate of NBT degradation by 50% is considered one SOD unit. The activity of SOD was determined as U g/FW using a UV spectrophotometer (Shimadzu UV-1800, Tokyo, Japan) with proper calibration.

To evaluate the activity of CAT, a reaction mixture was made by mixing 1.5 mL of a crude enzyme extract with 100 L, 20 mM of H_2O_2, and 50 mM of a pH 7 potassium phosphate buffer. The procedure began with the introduction of H_2O_2, and as a result of its decomposition, the absorbance at 240 nm at 25° C for 1 min decreased. CAT activity was measured as µmol H_2O_2/g FW.

APX activity was determined using a reaction mixture with a final volume of 1.5 mL. The mixture consisted of 20 µL of pure enzyme extract, 100 mM of potassium phosphate buffer (pH 7), 0.25 mM of ascorbic acid, 0.70 mM of H_2O_2, and 0.66 mM of sodium EDTA (pH 7). The process was initiated by introducing H_2O_2, and the ascorbic acid oxidation was evaluated by measuring the reduction at 290 nm. The activity of APX was expressed as µmol H_2O_2/g FW.

Polyphenol oxidase (PPO) activity was determined by incubating 1.5 mL of a final volume containing 500 mM of catechol in 100 mM of sodium phosphate buffer, pH 6.4, with 20 µL of a crude enzyme extract and observing the rise in absorbance at 398 nm. The molar difference in catechol-specific activity was expressed as µmol/g FW.

2.4. Measurement of H_2O_2 and Malondialdehyde (MDA) Content

To assay the hydrogen peroxide (H_2O_2) and MDA content, we followed the method according to Wang et al. [32]. In brief, 4 mL of 0.1% cold trichloroacetic acid was used to homogenize 0.5 g of the sample. After 20 min in an ice bath, the homogenate was centrifuged at 12,000× *g* and 4 °C for 20 min. The supernatants produced were collected for analysis.

For H_2O_2 content measurement, the reaction system was prepared by collecting 0.5 mL of supernatant, 1 mL of 1 M potassium iodide, and 0.5 mL of 10 mM potassium phosphate buffer (pH 7.0). The reaction system was incubated at 25 °C for one hour in the dark before determining the absorbance at 390 nm. H_2O_2 was used as a standard, and the H_2O_2 content was expressed as µmol H_2O_2/g FW.

To measure the MDA content, 0.4 g of the sample was homogenized in 8 mL of 10% cold trichloroacetic acid and centrifuged at 10,000× *g* and 4 °C for 20 min. The resulting 2 mL of supernatant was mixed with 2 mL of 20% trichloroacetic acid containing 0.5% thiobarbituric acid and incubated at 95 °C for 30 min. After quick cooling on ice, the reaction mixture was centrifuged at 10,000× *g* for 10 min at 4 °C. The absorbance of the system was measured at 450, 532, and 600 nm. The MDA content was measured using the equation below and expressed on a (μmol/g FW) basis:

$$\text{MDA content} = 6.452 \times (OD_{532} - OD_{600}) - 0.559 \times OD_{450} \quad (1)$$

2.5. Statistical Analysis

The data were analysed using one-way analysis of variance (ANOVA) with CoStat software, as described by Snedecor and Cochran [33]. The results were evaluated as the mean ± standard error (SE). Significance analysis was performed using Duncan's multiple range tests, with a p value < 0.05 considered significant.

3. Results

3.1. Effect of RT, LT, and MeJA on Weight Loss, Firmness, TSS, and TA

According to the results presented in Figure 1A,B, it is evident that there were significant differences between the treatments in terms of weight loss and berry firmness. In terms of weight loss, the application of MeJA (M_1 and M_2) resulted in significantly lower weight loss compared to both the RT and LT treatments across all storage intervals. Notably, there were no significant differences between the M_1 and M_2 treatments in terms of weight loss at any of the storage intervals. Moreover, the highest weight loss percentages were recorded for the RT group and the LT group during all storage intervals.

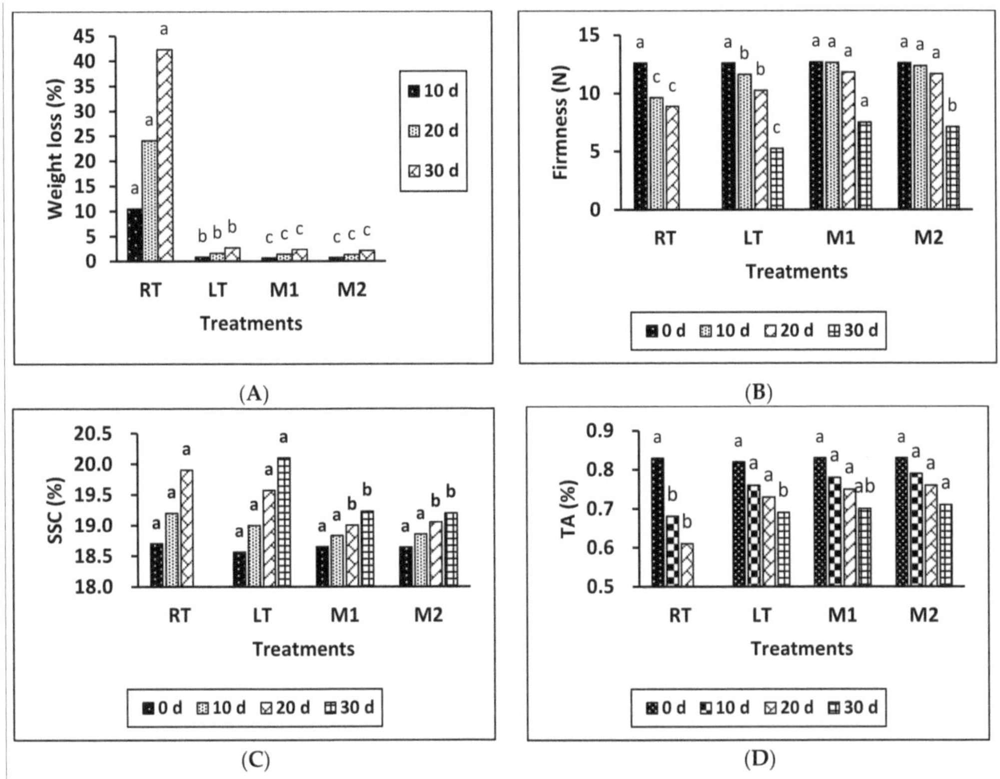

Figure 1. Effects of temperature and MeJA on weight loss (**A**), firmness (**B**), SSC (**C**), and TA content (**D**) in grapes during the storage period. The data presented are expressed as the mean ± (SE) of triplicate assays. Significant differences between the RT, LT, and MeJA samples within the same period were determined using Duncan's test (p < 0.05), indicated by different letters.

Considering berry firmness, the MeJA-treated samples exhibited significantly greater berry firmness compared to the untreated RT and LT samples. Additionally, berry firmness was consistently lower at the 10- and 20-day storage intervals for the RT treatment compared to the MeJA-treated samples.

In summary, the application of MeJA had a significant impact on reducing weight loss and maintaining berry firmness compared to both RT and LT treatments across the storage periods. Furthermore, there were no significant differences in weight loss or firmness between the M_1 and M_2 treatments at any of the storage intervals, except for the 30-day interval, where there was a significant difference in firmness between them.

According to the data provided in Figure 1C,D, the results indicate that the MeJA treatment effectively reduced the increase in SSC in comparison to samples stored at RT and LT, particularly over the period of 20 to 30 days of storage. Nevertheless, no notable disparities were detected among the various treatments over the 0- and 10-day storage periods. Furthermore, there was no discernible disparity observed between the M_1 and M_2 treatments after 20 and 30 days of storage. There was no discernible disparity between the RT and LT treatments during the storage period.

Furthermore, the TA percentage in all samples gradually decreased during storage. There were no significant differences between M_1, M_2, and LT treatments at 0, 10, and 20 days of storage, except for the 30-day storage interval, where a significant difference was observed between the M_2 and LT treatments. The RT treatment exhibited a much lower TA percentage compared to the other treatments. Therefore, the main result is that the MeJA treatment greatly slowed the rise in SSC, especially after 30 days of storage. It also had different effects on TA compared to the other treatments.

3.2. *Effect of RT, LT, and MeJA on SOD, CAT, APX, and PPO Activity*

According to the data shown in Figure 2A–C, the levels of SOD, CAT, and APX activity initially increased in all treatment groups and gradually declined toward the end of the storage period. The highest activity was observed on day 20 in the M_1, M_2, LT, and RT treatments. The application of MeJA resulted in higher values of SOD, CAT, and APX compared to all other treatments. Additionally, a significant difference was observed between LT and RT, with RT exhibiting the lowest values of SOD, CAT, and APX compared to the other treatments. Moreover, a significant difference in the content of SOD, CAT, and PPO was found between M_1 and M_2 at 20 and 30 days of storage. However, there was no clear effect observed in APX between the M_1 and M_2 treatments during all storage periods.

Based on the results provided, it can be concluded that the application of MeJA significantly increased the activity of SOD, CAT, and APX in all of the treatments. Furthermore, notable differences were observed in these enzyme activities between LT and RT, with the LT treatment exhibiting significantly higher values of SOD, CAT, and APX than the RT treatment at 10 and 20 days of storage. To sum up, adding MeJA increased the activity of SOD, CAT, and APX in all of the treatments. There were notable differences in these enzyme activities between LT and RT, as well as between M_1 and M_2 at different storage times. However, there was no clear effect on APX between M_1 and M_2 treatments during all storage periods.

According to the findings in Figure 2D, PPO content showed a gradual rise until it reached its peak on day 20 of storage, then experienced a slight decline near the end of the storage period. On day 20, there was a notable disparity between the LT and RT treatments. Furthermore, a significant difference was observed between the MeJA treatments and the LT treatment. There was a notable disparity observed in the MeJA treatments (M_1 and M_2), specifically during the 20- and 30-day storage periods. The M_1 treatment exhibited the highest values, followed by the M_2, LT, and RT treatments.

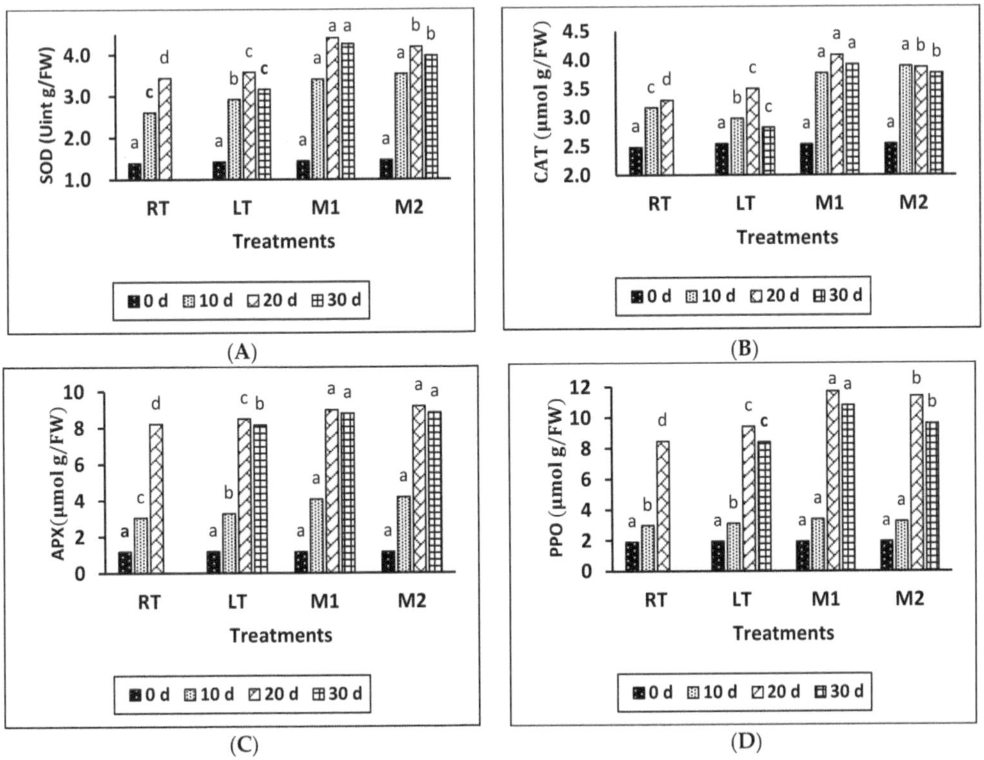

Figure 2. Effect of different temperatures and MeJA on SOD (**A**), CAT (**B**), APX (**C**), and PPO activity (**D**) in grapes during storage period. The data presented are expressed as the mean ± (SE) of triplicate assays. Significant differences between the RT, LT, and MeJA samples within the same period were determined using Duncan's test ($p < 0.05$), indicated by different letters.

3.3. *Effect of RT, LT, and MeJA on H_2O_2 and MDA Content*

The findings of Figure 3A demonstrate that the concentration of H_2O_2 in all treatments indicated a progressive rise, culminating at the conclusion of the storage duration. The LT samples exhibited the highest concentration of H_2O_2, indicating a substantial statistical disparity when compared to the RT samples and those treated with MeJA during storage. In addition, M_1 had the lowest H_2O_2 level compared to the other treatments during the storage period.

The analysis of the results presented in Figure 3B indicates a progressive increase in the content of MDA in all treated samples throughout the storage period. Notably, the MeJA treatments effectively preserved a lower content of MDA compared with the other treatments across the entire duration. Specifically, the M_1 treatment demonstrated the lowest MDA content during the storage period, followed by the M_2, RT, and LT treatments. A significant difference was observed between the LT and RT groups, as well as between the M_1 and M_2 treatments, during the storage period.

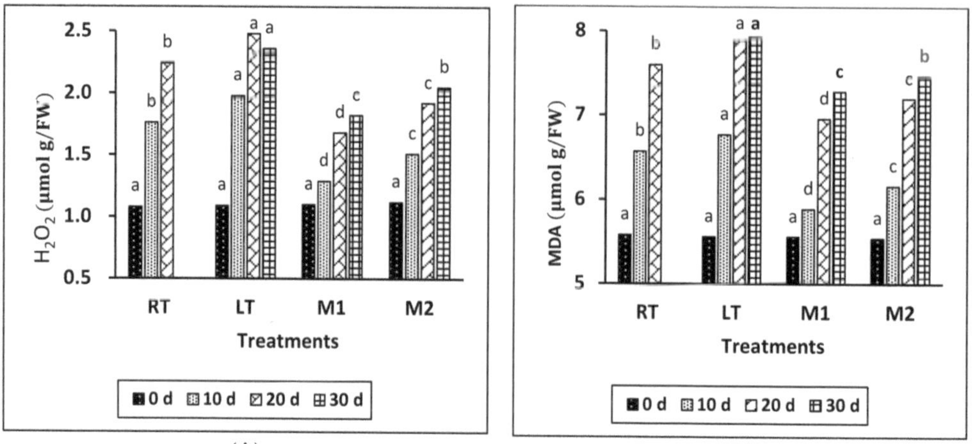

(A) (B)

Figure 3. Effect of different temperatures and MeJA treatment on H_2O_2 (**A**) and MDA content (**B**) in grapes during storage period. The data presented are expressed as the mean ± (SE) of triplicate assays. Significant differences between the RT, LT, and MeJA samples within the same period were determined using Duncan's test ($p < 0.05$), indicated by different letters.

4. Discussion

4.1. Effect of RT, LT, and MeJA on Weight Loss, Firmness, TSS, and TA

The weight loss of fruits is a critical parameter used to evaluate their quality and marketability [34]. As shown in Figure 1A, weight loss percentage increased in all treatments during the storage period. This increase in weight loss was due to increased respiratory intake and transpiration water loss [35]. Weight loss increased with storage temperature. Less weight was lost under LT settings compared to RT conditions at the conclusion of the storage period. These findings align with previous studies that have shown a significant reduction in weight loss percentage in fruits such as blueberry, mango, and winter jujube when stored under LT conditions compared to the control group stored under RT conditions. These results reveal that MeJA treatment could enhance and preserve the quality of grapes. The findings agreed with [25,36,37]. They indicated that MeJA could inhibit weight loss in mangosteen, tomato, dragon fruit, papaya, table grape, and wine grape during storage periods. The decrease in weight loss resulting from MeJA treatment is likely attributable to inhibition of the fruit's respiration process, which leads to a reduction in oxygen consumption and subsequent weight loss [38]. Furthermore, it improves the fruit's capacity to retain moisture and decreases the rate of evaporation, resulting in less weight loss during storage [39]. MeJA also impacts fruit components, specifically organic acids and flavonoids, resulting in enhanced fruit quality and delayed deterioration [40].

Table grape firmness is an important quality parameter for producers, since severe softening might result in postharvest decay or consumer rejection [3]. Based on the results of this research, it was observed that both MeJA treatment and LT storage contributed to delaying the loss of firmness in grapes. This delay in firmness loss helped in maintaining the crisp flavour and freshness of the fruit. This effect may be attributed to the ability of LT to prevent postharvest physiological metabolism, as shown in Figure 1B. These findings agreed with [34,41,42]. They indicated that storage at LT prevented firmness reduction in blueberry, winter jujube, and apricot, delaying fruit ripening and senescence. Additionally, the reduction in fresh weight loss percentage can be due to MeJA treatment reducing the respiration rate during storage, presumably due to the maintenance of firmness and fruit quality [43].

MeJA treatment has been shown to increase firmness in fruits, as reported in previous studies [44]. This increase in firmness can be due to cell wall integrity stimulation-related enzymes, such as chitin synthase and phenylalanine ammonia-lyase (PAL).

Previous studies have indicated that the application of MeJA has varying effects on the expression of PAL. Furthermore, the PAL expression pattern differs between the pulp and core tissues. However, it is uncertain if PAL directly contributes to lignin formation in kiwifruit [44]. In peach fruit, MeJA was found to enhance phospholipid remodelling, which promotes the integrity of the cell wall and reduces electrolyte leakage. This may be due to the effect of MeJA on the related enzymes of the cell wall [45]. Moreover, MeJA has been shown to delay the degradation of enzymes, such as cellulose and pectin methylesterase in mandarin fruit [26].

SSC and TA are key quality parameters that determine the sugar, acid content, and flavour of fruits. In grapes, these parameters play a critical role in determining the unique flavour of the grape [46]. As shown in Figure 1C, SSC in all the treatment samples increased gradually during the storage period. The increase in SSC can be attributed to the elevated proportion of water loss from the grape berry along with the ongoing conversion of starch into sugar during the ripening process, leading to a greater concentration of SSC [34]. The results of this research agree with previous findings that MeJA-treated blueberry fruits resulted in lower SSC values compared to the untreated, particularly during the 14- and 21-day storage periods [47].

In contrast, the content of TA in all samples reduced slightly and continuously during storage, as shown in Figure 1D. This can be due to the organic acid consumption in the respiration process, with a minor decrease shown in treated samples [48]. At the end of storage, the SSC was lower and the TA content was higher in MeJA samples than in RT and LT samples, indicating that MeJA delayed the increase in SSC and reduced the decrease in TA content to maintain a higher storage quality.

The TA content observed during storage could be attributed to the berries' metabolic activity during storage, as reported by Caleb et al. [49]. The results of this research are consistent with the results of El-Beltagi et al. [50], who indicated a decrease in TA content in pomegranate with increasing storage intervals under cold storage conditions, while MeJA treatment helped preserve the TA content compared to the control. Furthermore, it has been shown that blueberry fruits treated with MeJA exhibited a higher TA content compared to the control during storage, as reported by Huang et al. [47].

Briefly, the application of MeJA substantially suppressed the reduction of titratable acidity caused by storage-related damage while decreasing the sugars at the end of storage. The decrease in sugars is a result of the production and accumulation of phenolic compounds, which are driven by MeJA, as well as of the carbohydrates that play a crucial role as a material and energy source in the metabolism of phenolic compounds [51]. In this study, grape fruits preserved high quality criteria after 30 days of storage, although the SSC percentage was reduced.

4.2. *Effect of RT, LT, and MeJA on SOD, CAT, APX, and PPO Activity*

Antioxidant enzymes, including SOD, CAT, and APX, are essential components of the antioxidant system and play a main role in alleviating and eliminating ROS. These antioxidant enzymes also play a significant role during the fruit development stages, especially the ripening stage. According to Mittler [52], the antioxidants (CAT and APX) convert H_2O_2 to O_2 and H_2O, while SOD catalyses the dissociation process of O_2^- into H_2O_2 and O_2. Previous studies have shown that SOD, CAT, and APX activities can alleviate chilling injuries and extend the shelf life of postharvest fruits by maintaining membrane integrity [53]. MeJA treatment improves the quality retention of *Agaricus bisporus* by inhibiting the activities of PPO, increasing the activities of antioxidant enzymes such as CAT and SOD, and reducing the expression levels of genes that encode PPO during storage [54].

Based on the findings of this study, it was found that MeJA had a significant impact on the content of SOD, CAT, APX, and PPO enzymes (as shown in Figure 2). This increase in enzyme activity was linked to better fruit quality parameters compared to the RT and LT treatments [36,44]. As a result, the amount of ROS in the fruits after harvest decreased [55]. Similar outcomes were observed in wine grapes treated with MeJA [25]. The results indicate that one mechanism by which MeJA-treated grapes preserve their quality is through their increased SOD, CAT, and APX enzyme content. According to [55], MeJA treatments that preserved higher SOD activity may have lowered O_2^- by converting it to H_2O_2. For example, the H_2O_2 content of MeJA-treated grapes increased slightly during storage, which may have been due to an increase in CAT enzyme content. Higher CAT enzyme content has been demonstrated to be the cause of conversion of H_2O_2 to water and O_2 [56]. Treatment with 10 μM/L MeJA was more effective than treatment with 100 μM/L MeJA in increasing the levels of antioxidant enzymes during storage. The results of this study agreed with previous studies that demonstrated that a low concentration of MeJA can effectively suppress the biosynthesis of ethylene, a crucial hormone that plays a role in the ageing process of fruits and vegetables [57].

The postharvest treatment of grape fruits with MeJA has been observed to result in an increase in PPO enzyme activity during storage. This increase in activity is likely due to the enhanced activity of enzyme precursors and the regeneration of enzymes over time [58,59]. However, as enzyme levels drop toward the end of the storage period, there is a potential for the accumulation of reactive oxygen species (ROS), leading to damage to DNA and RNA, promotion of membrane peroxidation, and early senescence [60]. These effects can ultimately result in the loss of texture and quality, as well as impact the taste and nutritional value of the grape fruit. The data presented suggest that MeJA effectively modulates the stress response pathways, leading to the induction of PPO activity [61].

It is important to consider these findings when managing postharvest treatments for grape fruits in order to minimize the negative effects of PPO activity and ROS accumulation during storage. Strategies for controlling PPO activity and ROS accumulation, such as optimizing storage conditions and employing antioxidant treatments, may be beneficial in preserving the texture, quality, taste, and nutritional value of grape fruits during postharvest storage.

4.3. Effect of RT, LT, and MeJA on H_2O_2 and MDA Content

H_2O_2, a reactive oxygen species, can lead to oxidative stress, membrane peroxidation, and cell death in plants [62]. Based on the data provided in Figure 3A, the study indicates that the production of H_2O_2 and MDA content increased during storage, but following treatment with MeJA, the levels remained significantly lower compared to grapes treated with RT and LT. Furthermore, the activity of CAT and APX enzymes appeared to be linked to stress levels, as their increased activity might have contributed to a reduction in H_2O_2 accumulation in the tissue. This finding aligns with previous research on MeJA-treated wine grapes [25].

It is clear that the study suggests a potential role for MeJA in mitigating oxidative stress and H_2O_2 accumulation in grapes during storage. This aligns with previous research on other plant species [55]. The precise mechanism behind MeJA's impact on H_2O_2 and MDA content, as well as its effect on CAT and APX enzymes, would likely benefit from further investigation and clarification.

MDA, a by-product of lipid peroxidation, is utilized as a marker to assess membrane damage induced by oxidative stress in plants [63]. In the present study, it was observed that samples treated with MeJA exhibited lower levels of MDA compared to other treatments, possibly due to an increase in enzymatic antioxidants. The reduction in MDA formation attributed to MeJA treatment suggests it may inhibit peroxidation reactions [64]. Conversely, the research highlighted a significant increase in MDA content in grapes treated with LT during storage, indicating that LT may accelerate the breakdown of membrane lipids.

This effect could potentially impact cell compartmentalization and lead to a loss of cell integrity [65].

These findings suggest that MeJA treatment may play a role in reducing MDA content, potentially through the modulation of peroxidation reactions and the enhancement of enzymatic antioxidants. Additionally, the contrasting impact of LT on MDA content emphasizes the significance of environmental factors in influencing membrane lipid breakdown and cellular integrity during storage.

5. Conclusions and Prospects

Notably, this study presents compelling evidence for the efficacy of MeJA treatment, particularly at a concentration of 10 μmol/L, in preserving the quality attributes of grapes. This includes the reduction of weight loss, enhancement of firmness, and maintenance of titratable acidity (TA), alongside the mitigation of the increase in soluble solid content (SSC). Additionally, the application of MeJA led to elevated levels of crucial antioxidant enzymes such as superoxide dismutase (SOD), catalase (CAT), ascorbate peroxidase (APX), and polyphenol oxidase (PPO). Furthermore, it effectively reduced the increase in H_2O_2 and MDA content during cold storage. Hence, the application of MeJA as a stimulator of berry quality emerges as a promising strategy to enhance the physical, chemical, and physiological characteristics of grapes, ultimately improving berry quality.

According to this study, the application of MeJA as a chemical molecule in the wine-making process holds potential for decreasing the sugar levels in grapes, particularly in the delayed stages of storage. However, additional investigation is necessary to fully understand its effectiveness. It is vital to meticulously consider the suitable dosage in order to attain the desired outcome without jeopardising the excellence of the wine. Moreover, the use of MeJA might potentially influence the inherent taste of the wine, requiring careful calibration to maintain the ideal equilibrium among sugar, acidity, and flavour. Due to the insufficient state of study on this topic, it is crucial to perform further investigation and achieve thorough knowledge.

Author Contributions: Conceptualization, E.E., A.E. and J.F.; methodology, E.E., A.E., Y.X., L.S. (Li Shaonan), L.S. (Lu Suwen), D.T. and J.F.; software, E.E. and A.E.; validation, E.E. and J.F.; formal analysis, E.E., A.E. and Y.X.; investigation, E.E., L.S. (Li Shaonan), L.S. (Lu Suwen) and D.T.; resources, E.E. and L.S. (Lu Suwen); data curation, E.E. and J.F.; writing—original draft preparation, E.E.; writing—review and editing, E.E., A.E., Y.X., L.S. (Li Shaonan), L.S. (Lu Suwen), D.T. and J.F.; visualization, E.E.; supervision, J.F.; project administration, J.F.; funding acquisition, J.F. All authors have read and agreed to the published version of the manuscript.

Funding: This study was funded by National Natural Science Foundation of China (32272647), the Fundamental Research Funds for the Central Universities (YDZX2023018), and Jiangsu Agricultural Industry Technology System (JATS[2022]457).

Data Availability Statement: Data are contained within the article.

Conflicts of Interest: The authors declare no conflict of interest and the funders had no role in the design of the study; in the collection, analyses, or interpretation of data; in the writing of the manuscript; or in the decision to publish the results.

References

1. FAO. Food and Agriculture Organization of the United Nations. Last Update 21 March 2023. Available online: https://www.fao.org/faostat/en/#data/QCL (accessed on 21 March 2021).
2. Deng, Q.; Xia, H.; Lin, L.; Wang, J.; Yuan, L.; Li, K.; Zhang, J.; Lv, X.; Liang, D. SUNRED, a natural extract-based biostimulant, application stimulates anthocyanin production in the skins of grapes. *Sci. Rep.* **2019**, *9*, 2590. [CrossRef] [PubMed]
3. Romero, I.; Vazquez-Hernandez, M.; Maestro-Gaitan, I.; Escribano, M.I.; Merodio, C.; Sanchez-Ballesta, M.T. Table grapes during postharvest storage: A review of the mechanisms implicated in the beneficial effects of treatments applied for quality retention. *Int. J. Mol. Sci.* **2020**, *21*, 9320. [CrossRef] [PubMed]
4. Poni, S.; Gatti, M.; Palliotti, A.; Dai, Z.; Duchêne, E.; Truong, T.-T.; Ferrara, G.; Matarrese, A.M.S.; Gallotta, A.; Bellincontro, A. Grapevine quality: A multiple choice issue. *Sci. Hortic.* **2018**, *234*, 445–462. [CrossRef]

5. Ruiz-García, L.; Hellín, P.; Flores, P.; Fenoll, J. Prediction of Muscat aroma in table grape by analysis of rose oxide. *Food Chem.* **2014**, *154*, 151–157. [CrossRef] [PubMed]
6. Palou, L.; Serrano, M.; Martínez-Romero, D.; Valero, D. New approaches for postharvest quality retention of table grapes. *Fresh Prod.* **2010**, *4*, 103–110.
7. Zhang, L.; Wang, L.; Zeng, X.; Chen, R.; Yang, S.; Pan, S. Comparative transcriptome analysis reveals fruit discoloration mechanisms in postharvest strawberries in response to high ambient temperature. *Food Chem. X* **2019**, *2*, 100025. [CrossRef] [PubMed]
8. Wu, Z.; Tu, M.; Yang, X.; Xu, J.; Yu, Z. Effect of cutting and storage temperature on sucrose and organic acids metabolism in postharvest melon fruit. *Postharvest Biol. Technol.* **2020**, *161*, 111081. [CrossRef]
9. Ayala-Zavala, J.F.; Wang, S.Y.; Wang, C.Y.; González-Aguilar, G.A. Effect of storage temperatures on antioxidant capacity and aroma compounds in strawberry fruit. *LWT Food Sci. Technol.* **2004**, *37*, 687–695. [CrossRef]
10. Luo, Y.; Tang, H.; Zhang, Y. Production of reactive oxygen species and antioxidant metabolism about strawberry leaves to low temperatures. *J. Agric. Sci.* **2011**, *3*, 89. [CrossRef]
11. Zhang, Z.; Zhu, Q.; Hu, M.; Gao, Z.; An, F.; Li, M.; Jiang, Y. Low-temperature conditioning induces chilling tolerance in stored mango fruit. *Food Chem.* **2017**, *219*, 76–84. [CrossRef]
12. Zhao, H.; Liu, B.; Zhang, W.; Cao, J.; Jiang, W. Enhancement of quality and antioxidant metabolism of sweet cherry fruit by near-freezing temperature storage. *Postharvest Biol. Technol.* **2019**, *147*, 113–122. [CrossRef]
13. An, X.; Xu, Y.; Jiang, L.; Huan, C.; Yu, Z. Effects of postharvest temperature on apoptosis-related enzyme activity and gene expression in peach fruits (*Prunus persica* L. cv. Xiahui 8). *Sci. Hortic.* **2019**, *245*, 178–184. [CrossRef]
14. Bertini, L.; Palazzi, L.; Proietti, S.; Pollastri, S.; Arrigoni, G.; de Laureto, P.P.; Caruso, C. Proteomic analysis of MeJa-induced defense responses in rice against wounding. *Int. J. Mol. Sci.* **2019**, *20*, 2525. [CrossRef] [PubMed]
15. Faizy, A.H.; Ozturk, B.; Aglar, E.; Yıldız, K. Role of methyl jasmonate application regime on fruit quality and bioactive compounds of sweet cherry at harvest and during cold storage. *J. Food Process. Preserv.* **2021**, *45*, e15882. [CrossRef]
16. Yu, X.; Zhang, W.; Zhang, Y.; Zhang, X.; Lang, D.; Zhang, X. The roles of methyl jasmonate to stress in plants. *Funct. Plant Biol.* **2018**, *46*, 197–212. [CrossRef] [PubMed]
17. Rehman, M.; Singh, Z.; Khurshid, T. Methyl jasmonate alleviates chilling injury and regulates fruit quality in "Midknight" Valencia orange. *Postharvest Biol. Technol.* **2018**, *141*, 58–62. [CrossRef]
18. Chen, L.; Pan, Y.; Li, H.; Jia, X.; Guo, Y.; Luo, J.; Li, X. Methyl jasmonate alleviates chilling injury and keeps intact pericarp structure of pomegranate during low temperature storage. *Food Sci. Technol. Int.* **2021**, *27*, 22–31. [CrossRef]
19. Bari, R.; Jones, J.D. Role of plant hormones in plant defence responses. *Plant Mol. Biol.* **2009**, *69*, 473–488. [CrossRef]
20. Sangprayoon, P.; Supapvanich, S.; Youryon, P.; Wongs-Aree, C.; Boonyaritthongchai, P. Efficiency of salicylic acid or methyl jasmonate immersions on internal browning alleviation and physicochemical quality of *Queen pineapple* cv. "Sawi" fruit during cold storage. *J. Food Biochem.* **2019**, *43*, e13059. [CrossRef]
21. Seo, J.; Yi, G.; Lee, J.G.; Choi, J.H.; Lee, E.J. Seed browning in pepper (*Capsicum annuum* L.) fruit during cold storage is inhibited by methyl jasmonate or induced by methyl salicylate. *Postharvest Biol. Technol.* **2020**, *166*, 111210. [CrossRef]
22. Liu, H.; Meng, F.; Miao, H.; Chen, S.; Yin, T.; Hu, S.; Shao, Z.; Liu, Y.; Gao, L.; Zhu, C. Effects of postharvest methyl jasmonate treatment on main health-promoting components and volatile organic compounds in cherry tomato fruits. *Food Chem.* **2018**, *263*, 194–200. [CrossRef] [PubMed]
23. Habibi, F.; Ramezanian, A.; Guillén, F.; Serrano, M.; Valero, D. Blood oranges maintain bioactive compounds and nutritional quality by postharvest treatments with γ-aminobutyric acid, methyl jasmonate or methyl salicylate during cold storage. *Food Chem.* **2020**, *306*, 125634. [CrossRef]
24. Ozturk, A.; Yildiz, K.; Ozturk, B.; Karakaya, O.; Gun, S.; Uzun, S.; Gundogdu, M. Maintaining postharvest quality of medlar (*Mespilus germanica*) fruit using modified atmosphere packaging and methyl jasmonate. *LWT* **2019**, *111*, 117–124. [CrossRef]
25. Modesti, M.; Petriccione, M.; Forniti, R.; Zampella, L.; Scortichini, M.; Mencarelli, F. Methyl jasmonate and ozone affect the antioxidant system and the quality of wine grape during postharvest partial dehydration. *Food Res. Int.* **2018**, *112*, 369–377. [CrossRef] [PubMed]
26. Baswal, A.; Dhaliwal, H.; Singh, Z.; Mahajan, B.; Gill, K. Postharvest application of methyl jasmonate, 1-methylcyclopropene and salicylic acid extends the cold storage life and maintain the quality of 'Kinnow'mandarin (*Citrus nobilis* L. *X C. deliciosa* L.) fruit. *Postharvest Biol. Technol.* **2020**, *161*, 111064. [CrossRef]
27. Buraphaka, H.; Putalun, W. Stimulation of health-promoting triterpenoids accumulation in *Centella asiatica* (L.) Urban leaves triggered by postharvest application of methyl jasmonate and salicylic acid elicitors. *Ind. Crops Prod.* **2020**, *146*, 112171. [CrossRef]
28. Singh, U.B.; Malviya, D.; Singh, S.; Kumar, M.; Sahu, P.K.; Singh, H.; Kumar, S.; Roy, M.; Imran, M.; Rai, J.P. Trichoderma harzianum-and methyl jasmonate-induced resistance to Bipolaris sorokiniana through enhanced phenylpropanoid activities in bread wheat (*Triticum aestivum* L.). *Front. Microbiol.* **2019**, *10*, 1697. [CrossRef] [PubMed]
29. Asghari, M. Impact of jasmonates on safety, productivity and physiology of food crops. *Trends Food Sci. Technol.* **2019**, *91*, 169–183. [CrossRef]
30. Aghdam, M.S.; Bodbodak, S. Physiological and biochemical mechanisms regulating chilling tolerance in fruits and vegetables under postharvest salicylates and jasmonates treatments. *Sci. Hortic.* **2013**, *156*, 73–85. [CrossRef]

31. AOAC. *Official Methods of Analysis of the Association of Analytical Chemists International, Official Methods*; Association of Officiating Analytical Chemists: Gaithersburg, MD, USA, 2005.
32. Wang, Y.; Gao, L.; Wang, Q.; Zuo, J. Low temperature conditioning combined with methyl jasmonate can reduce chilling injury in bell pepper. *Sci. Hortic.* **2019**, *243*, 434–439. [CrossRef]
33. Snedecor, G.W.; Cochran, W.G. *Statistical Methods*, 8th ed.; Iowa State University Press: Ames, IA, USA, 1989; p. 1191.
34. Sang, Y.; Yang, W.; Liu, Y.; Zhang, W.; Guo, T.; Shen, P.; Tang, Y.; Guo, M.; Chen, G. Influences of low temperature on the postharvest quality and antioxidant capacity of winter jujube (*Zizyphus jujuba Mill.* cv. *Dongzao*). *LWT* **2022**, *154*, 112876. [CrossRef]
35. Ozturk, B.; Yildiz, M.; Yildiz, K.; Gun, S. Maintaining the postharvest quality and bioactive compounds of jujube (*Ziziphus jujuba Mill.* Cv. "Li") fruit by applying 1-methylcyclopropene. *Sci. Hortic.* **2021**, *275*, 109671. [CrossRef]
36. Mustafa, M.A.; Ali, A.; Seymour, G.; Tucker, G. Treatment of dragonfruit (*Hylocereus polyrhizus*) with salicylic acid and methyl jasmonate improves postharvest physico-chemical properties and antioxidant activity during cold storage. *Sci. Hortic.* **2018**, *231*, 89–96. [CrossRef]
37. Mustafa, M.A.; Ali, A.; Seymour, G.; Tucker, G. Delayed pericarp hardening of cold stored mangosteen (*Garcinia mangostana* L.) upon pre-treatment with the stress hormones methyl jasmonate and salicylic acid. *Sci. Hortic.* **2018**, *230*, 107–116. [CrossRef]
38. Ozturk, B.; Altuntas, E.; Yildiz, K.; Ozkan, Y.; Saracoglu, O. Effect of methyl jasmonate treatments on the bioactive compounds and physicochemical quality of "Fuji" apples. *Cienc. Investig. Agrar.* **2013**, *40*, 201–221. [CrossRef]
39. Asghari, M.; Hasanlooe, A.R. Interaction effects of salicylic acid and methyl jasmonate on total antioxidant content, catalase and peroxidase enzymes activity in "Sabrosa" strawberry fruit during storage. *Sci. Hortic.* **2015**, *197*, 490–495. [CrossRef]
40. Saavedra, G.M.; Figueroa, N.E.; Poblete, L.A.; Cherian, S.; Figueroa, C.R. Effects of preharvest applications of methyl jasmonate and chitosan on postharvest decay, quality and chemical attributes of Fragaria chiloensis fruit. *Food Chem.* **2016**, *190*, 448–453. [CrossRef]
41. Fan, X.; Xi, Y.; Zhao, H.; Liu, B.; Cao, J.; Jiang, W. Improving fresh apricot (*Prunus armeniaca* L.) quality and antioxidant capacity by storage at near freezing temperature. *Sci. Hortic.* **2018**, *231*, 1–10. [CrossRef]
42. Ge, Y.; Tang, Q.; Li, C.; Duan, B.; Li, X.; Wei, M.; Li, J. Acibenzolar-S-methyl treatment enhances antioxidant ability and phenylpropanoid pathway of blueberries during low temperature storage. *LWT* **2019**, *110*, 48–53. [CrossRef]
43. El-Mogy, M.M.; Ali, M.R.; Darwish, O.S.; Rogers, H.J. Impact of salicylic acid, abscisic acid, and methyl jasmonate on postharvest quality and bioactive compounds of cultivated strawberry fruit. *J. Berry Res.* **2019**, *9*, 333–348. [CrossRef]
44. Li, H.; Suo, J.; Han, Y.; Liang, C.; Jin, M.; Zhang, Z.; Rao, J. The effect of 1-methylcyclopropene, methyl jasmonate and methyl salicylate on lignin accumulation and gene expression in postharvest "Xuxiang" kiwifruit during cold storage. *Postharvest Biol. Technol.* **2017**, *124*, 107–118. [CrossRef]
45. Chen, M.; Guo, H.; Chen, S.; Li, T.; Li, M.; Rashid, A.; Xu, C.; Wang, K. Methyl jasmonate promotes phospholipid remodeling and jasmonic acid signaling to alleviate chilling injury in peach fruit. *J. Agric. Food Chem.* **2019**, *67*, 9958–9966. [CrossRef] [PubMed]
46. Zhao, Y.; Zhu, X.; Hou, Y.; Pan, Y.; Shi, L.; Li, X. Effects of harvest maturity stage on postharvest quality of winter jujube (*Zizyphus jujuba Mill. cv. Dongzao) fruit during cold storage. Sci. Hortic.* **2021**, *277*, 109778.
47. Huang, X.; Li, J.; Shang, H.; Meng, X. Effect of methyl jasmonate on the anthocyanin content and antioxidant activity of blueberries during cold storage. *J. Sci. Food Agric.* **2015**, *95*, 337–343. [CrossRef]
48. Fagundes, C.; Moraes, K.; Pérez-Gago, M.B.; Palou, L.; Maraschin, M.; Monteiro, A. Effect of active modified atmosphere and cold storage on the postharvest quality of cherry tomatoes. *Postharvest Biol. Technol.* **2015**, *109*, 73–81. [CrossRef]
49. Caleb, O.J.; Opara, U.L.; Mahajan, P.V.; Manley, M.; Mokwena, L.; Tredoux, A.G. Effect of modified atmosphere packaging and storage temperature on volatile composition and postharvest life of minimally-processed pomegranate arils (cvs. "Acco" and "Herskawitz"). *Postharvest Biol. Technol.* **2013**, *79*, 54–61. [CrossRef]
50. El-Beltagi, H.S.; Al-Otaibi, H.H.; Ali, M.R. A new approach for extending shelf-life of pomegranate arils with combined application of salicylic acid and methyl jasmonate. *Horticulturae* **2023**, *9*, 225. [CrossRef]
51. Li, X.; Li, M.; Wang, J.; Wang, L.; Han, C.; Jin, P.; Zheng, Y. Methyl jasmonate enhances wound-induced phenolic accumulation in pitaya fruit by regulating sugar content and energy status. *Postharvest Biol. Technol.* **2018**, *137*, 106–112. [CrossRef]
52. Mittler, R. Oxidative stress, antioxidants and stress tolerance. *Trends Plant Sci.* **2002**, *7*, 405–410. [CrossRef]
53. Cao, S.; Zheng, Y.; Wang, K.; Jin, P.; Rui, H. Methyl jasmonate reduces chilling injury and enhances antioxidant enzyme activity in postharvest loquat fruit. *Food Chem.* **2009**, *115*, 1458–1463. [CrossRef]
54. Meng, D.; Song, T.; Shen, L.; Zhang, X.; Sheng, J. Postharvest application of methyl jasmonate for improving quality retention of Agaricus bisporus fruit bodies. *J. Agric. Food Chem.* **2012**, *60*, 6056–6062. [CrossRef] [PubMed]
55. Dong, Y.; Zhi, H.; Xu, J.; Zhang, L.; Liu, M.; Zong, W. Effect of methyl jasmonate on reactive oxygen species, antioxidant systems, and microstructure of Chinese winter jujube at two major ripening stages during shelf life. *J. Hortic. Sci. Biotechnol.* **2016**, *91*, 316–323. [CrossRef]
56. Apel, K.; Hirt, H. Reactive oxygen species: Metabolism, oxidative stress, and signal transduction. *Annu. Rev. Plant Biol.* **2004**, *55*, 373–399. [CrossRef] [PubMed]
57. Reyes-Díaz, M.; Lobos, T.; Cardemil, L.; Nunes-Nesi, A.; Retamales, J.; Jaakola, L.; Alberdi, M.; Ribera-Fonseca, A. Methyl jasmonate: An alternative for improving the quality and health properties of fresh fruits. *Molecules* **2016**, *21*, 567. [CrossRef] [PubMed]

58. Glowacz, M.; Rees, D. Using jasmonates and salicylates to reduce losses within the fruit supply chain. *Eur. Food Res. Technol.* **2016**, *242*, 143–156. [CrossRef]
59. Christopoulos, M.V.; Tsantili, E. Participation of phenylalanine ammonia-lyase (PAL) in increased phenolic compounds in fresh cold stressed walnut (*Juglans regia* L.) kernels. *Postharvest Biol. Technol.* **2015**, *104*, 17–25. [CrossRef]
60. Hodges, D.M.; Lester, G.E.; Munro, K.D.; Toivonen, P. Oxidative stress: Importance for postharvest quality. *HortScience* **2004**, *39*, 924. [CrossRef]
61. Boughton, A.J.; Hoover, K.; Felton, G.W. Impact of chemical elicitor applications on greenhouse tomato plants and population growth of the green peach aphid. *Myzus Persicae Entomol. Exp. Appl.* **2006**, *120*, 175–188. [CrossRef]
62. Xia, X.-J.; Zhou, Y.-H.; Shi, K.; Zhou, J.; Foyer, C.H.; Yu, J.-Q. Interplay between reactive oxygen species and hormones in the control of plant development and stress tolerance. *J. Exp. Bot.* **2015**, *66*, 2839–2856. [CrossRef]
63. Hu, Z.; Weijian, L.; Yali, F.; Huiquan, L. Gibberellic acid enhances postharvest toon sprout tolerance to chilling stress by increasing the antioxidant capacity during the short-term cold storage. *Sci. Hortic.* **2018**, *237*, 184–191. [CrossRef]
64. Flores, G.; Blanch, G.P.; Del Castillo, M.L.R. Effect of postharvest methyl jasmonate treatment on fatty acid composition and phenolic acid content in olive fruits during storage. *J. Sci. Food Agric.* **2017**, *97*, 2767–2772. [CrossRef]
65. Wang, J.W.; Zhou, X.; Zhou, Q.; Liu, Z.Y.; Sheng, L.; Wang, L.; Cheng, S.C.; Ji, S.J. Proteomic analysis of peel browning of "Nanguo" pears after low-temperature storage. *J. Sci. Food Agric.* **2017**, *97*, 2460–2467. [CrossRef]

horticulturae

MDPI

Article

Changes in Physicochemical Characteristics, Peel Color, and Juice Attributes of 'Moro' Blood Orange Fruit Treated with Glycine Betaine and Methyl Salicylate during Cold Quarantine Storage

Fariborz Habibi [1,*], Ali Sarkhosh [1], Fabián Guillén [2], María Serrano [3] and Daniel Valero [2]

1 Horticultural Sciences Department, University of Florida, Gainesville, FL 32611, USA
2 Department of Food Technology, University Miguel Hernández, Ctra. Beniel km. 3.2, 03312 Orihuela, Alicante, Spain
3 Department of Applied Biology, University Miguel Hernández, Ctra. Beniel km. 3.2, 03312 Orihuela, Alicante, Spain
* Correspondence: fhabibi@ufl.edu or fariborz_h659@yahoo.com

Abstract: Cold quarantine storage is the practice of subjecting citrus fruit to low temperatures after harvesting to comply with stringent international phytosanitary standards for export, but fruit quality can be affected during storage. Therefore, this study investigated the effects of glycine betaine (GB) and/or methyl salicylate (MeSA) on physicochemical changes, chemical attributes of juice, and peel color of 'Moro' blood orange at cold quarantine storage (2 °C) for 60 days. Fruit were treated with GB (15 and 30 mM) by vacuum infiltration at 30 kPa for 8 min and vapor treatment of MeSA (100 μM) for 18 h as well as the combination of both GB concentrations with MeSA. The key findings of this research revealed that the combined treatment of 30 mM GB and 100 μM MeSA significantly mitigated weight and firmness losses in 'Moro' blood orange fruit during the cold quarantine period. Furthermore, there was a decrease in titratable acidity (TA) across all treatments, with the highest TA recorded for the 30 mM GB + 100 μM MeSA combination. Conversely, total soluble solids (TSS), TSS/TA ratio, and juice pH increased in all treatments, with the control treatment displaying the highest values. Regarding peel color parameters, which encompass L^* (lightness), b^*, hue angle ($h°$), chroma (C^*), and a^*, as well as the citrus color index (*CCI*), these exhibited characteristic changes during cold quarantine storage. However, the application of GB and MeSA, especially at the 30 mM GB + 100 μM MeSA level, noticeably delayed these peel color variations. Overall, GB and MeSA treatments offer significant advantages in preserving the physicochemical characteristics and chemical attributes of 'Moro' blood oranges during cold quarantine storage. These findings underscore the potential of GB and MeSA treatments for maintaining the quality of 'Moro' blood oranges during cold quarantine storage, with a noteworthy synergistic effect between MeSA and GB in preserving fruit quality.

Keywords: chroma; firmness; juice content; titratable acidity; weight loss

Citation: Habibi, F.; Sarkhosh, A.; Guillén, F.; Serrano, M.; Valero, D. Changes in Physicochemical Characteristics, Peel Color, and Juice Attributes of 'Moro' Blood Orange Fruit Treated with Glycine Betaine and Methyl Salicylate during Cold Quarantine Storage. *Horticulturae* **2023**, *9*, 1103. https://doi.org/10.3390/horticulturae9101103

Academic Editor: Antonio Ferrante

Received: 4 September 2023
Revised: 21 September 2023
Accepted: 4 October 2023
Published: 5 October 2023

1. Introduction

Blood oranges (*Citrus sinensis* L. Osbeck) are a desirable citrus crop due to their high content of bioactive compounds, particularly anthocyanin pigments, known for their health-related properties. This makes blood oranges popular among consumers who value their nutritional benefits. The quality attributes of blood oranges, including anthocyanin, color, flavor, texture, aroma, juiciness, sugar content, seedlessness, size, freshness, and nutritional value, all play a crucial role in promoting consumer acceptance and satisfaction with blood oranges [1].

Postharvest life extension and maintenance of citrus fruit quality can be achieved by storing them at temperatures ranging from 1 to 4 °C. For the disinfestation of citrus fruit, cold quarantine storage at temperatures between 0.5 and 2 °C is widely used and accepted by regulatory agencies of most importing countries. This method mainly controls the Mediterranean fruit fly and can be implemented on a commercial scale. Chemical treatments for the disinfestation of fruit flies are not permitted, and physical treatments such as controlled atmosphere, dielectric heating, heat treatments, irradiation, and ultrasound treatments can be expensive and difficult to implement due to the fruit's threshold tolerance to treatment. However, it is essential to note that while cold quarantine storage is widely accepted, this method may cause cold damage. Therefore, protecting citrus fruit from such damage needs to be clearly defined and addressed [2].

Elicitors are substances that can induce or enhance the production of bioactive compounds and nutritional quality in fruit and vegetables. In recent times, elicitor compounds have gained popularity as a means of preserving fruit quality during cold storage. One such treatment involves the use of methyl salicylate (MeSA), a natural volatile compound, which has been found to effectively maintain the quality of fruits like pomegranate [3] and sweet cherry [4] during cold storage. In addition, the application of postharvest glycine betaine (GB) treatment, which is a quaternary ammonium compound, has become prevalent in the treatment of vegetables during storage including sweet pepper [5] and zucchini [6], climacteric fruits including peach [7–10], loquat [11], hawthorn [12], pear [13–15], plum [16], banana [17], and jujube [18], as well as non-climacteric fruits including pomegranate [19] and strawberry [20].

Exogenous application of elicitor compounds on blood oranges can have beneficial effects on reducing chilling injury (CI), increasing bioactive compounds, and antioxidant activity during cold storage. On the other hand, postharvest elicitor treatments are a promising approach to enhance the postharvest life of blood oranges by mitigating physiological disorders, preserving quality attributes, and boosting healthy phytochemicals [1]. The mechanisms through which postharvest treatments of GB and MeSA can reduce postharvest losses and control CI in blood oranges involve several processes. These include enhancing antioxidant enzyme activities such as catalase, ascorbate peroxidase, and superoxide dismutase, promoting proline accumulation, reducing electrolyte leakage, malondialdehyde, and hydrogen peroxide content, increasing phenylalanine ammonia-lyase activity, and decreasing polyphenol oxidase activity. Additionally, these mechanisms suppress polyphenol oxidase and peroxidase activities. Together, these processes inhibit lipid peroxidation and maintain cell membrane integrity, ultimately preventing fruit losses and chilling symptoms [21].

The cold temperatures can prolong the storage life of blood oranges by reducing fruit metabolism and other processes. Previous research has indicated that MeSA or GB can effectively mitigate CI and preserve the quality of blood oranges stored at 3 °C [21,22]. However, there is currently no scientific evidence on the combined effects of organic osmolyte (GB) and plant growth regulators (MeSA) on blood oranges during cold quarantine storage, and their potential roles have not been explored. Therefore, this study aimed to assess the impact of GB and/or MeSA treatments on the physicochemical attributes of 'Moro' blood oranges during cold quarantine storage.

2. Materials and Methods

2.1. Fruit Treatments and Storage Conditions

Blood oranges (cv. Moro) were harvested at commercial maturity based on the ratio of total soluble solids (TSS) to titratable acidity (TA) (TSS/TA ≅ 10). They were then transported to the postharvest laboratory, where they were checked for uniform size and the absence of any defects or rind injuries. The fruit were subjected to a vacuum infiltration treatment with an aqueous solution of 15 and 30 mM GB at 30 kPa for 8 min. Vapor treatment of 100 µM MeSA was applied to the fruit by placing them in 20 L plastic containers for 18 h. The combination of GB with MeSA (15 mM GB + 100 µM MeSA, 30 mM GB + 100 µM MeSA) was also performed under the same conditions. The optimal concentrations of GB and MeSA were determined based on previous study [23]. The control fruit received no treatment. Each treatment was conducted on three replicates of five fruit, which were then placed in polyethylene bags with 16 holes and stored for 60 days at 2 °C with 90% relative humidity (RH). The following parameters were measured after 1, 30, and 60 days of cold storage plus shelf-life conditions (20 °C) for two days.

2.2. Weight Loss

The fruit weight was measured individually before storage ($W1$) and at each sampling time ($W2$) using a digital balance (HL-i, A&D, Tokyo, Japan) with an accuracy of 0.001, to determine the fruit weight loss (WL). The percentage of weight loss was calculated using the following formula [24]:

$$Weigt\ loss\ (\%) = \frac{W1-W2}{W1} \times 100 \quad (1)$$

2.3. Firmness

To determine the firmness of each treatment, a texture analyzer (TA-XT2, Surrey, UK) was used with a 3.5 mm diameter probe to compress the equatorial area of the fruit at a rate of 10%. The measured data were reported in newton [21].

2.4. Chemical Attributes of Juice

The TSS were measured using a portable refractometer (RBX0032A, Petro Centre, Singapore) and recorded as a percentage (%). TA was measured by titration with NaOH 0.1 N to pH 8.2 as an endpoint with a pH meter (Jenway 351, London, UK). The maturity index was calculated by dividing the TSS value by the TA value. The pH of the juice was measured using a pH meter [24].

2.5. Juice Content

The juice content of the fruit was determined by first weighing the fruit using a digital balance (HL-i, A&D, Tokyo, Japan). Next, the juice from each treatment was manually squeezed from hand-peeled fruit and weighed. The fruit juice content was then calculated using the following formula [24]:

$$Juice\ content\ (\%) = \frac{Weigt\ of\ juice}{Weigt\ of\ fruit} \times 100 \quad (2)$$

2.6. Peel Color

The color of the peel of each fruit was measured using a colorimeter (CR400/4P, Minolta Camera Co., Osaka, Japan) on two opposite sides of the equatorial area of 15 fruit. The L^* value represented lightness (0 = black to 100 = white), while a^* and b^* values represented green (−) to red (+) and blue (−) to yellow (+), respectively. Hue angle ($h°$), chroma (C^*), and citrus color index (CCI) were then calculated using the following formula [24]:

$$Chroma = \sqrt{a^{*2} + b^{*2}} \quad (3)$$

$$h° = Arctan\frac{b^*}{a^*} \quad (4)$$

$$CCT = 1000 \times \frac{a^*}{L^* \times b^*} \quad (5)$$

2.7. Statistical Analysis

The experiment was carried out using a completely randomized design (CRD) with three replications, and the collected data were subjected to a two-factor analysis of variance (ANOVA) to analyze the effects of both treatments and storage times. The SAS software (version 9.4) for Windows was used to perform the data analysis. Mean comparisons were carried out using the least significant difference (LSD) test, and the standard errors (SE) of means were considered with a significance level of $p < 0.05$.

3. Results

3.1. Weight Loss

All treatments resulted in an increase in weight loss during cold quarantine storage, with control samples displaying the highest weight loss (Figure 1). However, treatments led to a significant reduction in weight loss at all sampling times. The combined treatment of GB and MeSA, especially at 30 mM GB + 100 µM MeSA, had the most positive effect on weight loss reduction.

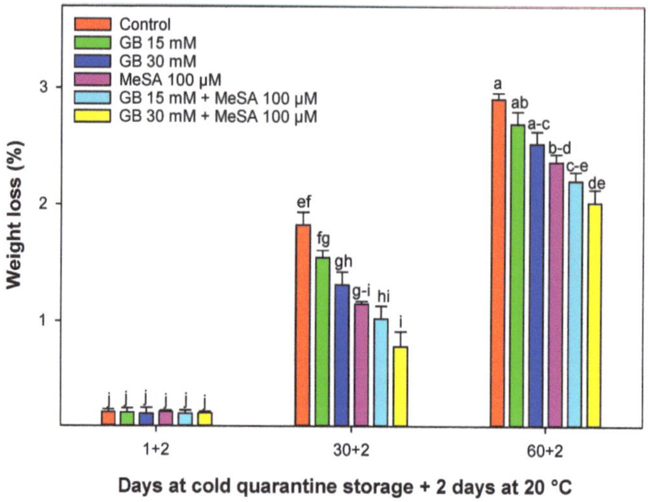

Figure 1. Changes in weight loss in control and treated fruit during cold quarantine storage. Vertical bars represent ± standard errors (SE) of means. Different letters above the bars on columns indicate significant difference at $p < 0.05$ level of probability based on LSD test.

3.2. Firmness

All treatments showed a decrease in firmness during cold quarantine storage, with the control samples exhibiting the most significant loss of firmness (Figure 2). All treatments were effective in reducing fruit firmness loss at all sampling times, but GB treatments displayed a synergistic effect in combination with MeSA displaying fruit firmness values of 27% and 34% higher when 15 mM GB and 30 mM GB, respectively, were applied as a combined treatment with MeSA respectively as compared to control fruit.

3.3. Chemical Attributes of Juice

Chemical attributes of juice (TSS, TA, TSS/TA, and juice pH) changed during cold quarantine storage in all treatments (Figure 3). TSS, TSS/TA, and juice pH increased in all treatments and the highest and lowest values were observed in control and 30 mM GB + 100 µM MeSA treatment, respectively. TA significantly decreased in all treatments and treated fruit had the highest TA, especially for 30 mM GB + 100 µM MeSA during cold quarantine while the lowest TA was observed in control samples. On the other hand, a dose-dependent effect related to GB concentrations applied was observed to delay the evolution of these parameters, applied alone or combined with MeSA treatments.

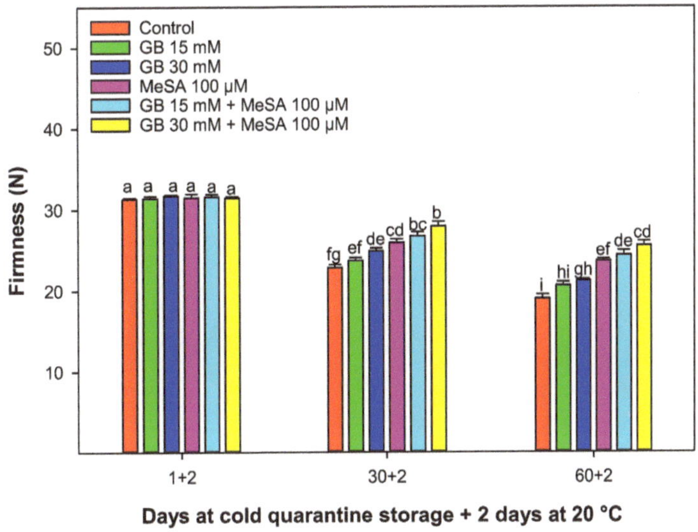

Figure 2. Changes in firmness in control and treated fruit during cold quarantine storage. Vertical bars represent ± standard errors (SE) of means. Different letters above the bars on columns indicate significant difference at $p < 0.05$ level of probability based on LSD test.

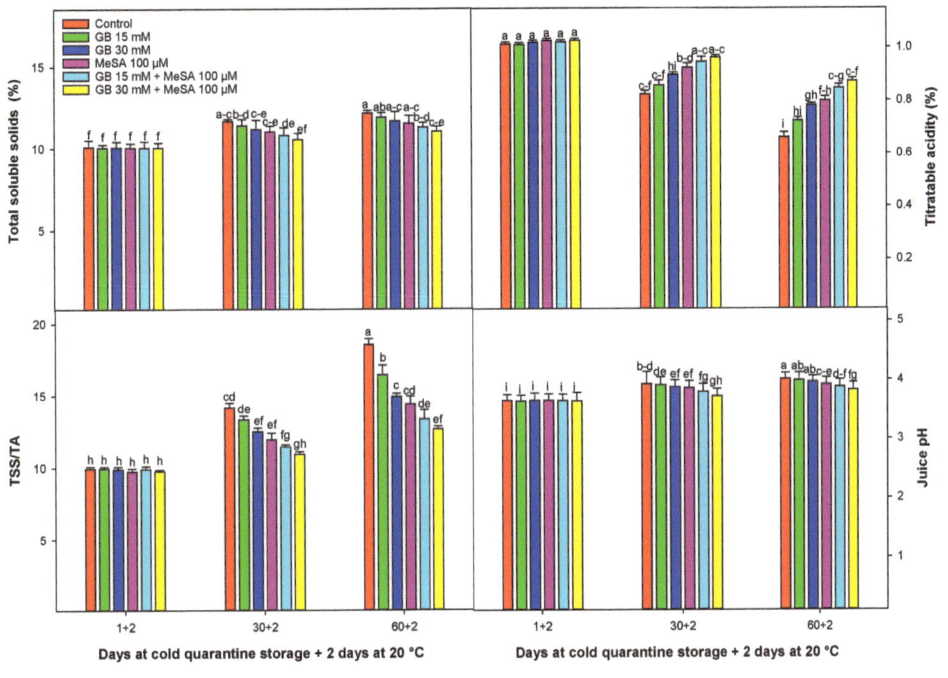

Figure 3. Changes in total soluble solids (TSS), titratable acidity (TA), TSS/TA, juice pH in control and treated fruit during cold quarantine storage. Vertical bars represent ± standard errors (SE) of means. Different letters above the bars on columns indicate significant difference at $p < 0.05$ level of probability based on LSD test.

3.4. Juice Content

Juice content decreased during cold quarantine storage and control samples revealed the most significant loss of juice content (Figure 4). All treatments were effective in reducing juice content loss at all sampling times, but the most significant juice content loss reduction was observed with the combined treatment of GB and MeSA, specifically with 30 mM GB + 100 μM MeSA.

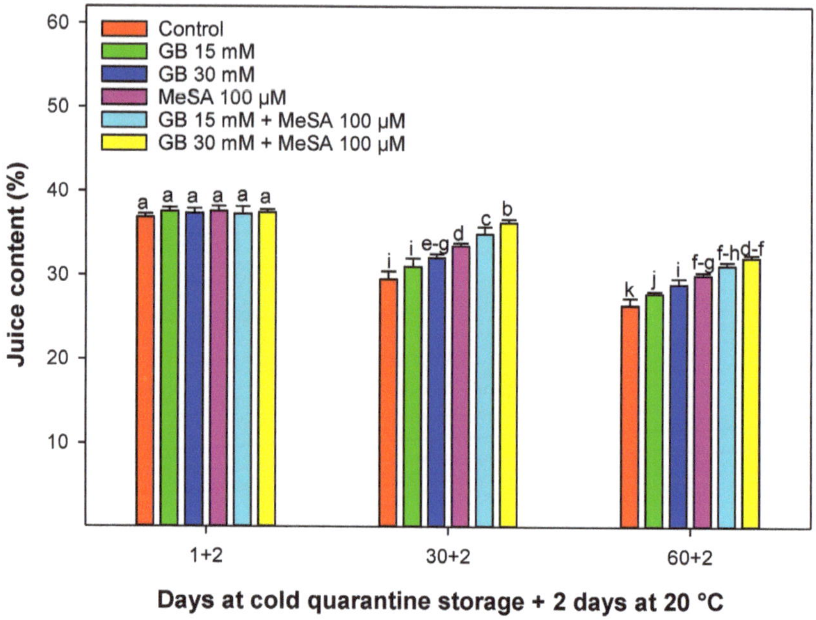

Figure 4. Changes in juice content in control and treated fruit during cold quarantine storage. Vertical bars represent ± standard errors (SE) of means. Different letters above the bars on columns indicate significant difference at $p < 0.05$ level of probability based on LSD test.

3.5. Peel Color

Color parameters, L^*, a^*, b^*, h°, C^*, and *CCI* changed during cold quarantine storage (Figure 5). In this study, L^*, b^*, h°, and C^* decreased in all treatments during cold quarantine storage, while a^* and CCI increased. The L^* value decreased with the same trends for all treatments along storage and the reduction of L^* in control samples was higher than in treated fruit. The b^* and h° values decreased in all treatments and this reduction was more pronounced in control samples. *CCI* increased during cold storage in all treatments during cold quarantine storage. Overall, the combined treatment of GB and MeSA, especially at 30 mM GB + 100 μM MeSA, had the most positive effect on color parameter changes.

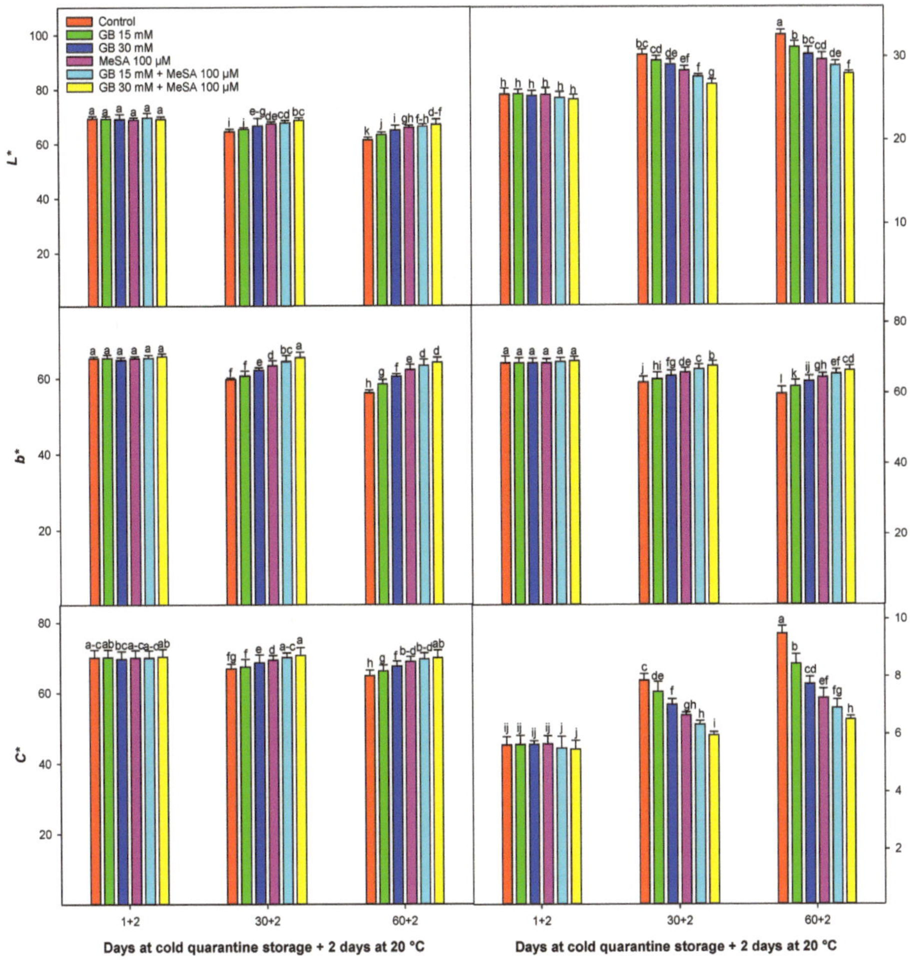

Figure 5. Changes in L^*, a^*, b^*, hue angle ($h°$), chroma (C^*), and citrus color index (*CCI*) in control and treated fruit during cold quarantine storage. Vertical bars represent ± standard errors (SE) of means. Different letters above the bars on columns indicate significant difference at $p < 0.05$ level of probability based on LSD test.

4. Discussion

Cold quarantine storage involves maintaining citrus fruit at low temperatures postharvest to meet strict phytosanitary requirements for international export [2,25]. While cold quarantine storage can have some benefits for the exportation of citrus fruit, prolonged storage at low temperatures can affect the texture, juice content, and peel color changes [24]. Therefore, postharvest elicitor treatments can help reduce these negative effects and ensure the highest possible quality of citrus fruit. In this study, the effects of postharvest treatment with GB and MeSA on 'Moro' blood orange were evaluated to determine their efficiency in mitigating weight loss, preserving firmness, maintaining the chemical attributes of juice, retaining juice content, and preserving peel color during cold quarantine storage.

Blood oranges can lose weight during cold storage due to a combination of factors, including respiration, water loss, chemical changes, transpiration, and cell-wall degradation [21]. Respiration consumes stored energy, leading to the breakdown of cells, consumption of stored energy, and water, resulting in weight loss. This can lead to weight loss and affect the fruit's quality and texture [26]. Although the respiration rate slows down during cold storage, it still continues, leading to a continuous loss of weight, as well as a decrease in size and volume of the fruit. Proper cold storage conditions can help minimize weight loss and maintain fruit quality. In this study, the applied treatments resulted in a noteworthy reduction in weight loss. Particularly, the combined application of GB and MeSA exhibited the most pronounced positive impact on curtailing weight loss. This observation suggests that GB and MeSA played a significant role in regulating water loss, transpiration, and the rate of respiration in the treated fruit. Consequently, the treated fruit displayed lower weight loss compared to the control group, as observed in this study.

Fruit acceptability can be determined by fruit firmness, which is influenced by temperature. Lower temperatures can reduce metabolic activity and the activity of cell-wall degrading enzymes, resulting in the softening of the fruit. The softening process occurs due to the breakdown of pectin, a key component of the cell wall, through the activity of several enzymes, including polygalacturonase, pectin lyase, pectin methylesterase, and cellulose. When these enzymes are active, pectin depolymerizes, leading to softening [26]. These changes in the cell-wall composition are responsible for the decrease in citrus fruit firmness during cold storage. In this study, the treated fruit exhibited a decrease in firmness loss compared to the control fruit, with the most pronounced effect observed in the combination treatment of GB and MeSA during cold quarantine storage. This combination appeared to synergistically mitigate cell-wall degradation, resulting in the preservation of turgor pressure and the avoidance of fruit tissue collapse. Consequently, this mechanism contributed to the maintenance of fruit firmness [21].

During cold storage, citrus fruit undergo glycolysis and the Krebs cycle to produce cellular energy. However, these processes can also alter the levels of sugars and organic acids in the citrus fruit through biosynthesis and catabolism [24]. The TSS of citrus fruit juice is a reliable measure of its sugar content, which makes up around 80% of the TSS, mainly from sucrose, glucose, and fructose, while 10% consists of organic acids, such as citric, malic, and ascorbic acids, as well as other components like vitamins, proteins, free amino acids, and glucosides. The conversion of organic acids to sugars by glycolytic enzymes can lead to an increase in TSS [21]. However, prolonged cold storage can also cause a decrease in TSS due to the breakdown of sugars and other soluble constituents [24].

The TA in citrus fruit juice is closely related to the amount of organic acids present. Organic acids are essential components of citrus fruit juice, and blood oranges are known to have high levels of citric and malic acids. During cold storage, the organic acids in the fruit can be consumed for the synthesis of sugars and ATP production, which can result in a reduction of TA [23]. In this study, the treated fruit, particularly in the case of 30 mM GB + 100 μM MeSA treatment, exhibited elevated TA, which could be attributed to the potential prevention of alcoholic fermentation during cold quarantine storage [24]. The applications of GB and MeSA may have played a role in preserving organic acids, thereby delaying their breakdown and conversion into sugars. The correlation between TA and TSS was negative. Consequently, these treatments contributed to the preservation of higher TA levels in the juice of treated fruit as compared to control samples.

The TSS/TA ratio exhibited differences among the treatments in this study, and it was observed to increase during cold storage. The ratio of sugars to organic acids is not only an indicator of the harvest maturity of citrus fruit, but it also plays a crucial role in determining the taste and flavor of the fruit during cold storage [24]. Citric and malic acids are primarily responsible for the acidity of citrus fruit, and changes in their levels due to the conversion of organic acids during fruit respiration can significantly alter the TSS/TA ratio, resulting in varying qualities among different cultivars of citrus fruit [21]. As a result, the varying changes in the levels of sugars and organic acids can be associated with the differences in the TSS/TA ratio among the treatments, which exhibit a different range of values. These values were affected by lower TSS and higher TA in fruit treated with MeSA and GB alone or in combination, as compared to control fruit, displaying a delayed metabolism, as observed in different vegetal species treated with these elicitors [5,6]. The treatments involving MeSA and GB, particularly when combined, appear to exert a positive influence on maintaining the delicate balance between sugars and organic acids throughout the cold quarantine storage period. This effect contributes substantially to enhancing the overall quality and flavor profile of the blood orange fruit, as evidenced by the MeSA and GB combination.

The results of this study indicated that the pH of the citrus fruit juice increased during storage, which can be attributed to the biochemical activity occurring within the fruit. This activity leads to the conversion of organic acids into sugar products, ultimately resulting in a shift towards a higher pH [24]. As the concentration of glucose and other sugars in the fruit increases, it can reduce the rate of respiration, which can also contribute to the observed increase in pH. Therefore, the increase in the pH of the fruit juice can serve as an indicator of the consumption of organic acids during cold storage. Furthermore, the reduction in TA can also be associated with an increase in juice pH, which is consistent with the observed results of this study [21].

The juice content of citrus fruit is an important factor in determining the overall quality of citrus fruit. The results of this study showed that the juice content decreased across all treatments during cold quarantine storage. The variation in the extent of juice content reduction among treatments could be attributed to the ability of the treatment to maintain cell membrane integrity [24]. The maintenance of membrane integrity is essential in preventing the drying of juice vesicles by protecting the sac-juice membranes from damage under low-temperature conditions [27]. Therefore, fruits that exhibit better cell membrane integrity are expected to have a lower reduction in juice content during cold storage, which is in consonance with the weight losses observed in this study.

Peel color is a crucial aspect of citrus fruit quality that greatly influences consumer preference. The color of the peel can be evaluated primarily based on three major types of pigments: chlorophylls, carotenoids, and anthocyanins. However, for a commercial perspective, it may also be important to consider factors such as visual quality and market acceptability. While chlorophylls are responsible for the green color of citrus fruit before ripening, carotenoids and anthocyanins play a significant role in determining the final color of the peel. Anthocyanins are particularly responsible for the red color of the flesh in blood oranges, but some cultivars also have them present in the peel [28]. When citrus fruits are exposed to cold temperatures, the color parameters of their peel may be affected. Specifically, the color may become darker, browner, and less orange. This is because the low temperature can cause the breakdown of chlorophyll in the peel, leading to changes in color [29]. Even though citrus fruits are not climacteric and have low ethylene production and respiration rates during maturity and ripening, color change of their peel can still occur after harvesting and during cold storage. In the present study, changes in peel color were observed in all treatments during cold quarantine storage of fruits. The lightness (L^*) value, which represents the amount of light reflected by an object on a scale from black (0) to white (100), decreased in all treatments, with control samples having the lowest L^* value. More specifically, a decrease in the L^* value indicates that the color is becoming darker. The a^* value indicates a shift towards a reddish hue. The a^* value, which ranges from green (−) to red (+), had a positive value in blood orange cultivars and increased in all treatments during cold quarantine storage. The b^* value indicates that the color is becoming less yellow and browner. The b^* value, which ranges from blue (−) to yellow (+), decreased in all treatments, possibly due to the increase in the a^* value or fruit senescence after prolonged storage [30]. Therefore, it seems that the effect of delaying color parameters was related to a delay in senescence, as it has been observed in other species treated with MeSA [3,4] and GB [7,11,18].

During cold storage in this study, the h° value, which represents the actual perceived color of orange or green and is a primary variable in changes in orange color, decreased in all treatments. The h° value ranges from 0 or 360° for red, 90° for yellow, 180° for green, and 270° for blue colors [24]. The h° values in this study were in the range of 59−67°, which corresponds to orange-yellow to yellow colors in all samples. Control samples had the lowest h° value during cold quarantine storage. Chroma (C^*) or saturation index, which quantifies the intensity or saturation of color, decreased during cold storage [30]. In this study, CCI increased in all treatments. A comparative study on blood orange cultivars stored at 2 and 5 °C showed that changes in peel color parameters at 5 °C were higher than at 2 °C. The reason for this may be that lower temperatures reduced the biosynthesis or degradation of blood orange peel pigments [24]. The results indicate that the application of GB and MeSA treatments can influence these color changes, with the combined treatment showing the most promising results. These findings provide valuable insights into the potential strategies for maintaining fruit color quality during extended storage periods. Specifically, the combined treatment of GB and MeSA, particularly at the concentration of 30 mM GB + 100 μM MeSA, exhibited the most positive effect on color parameter changes. This finding suggests that the interaction between these two treatments could synergistically impact color preservation during cold quarantine storage.

5. Conclusions

In conclusion, the application of GB and MeSA treatments on 'Moro' blood oranges during cold quarantine storage showed significant preservation in physicochemical changes and chemical attributes of the fruit juice. The combination of 30 mM GB and 100 μM MeSA proved to be the most effective treatment, reducing weight and firmness losses while also delaying changes in peel color parameters such as L^*, b^*, h°, and C^*. Furthermore, the treated fruit exhibited higher TA, TSS, and TSS/TA compared to the control samples. These findings highlight the potential of GB and MeSA treatments in maintaining the physicochemical attributes of 'Moro' blood oranges during cold quarantine storage, with MeSA showing a synergistic effect when combined with GB in preserving the fruit's quality. Further research is needed to better understand how GB and MeSA interact with fruit physiologically. This should involve studying effects of cold quarantine storage through sensory, nutritional, and molecular analyses.

Author Contributions: Conceptualization, F.H. and D.V.; methodology, F.H.; software, F.H.; validation, F.H., A.S. and F.G.; formal analysis, F.H.; investigation, F.H.; resources, D.V.; data curation, F.H.; writing—original draft preparation, F.H.; writing—review and editing, A.S., D.V., M.S. and F.G; visualization, F.H.; supervision, D.V.; project administration, F.H.; funding acquisition, D.V. All authors have read and agreed to the published version of the manuscript.

Funding: This research received no external funding.

Data Availability Statement: Data are contained within the article.

Acknowledgments: The authors would like to thank University Miguel Hernández for the sabbatical opportunity for Fariborz Habibi.

Conflicts of Interest: The authors declare no conflict of interest.

References

1. Habibi, F.; García-Pastor, M.E.; Puente-Moreno, J.; Garrido-Auñón, F.; Serrano, M.; Valero, D. Anthocyanin in blood oranges: A review on postharvest approaches for its enhancement and preservation. *Crit. Rev. Food Sci. Nutr.* **2022**, 1–13. [CrossRef] [PubMed]
2. Carmona, L.; Zacarías, L.; Rodrigo, M.J. Stimulation of coloration and carotenoid biosynthesis during postharvest storage of 'Navelina' orange fruit at 12 C. *Postharvest Biol. Technol.* **2012**, *74*, 108–117. [CrossRef]
3. Sayyari, M.; Babalar, M.; Kalantari, S.; Martínez-Romero, D.; Guillén, F.; Serrano, M.; Valero, D. Vapour treatments with methyl salicylate or methyl jasmonate alleviated chilling injury and enhanced antioxidant potential during postharvest storage of pomegranates. *Food Chem.* **2011**, *124*, 964–970. [CrossRef]
4. Giménez, M.J.; Valverde, J.M.; Valero, D.; Zapata, P.J.; Castillo, S.; Serrano, M. Postharvest methyl salicylate treatments delay ripening and maintain quality attributes and antioxidant compounds of 'Early Lory' sweet cherry. *Postharvest Biol. Technol.* **2016**, *117*, 102–109. [CrossRef]
5. Wang, Q.; Ding, T.; Zuo, J.; Gao, L.; Fan, L. Amelioration of postharvest chilling injury in sweet pepper by glycine betaine. *Postharvest Biol. Technol.* **2016**, *112*, 114–120. [CrossRef]
6. Yao, W.; Xu, T.; Farooq, S.U.; Jin, P.; Zheng, Y. Glycine betaine treatment alleviates chilling injury in zucchini fruit (*Cucurbita pepo* L.) by modulating antioxidant enzymes and membrane fatty acid metabolism. *Postharvest Biol. Technol.* **2018**, *144*, 20–28. [CrossRef]

7. Shan, T.; Jin, P.; Zhang, Y.; Huang, Y.; Wang, X.; Zheng, Y. Exogenous glycine betaine treatment enhances chilling tolerance of peach fruit during cold storage. *Postharvest Biol. Technol.* **2016**, *114*, 104–110. [CrossRef]
8. Wang, L.; Bokhary, S.U.F.; Xie, B.; Hu, S.; Jin, P.; Zheng, Y. Biochemical and molecular effects of glycine betaine treatment on membrane fatty acid metabolism in cold stored peaches. *Postharvest Biol. Technol.* **2019**, *154*, 58–69. [CrossRef]
9. Wang, L.; Shan, T.; Xie, B.; Ling, C.; Shao, S.; Jin, P.; Zheng, Y. Glycine betaine reduces chilling injury in peach fruit by enhancing phenolic and sugar metabolisms. *Food Chem.* **2019**, *272*, 530–538. [CrossRef]
10. Jia, Z.; Wang, Y.; Wang, L.; Zheng, Y.; Jin, P. Amino acid metabolomic analysis involved in flavor quality and cold tolerance in peach fruit treated with exogenous glycine betaine. *Food Res. Int.* **2022**, *157*, 111204. [CrossRef]
11. Zhang, Y.; Jin, P.; Huang, Y.; Shan, T.; Wang, L.; Li, Y.; Zheng, Y. Effect of hot water combined with glycine betaine alleviates chilling injury in cold-stored loquat fruit. *Postharvest Biol. Technol.* **2016**, *118*, 141–147. [CrossRef]
12. Razavi, F.; Mahmoudi, R.; Rabiei, V.; Aghdam, M.S.; Soleimani, A. Glycine betaine treatment attenuates chilling injury and maintains nutritional quality of hawthorn fruit during storage at low temperature. *Sci. Hortic.* **2018**, *233*, 188–194. [CrossRef]
13. Luo, M.; Zhou, X.; Sun, H.; Zhou, Q.; Wei, B.; Cheng, S.; Ji, S.-j. Glycine betaine treatment alleviates loss of aroma-related esters in cold-stored 'Nanguo' pears by regulating the lipoxygenase pathway. *Food Chem.* **2020**, *316*, 126335. [CrossRef] [PubMed]
14. Sun, H.; Luo, M.; Zhou, X.; Zhou, Q.; Sun, Y.; Ge, W.; Wei, B.; Cheng, S.; Ji, S. Exogenous glycine betaine treatment alleviates low temperature-induced pericarp browning of 'Nanguo' pears by regulating antioxidant enzymes and proline metabolism. *Food Chem.* **2020**, *306*, 125626. [CrossRef] [PubMed]
15. Wang, J.; Lv, M.; He, H.; Jiang, Y.; Yang, J.; Ji, S. Glycine betaine alleviated peel browning in cold-stored 'Nanguo' pears during shelf life by regulating phenylpropanoid and soluble sugar metabolisms. *Sci. Hortic.* **2020**, *262*, 109100. [CrossRef]
16. Mahmoudi, R.; Razavi, F.; Rabiei, V.; Gohari, G.; Palou, L. Application of glycine betaine coated chitosan nanoparticles alleviate chilling injury and maintain quality of plum (*Prunus domestica* L.) fruit. *Int. J. Biol. Macromol.* **2022**, *207*, 965–977. [CrossRef]
17. Chen, L.-L.; Shan, W.; Cai, D.-L.; Chen, J.-Y.; Lu, W.-J.; Su, X.-G.; Kuang, J.-F. Postharvest application of glycine betaine ameliorates chilling injury in cold-stored banana fruit by enhancing antioxidant system. *Sci. Hortic.* **2021**, *287*, 110264. [CrossRef]
18. Zhang, J.; Wang, C.; Chen, C.; Zhang, S.; Zhao, X.; Wu, C.; Kou, X.; Xue, Z. Glycine betaine inhibits postharvest softening and quality decline of winter jujube fruit by regulating energy and antioxidant metabolism. *Food Chem.* **2023**, *410*, 135445. [CrossRef] [PubMed]
19. Molaei, S.; Rabiei, V.; Soleimani, A.; Razavi, F. Exogenous application of glycine betaine increases the chilling tolerance of pomegranate fruits cv. Malase Saveh during cold storage. *J. Food Process. Preserv.* **2021**, *45*, e15315. [CrossRef]
20. Huang, S.; Lim, S.Y.; Lau, H.; Ni, W.; Li, S.F.Y. Effect of glycinebetaine on metabolite profiles of cold-stored strawberry revealed by 1H NMR-based metabolomics. *Food Chem.* **2022**, *393*, 133452. [CrossRef]
21. Habibi, F.; Guillén, F.; Serrano, M.; Valero, D. Postharvest treatment with glycine betaine enhances chilling tolerance of blood orange fruit by increasing antioxidant defence systems and osmoregulation during cold storage. *Sci. Hortic.* **2022**, *305*, 111352. [CrossRef]
22. Habibi, F.; Ramezanian, A.; Guillén, F.; Serrano, M.; Valero, D. Blood oranges maintain bioactive compounds and nutritional quality by postharvest treatments with γ-aminobutyric acid, methyl jasmonate or methyl salicylate during cold storage. *Food Chem.* **2020**, *306*, 125634. [CrossRef] [PubMed]
23. Habibi, F.; Sarkhosh, A.; Guillén, F.; Serrano, M.; Valero, D. Postharvest treatments with methyl salicylate and glycine betaine synergistically enhanced chilling tolerance and maintained bioactive compounds of blood orange fruit subjected to cold quarantine storage. *LWT* **2023**, *185*, 115141. [CrossRef]
24. Habibi, F.; Guillén, F.; Serrano, M.; Valero, D. Physicochemical changes, peel colour, and juice attributes of blood orange cultivars stored at different temperatures. *Horticulturae* **2021**, *7*, 320. [CrossRef]
25. Sdiri, S.; Navarro, P.; Monterde, A.; Benabda, J.; Salvador, A. Effect of postharvest degreening followed by a cold-quarantine treatment on vitamin C, phenolic compounds and antioxidant activity of early-season citrus fruit. *Postharvest Biol. Technol.* **2012**, *65*, 13–21. [CrossRef]
26. Valero, D.; Serrano, M. *Postharvest Biology and Technology for Preserving Fruit Quality*, 1st ed.; CRC Press: Boca Raton, FL, USA, 2010. [CrossRef]
27. Cao, J.; Kang, C.; Chen, Y.; Karim, N.; Wang, Y.; Sun, C. Physiochemical changes in *Citrus reticulata* cv. Shatangju fruit during vesicle collapse. *Postharvest Biol. Technol.* **2020**, *165*, 111180. [CrossRef]
28. Chen, C. Pigments in Citrus Fruit: Mutants, Compounds, Genes, and Beyond. In *The Citrus Genome. Compendium of Plant Genomes*; Gentile, A., La Malfa, S., Deng, Z., Eds.; Springer: Cham, Switzerland, 2020; pp. 195–209. [CrossRef]
29. Habibi, F.; Ramezanian, A.; Guillén, F.; Martínez-Romero, D.; Serrano, M.; Valero, D. Susceptibility of blood orange cultivars to chilling injury based on antioxidant system and physiological and biochemical responses at different storage temperatures. *Foods.* **2020**, *9*, 1609. [CrossRef] [PubMed]
30. Van Wyk, A.A.; Huysamer, M.; Barry, G.H. Extended low-temperature shipping adversely affects rind colour of 'Palmer Navel' sweet orange [*Citrus sinensis* (L.) Osb.] due to carotenoid degradation but can partially be mitigated by optimising post-shipping holding temperature. *Postharvest Biol. Technol.* **2009**, *53*, 109–116. [CrossRef]

horticulturae

Article

Effect of Oxalic Acid Treatments and Modified Atmosphere Packaging on the Quality Attributes of Rocket Leaves during Different Storage Temperatures

Derya Erbaş

Department of Horticulture, Faculty of Agriculture, Isparta University of Applied Sciences, Isparta 32260, Türkiye; deryaerbas@isparta.edu.tr; Tel.: +90-246-214-63-65

Abstract: The effects of combinations of oxalic acid (OA) treatment with modified atmosphere packaging on the quality and biochemical content changes of rocket (*Eruca sativa* Mill. cv. Bengi) leaves were examined. After harvest, selected leaves were dipped into an aqueous solution containing different concentrations of oxalic acid (0-control, 0.25 mM, 0.5 mM, and 1 mM) for 1 min. Treated samples were dried and placed in modified atmosphere packages. Treated rockets were stored at two different temperatures (0 °C and 10 °C) and 90 ± 5% relative humidity conditions for 10 days. Leaves were analyzed at 2-day intervals for some quality and biochemical parameters during storage. OA-treated leaves were greener than those of the control group. At the end of the storage, high doses (1 mM) of OA applications successfully suppressed the respiration rate (0 °C: 63.12 mL CO_2 kg^{-1} h^{-1}, 10 °C: 78.09 mL CO_2 kg^{-1} h^{-1}) and retarded the weight loss (0 °C: 0.14%, 10 °C: 0.49%) and color discoloration (0 °C: ΔE 7.23, 10 °C: ΔE 8.34) of rocket leaves. In addition, OA treatments decreased the vitamin C losses and chlorophyll degradation. In conclusion, rocket leaves could be stored at 0 °C for 8–9 days with 1 mM OA treatment and 6 days with the control treatment and at 10 °C for 6–7 days with 1 mM OA treatment and 4 days with the control (C) treatment with a minimum quality loss under MAP conditions.

Keywords: *Eruca sativa*; oxalic acid; quality; storage; yellowing

Citation: Erbaş, D. Effect of Oxalic Acid Treatments and Modified Atmosphere Packaging on the Quality Attributes of Rocket Leaves during Different Storage Temperatures. *Horticulturae* **2023**, *9*, 718. https://doi.org/10.3390/horticulturae9060718

Academic Editors: Wenzhong Hu, Tian Zhong and Xiuxiu Sun

Received: 11 May 2023
Revised: 8 June 2023
Accepted: 15 June 2023
Published: 19 June 2023

1. Introduction

Vegetables undergo numerous biological and physiological changes following harvest. As a result of the acceleration of biochemical reactions, including respiration rate and ethylene production in vegetables after harvest, the concentrations of some substances (including color substances, vitamins, organic acids, and oils/fats) that determine the quality and nutritional content of the products change, water loss increases, and color changes take place [1]. Therefore, vegetables cannot be stored for very long following harvest [2,3]. The best way to preserve postharvest quality by delaying the senescence or deterioration of products is to minimize respiration, and the most effective way to do this is to reduce the ambient temperature. In addition to cold storage, heat treatments, edible coatings, ozone applications, modified atmosphere packaging, controlled atmosphere storage, and chemical applications have been widely adopted to increase vegetables' storage and shelf life and maintain their quality for extended periods [4]. However, the fact that the toxic effects of chemical applications do not entirely disappear in products with short storage periods, including vegetables, resulted in a preference for natural applications in the post-harvest period [5].

Oxalic acid (OA) is a natural organic acid with the formula $H_2C_2O_4$. It is abundantly found in many plants (sorrel, rocket, peas, tomatoes, spinach, etc.) [6]. As the final metabolic product in plants, OA is involved in many vital functions, such as the response to environmental stressors and resistance [7]. In recent studies, exogenously applied OA at non-toxic concentrations has been reported to reduce enzymatic browning in litchi [8], control diseases and rot in peaches and kiwifruit, reduce respiration rate and ethylene production and maintain nutritional quality [9,10], extend shelf life in bananas and plums [11,12], maintain fruit quality in cherries [13], stimulate the antioxidant system and phenolic content of lemons [14], and regulate ethanol fermentation in kiwifruit [15]. However, such studies conducted with vegetables have been very limited. Exogenously applied OA has been reported to delay quality losses and extend shelf life in tomatoes [16], artichokes [17], asparagus [18], lettuce, and rocket [19].

Vegetables play an essential role in human nutrition thanks to their vitamin and mineral content. Nutritionists state that consuming fruit and vegetables reduces the risk of many diseases and contributes significantly to human health [20]. Therefore, in line with the increase in people's interest in health and quality of life in recent years, there has been an increase in the demand for healthy and fresh products [21]. Particularly, sorted and washed fresh vegetables have been increasingly attracting consumers' attention, especially working people, due to their ease of use. In Turkey, the production of parsley, mint, dill, cress, and rocket, among the vegetables of which the leaves are eaten, has increased significantly in recent years. Between 2012 and 2022, the increase in production volume was 86.5% for parsley, 113.6% for mint, 128.4% for cress, 252.6% for dill, and 429.0 % for rocket. Rocket production in Turkey was 7689 tons in 2012 and increased to 40,674 tons in 2022 [22]. Rocket (*Eruca sativa* Mill.) belongs to the *Brassicaceae* family [23] and can grow naturally in the Mediterranean Basin. Rocket is gaining popularity worldwide due to its health benefits. Glucosinolates give rocket, like other members of this family, a unique and rich aroma [24]. However, due to the high metabolic activity in vegetables of which the leaves are consumed, such as rocket, yellowing and water loss occur faster, shortening their shelf life [25,26]. Yellowing due to chlorophyll breakdown, especially during storage, reduces the marketability of rocket [27,28]. Considering the nutritional value and health value of rocket, its aroma, and the increase in production, there is a need to investigate the correct storage conditions and post-harvest practices to minimize the losses that may occur after harvest.

In line with this information, this study aimed to investigate the effects of post-harvest OA application on quality changes and biochemical contents of rocket leaves at different storage temperatures.

2. Materials and Methods

2.1. Reagents

Oxalic acid (≥97%), Tween® 20, methanol (99.9%), Trolox standard, and acetone were obtained from Sigma Chemical Co. (St. Louis, MO, USA). Folin-Ciocalteu's and gallic acid standard were purchased from Merck Co. (Rahway, NJ, USA). The chemicals used in the preparation of the FRAP solution (2, 4, 6-Tri(2-pyridyl)-s-triazine, iron (3) chloride hexahydrate, and sodium acetate) were also obtained from Merck Co.

2.2. Plant Material

Rockets were grown with a photoperiod of 16 h, an ambient temperature of 21–22 °C, and a relative humidity of 60–70% under green greenhouse conditions. Commercially grown rockets (*Eruca sativa* Mill. cv. Bengi) were harvested (~18–20 cm length) in Antalya-Turkey in the early morning and immediately transported to the Laboratory of the Postharvest Physiology. In the laboratory, the roots were cut with sharp scissors, and foreign materials and discolored or bruising leaves were discarded by hand at 5 °C. The fully green leaves without roots were used as plant material. Leaves were dipped in tap water (at ~5–6°C) to remove soil particles for 1 min.

2.3. Oxalic Acid Treatments and Storage Conditions

Selected leaves were randomly divided into 8 groups for the following treatments (T) (3500 ± 150 g leaves for each group):

T1-T2-T3-T4: Leaves were dipped in aqueous solutions (pH 6.5 and at 5 °C) of OA at different doses for 1 min: 0 (control), 0.25 mM, 0.5 mM, and 1 mM, stored at 0 ± 1 °C.

T5-T6-T7-T8: Leaves were dipped in aqueous solutions (pH 6.5 and at 5 °C) of OA at different doses for 1 min: 0 (control), 0.25 mM, 0.5 mM, and 1 mM, stored at 10 ± 1 °C.

Control groups were dipped in distilled water (5 °C) for 1 min. Dipping time (1 min) was determined according to a previous study about rocket plants [19]. Storage temperatures were chosen to simulate shelf (market) conditions (10 ± 1 °C) and normal cold storage temperatures (0 ± 1 °C) for rocket leaves. Tween 20 (0.1%) was also added to enhance infiltration of all aqueous solutions as a surfactant. After dipping treatments, the leaves were spun in a salad spinner to dry for about 2 min and placed in modified atmosphere packages (LDPE) (about 200 ± 50 g per package) at 5 °C. For all treatments, 120 packages (3 replicates × 8 treatments × 5 storage periods − day 0 excluded) were stored for 10 days with 90 ± 5% relative humidity (RH). About 400–450 g of leaves were selected for initial (day 0) analysis. Initially and at 2-day intervals during storage, leaves were analyzed to determine weight loss, respiration rate, leaf color, antioxidant activity, total chlorophyll content, total phenol content, ascorbic acid (vitamin C) content, yellowing, and external appearance.

2.4. *Physical and Chemical Analysis*

2.4.1. Weight Loss and Respiration Rate

The individual modified atmosphere packages (each package was considered as a replicate, 3 replicates) were weighed at the beginning of the storage (day 0) and placed in the cold storage rooms. The analysis days (2, 4, 6, 8, and 10 days) MAPs were weighed again and calculated according to the following Equation (1):

$$\text{Weight loss } (\%) = [(\text{First weight} - \text{Last weight})/\text{First weight}] \times 100 \quad (1)$$

The respiration rate of the leaves was determined by placing (weight 75–80 g leaves) in a 1 L glass jar hermetically sealed for 30 min [17]. Afterwards, the gas sample was taken and injected into a gas chromatograph (Agilent 6890N, Palo Alto, CA, USA) equipped with a thermal conductivity detector. Measurements were performed with 3 replicates, and results were expressed as mL CO_2 kg^{-1} h^{-1}.

2.4.2. Leaf Color, Yellowing, and External Appearance

Leaf color (L*, a* and b*) was determined at two points on the leaf surface with a colorimeter (Minolta CR-200, Ramsey, NJ, USA). The colorimeter was calibrated with a white plate. Color changes (ΔE) at each sampling day were calculated from L*, a*, and b* values similar to Cefola and Pace [19]. Measurements were performed with 3 replicates, and each replicate contained 20 leaves. ΔE was calculated according to the following Equation (2):

$$\Delta E = [(L^*_0 - L^*)^2 + (a^*_0 - a^*)^2 + (b^*_0 - b^*)^2]^{1/2} \quad (2)$$

Yellowing was determined based on a five-point hedonic scale (1: dark green, 2: light green, 3: yellowish-green, 4: greenish-yellow, 5: yellow). When the leaves' scores reached 3, they were noted as unmarketable and at the end of their shelf life [25]. The degree of external appearance has a nine-point hedonic scale (≤1–4: poor, 9: excellent). Evaluations were made with 3 replicates. The yellowing and external appearance evaluation panel consisted of 7 members of the research staff (Horticulture Department) who were experienced in sensory analysis of horticultural crops.

2.4.3. Total Phenolic Content and Antioxidant Activity

Extraction for total phenolic content and antioxidant activity: The leaf samples (5 g) were placed in a tube (50 mL) and 80% methanol (5 mL) was added. The samples with methanol were crushed by using a homogenizer. The homogenized samples were placed in dark conditions (14–16 h at 4 °C), and supernatants were stored at −20 °C until the day of analysis (up to a week) [29]. Total phenolic content was determined using the Folin-Ciocalteu method as described and modified by Lola-Luz et al. [30]. The standard curve was developed using gallic acid standard. The absorbance was read at 725 nm using a spectrophotometer (Varian Cary Bio 100, Mulgrave, Victoria, Australia). Measurements were performed with 3 replicates. Results were calculated as mg of gallic acid equivalent (GAE) per g fresh weight (FW).

The ferric reducing antioxidant power (FRAP) assay was used to evaluate the antioxidant capacity of rocket leaves. All the equipment and chemicals and the FRAP method were chosen and determined according to the method stated by Gutiérrez et al. [31]. The absorbance was read at 593 nm using a spectrophotometer. Measurements were performed with 3 replicates. The calibrating curve was developed using the Trolox standard. Results were calculated as Trolox equivalents (Trolox Eq) in g kg^{-1} FW.

2.4.4. Vitamin C and Total Chlorophyll Content

Vitamin C analysis of the leaves was determined by high-performance liquid chromatography (HPLC) (Agilent Technologies Inc.). The leaf sample (10 g) with the extraction medium (20 mL) was homogenized. The HPLC conditions and method were chosen according to the procedure reported by Martínez-Sánchez et al. [32]. Measurements were performed with 3 replicates. The results were expressed in mg per kg of FW.

For total chlorophyll content analysis, chopped leaves (5 g) were extracted in 80% acetone (Sigma–Aldrich, Steinheim, Germany) with a homogenizer. The extracted samples were centrifuged, and the supernatants were collected. Measurements were performed with 3 replicates. The absorbency of acetone extracts was measured at 663 and 645 nm with the spectrophotometer [33].

2.5. Statistical Analysis

The study was set up according to a completely factorial randomized design. The measurements were made in triplicate. All data obtained from this study were statistically evaluated by the SPSS 19.0 package program. Main effects (treatments and storage periods) and interactions (treatments × storage periods) were analyzed, and means were compared by Tukey's test at a significance level of 0.05. The correlation matrix and principal components analysis (PCA) were obtained from the R-core program.

3. Results and Discussion

3.1. Weight Loss and Respiration Rate

The weight loss (%) of rocket leaves increased steadily at both storage temperatures (Figure 1A). However, in general, it can be argued that these increases were minimal due to the high moisture content in MAP and relatively low water loss. Similarly, Manolopoulou and Mallidis [34] reported that high CO_2, low O_2, and high humidity in MAP with prolonged storage time effectively reduced water loss from the products and thus weight loss. In the study, on day 2 of storage, weight loss ranged between 0.08% and 0.11% at 0 °C and between 0.03% and 0.08% at 10 °C. At the end of storage on day 10, weight loss values were between 0.14% and 0.46% at 0 °C and between 0.47% and 1.02% at 10 °C. It was determined that the weight loss of rocket leaves stored at 10 °C was higher than those stored at 0 °C (Figure 1A). This can be explained by the fact that respiration is better suppressed at low temperatures, and the metabolic rate is slowed down. Erbaş and Koyuncu [12] reported that the weight loss in the products increased due to the removal of water from the tissues during respiration; thus, the respiration rate was effective in weight loss. This is supported by the fact that OA treatments (especially 1 mM OA) were more effective than the control treatment at both storage temperatures in terms of reducing weight loss and suppressing respiration rate.

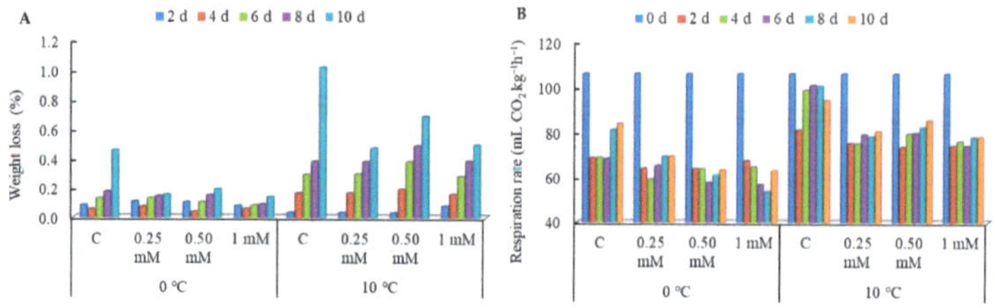

Figure 1. The effects of different OA treatments and storage temperatures on the weight loss (**A**) and respiration rate (**B**) of rocket leaves in MAP during storage. C: Control, d: days.

Respiration rate is one of the most critical factors affecting the quality and shelf life of products after harvest. Therefore, it is essential to suppress or slow down the respiration rate of products. As shown in Figure 1B, the respiration rate of the rocket leaves was measured as 106.04 CO_2 kg h^{-1} at the beginning, and on the second day of storage, the respiration rate decreased compared to the beginning and was between 63.92 mL CO_2 kg h^{-1} (0.5 mM, 0 °C) and 81.11 mL CO_2 kg h^{-1} (C, 10 °C). The decrease compared to the initial values can be explained by the action of lowering the ambient temperature and placing the products under MAP conditions. As expected, since the metabolic activity of rocket leaves stored at low temperatures slowed down, respiration rates were also lower. At the end of storage on day 10, the lowest respiration values (63.12 mL CO_2 kg h^{-1}, 0 °C and 78.09 mL CO_2 kg h^{-1}, 10 °C) were obtained from 1 mM OA treatment at both storage temperatures. All OA treatments were effective in suppressing the respiration rate compared to the control treatment (Table 1). In particular, the 1 mM OA dose was the best suppressing treatment for respiration rate at both storage temperatures (Figure 1B). This may be explained by the fact that OA application slows down the metabolic activity of the products. Our results are in accordance with previous works on vegetables [17,19] and fruit [12,35] treated with OA, which has been linked to a lower metabolic activity induced by OA.

Table 1. *p* values for storage period; treatments and their interactions for rocket leaves.

	0 °C			10 °C		
Parameters	**SP**	**T**	**SP × T**	**SP**	**T**	**SP × T**
WL	**	*	ns	**	**	**
RR	**	**	ns	**	**	**
ΔE	**	**	**	**	**	ns
Yellowing	**	**	*	**	**	**
EE	**	**	**	**	**	**
TPC	*	**	ns	**	*	ns
AOA	*	*	ns	*	*	ns
Vitamin C	*	**	ns	**	**	ns
TC	**	*	**	**	**	**

SP: Storage period, T: Treatments, WL: Weight loss, RR: Respiration rate, ΔE: Leaf color, EE: External appearance, TPC: Total phenolic content, AOA: Antioxidant activity, TC: Total chlorophyll content. Ns represents non significance at $p < 0.05$, * represents significance at the 0.05 level, ** represents significance at the 0.01 level.

3.2. Leaf Color, Yellowing, and External Appearance

Yellowing is the biggest ripening symptom in green leafy vegetables [26]. The rocket leaves' color and appearance were affected by treatments and storage temperature (Table 1). In line with the prolonged storage period, yellowing increased in the leaves, resulting in color and external appearance deterioration. The color change in the rocket leaves started on the second day, and the highest color change (ΔE: 8.48, 0 °C and 11.74, 10 °C) was detected in the control group at both storage temperatures on day 10 at the end of storage. At the end of storage, the 1 mM OA dose (ΔE: 7.23, 0 °C and 8.34, 10 °C) was the most effective treatment in color preservation at both storage temperatures (Figure 2A). In line with the color change, yellowing and external appearance scores also increased during storage. On the 10th day of storage, the rocket leaves stored at 10 °C had the highest score in terms of yellowing, while the lowest scores were obtained from those stored at 0 °C. At the end of storage, the control group (4.50 score: 0 °C, 4.83 score: 10 °C) was determined to have the highest yellowing scores at both storage temperatures. The lowest yellowing scores were found in 1 mM OA treatment (3.50 score: 0 °C, 4.33 score: 10 °C) at both temperatures (Figure 2B). At the end of storage, all OA treatments were more effective than the control treatments in terms of yellowing and external appearance (Figure 2B,C). The external appearance scores of the rocket leaves were not significantly changed in the initial 4 days of storage, but after the 4th day, the differences between the treatments became more apparent. At the end of storage, the lowest external appearance scores (4.50: 0 °C, 2.83 score: 10 °C) were obtained from the control groups. As expected, in parallel with the increase in temperature, the external appearance scores of rocket leaves stored at 0 °C were higher than those stored at 10 °C. During storage, the highest external appearance scores were generally found in the 1 mM OA treatment at both storage temperatures (Figure 2C). It was thought that the effect of OA on color change, yellowing, and external appearance could be explained by a slowdown of chlorophyll synthesis and lipid peroxidation, which is known to have an impact on quality. The positive effects of OA on color and appearance have also been reported in studies conducted with different vegetable and fruit species [15,19,36].

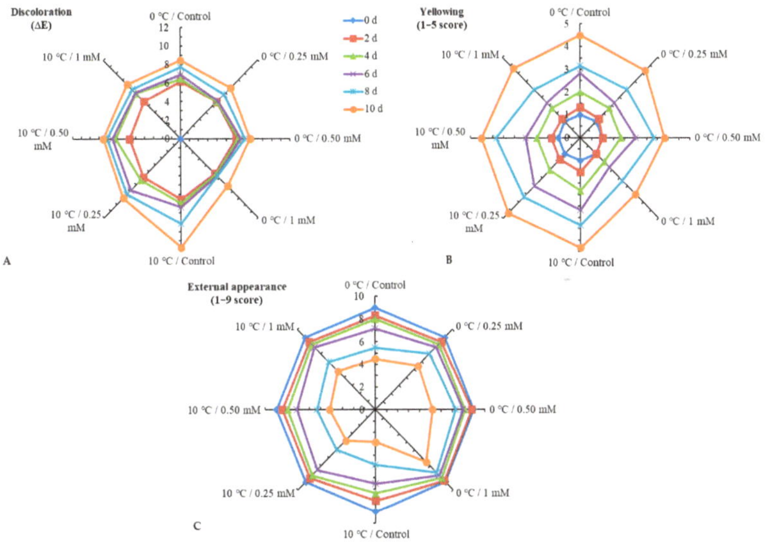

Figure 2. The effects of different OA treatments and storage temperatures on the leaf color (discoloration) (**A**), yellowing (**B**), and external appearance (**C**) of rocket leaves in MAP during storage. d: days.

3.3. Total Phenolic Content and Antioxidant Activity.

Phenolic compounds, one of the essential antioxidant substances in human nutrition [37], are known as secondary metabolism products of plants and are involved in the formation of sensory properties and coloration of products [38]. It has been known that changes in the phenolic content of fruit or vegetables after harvest can be affected by many factors, including species and variety, ripeness, and harvest time. The TPC values of rocket leaves, albeit not regularly, increased in parallel with the prolonged storage period. The increases were rapid from the beginning of storage until day 6 and remained stable on days 8 and 10. The TPC value was 126.5 mg GA 100 g^{-1} fw at the beginning of storage, 119.6 (10 °C, C) to 172.0 mg GA 100 g^{-1} fw (0 °C, 1 mM OA) on day 2, and 198.7 (10 °C, C) to

207.6 mg GA 100 g^{-1} fw (0 °C, 1 mM OA) at the end of storage on day 10. The highest increase was detected in the OA group samples stored at 0 °C (Figure 3A), and the effects of treatments on the TPC were statistically ($p < 0.05$) significant (Table 1).

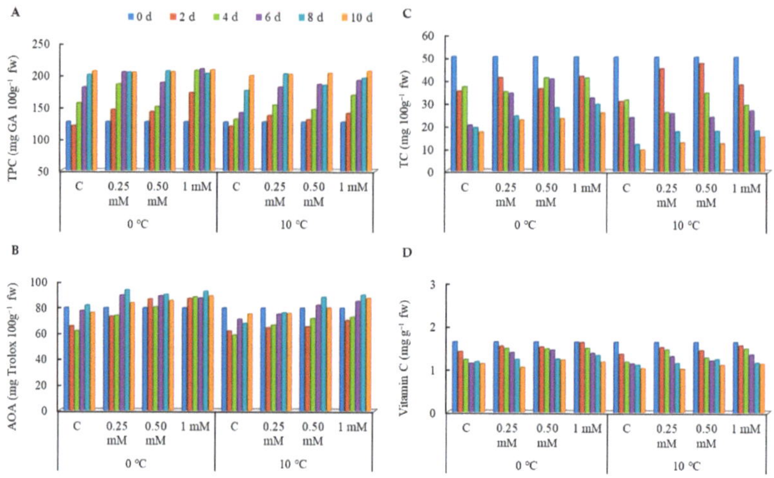

Figure 3. The effects of different OA treatments and storage temperatures on the total phenolic content (TPC) (**A**), antioxidant activity (AOA) (**B**), total chlorophyll content (TC) (**C**), and vitamin C (**D**) of rocket leaves in MAP during storage. C: Control, d: days.

Similar to the TPC values, the AOA values of rocket leaves fluctuated during storage; however, in general, they increased at the end of storage. The highest increases were detected in the OA-treated groups. The initial value measured as 79.0 mg 100 g^{-1} fw was between 74.4 (10 °C, C) and 88.4 mg 100 g^{-1} fw (0 °C, 1 mM OA) at the end of storage. OA applications were especially influential in terms of the preservation of AOA at 0 °C (Figure 3B). In studies conducted in different species, exogenously applied OA was reported to protect membrane integrity by preventing oxidation of phenolic substances and lipid peroxidation and increasing AOA [36,39,40].

3.4. Vitamin C and Total Chlorophyll Content

The total chlorophyll content of the rocket leaves decreased steadily throughout storage in all treatments and at both storage temperatures. The initial TC value measured as 50.2 mg 100 g^{-1} fw was between 17.1 mg 100 g^{-1} fw (C) and 25.7 mg 100 g^{-1} fw (1 mM OA) at 0 °C and between 9.5 mg 100 g^{-1} fw (C) and 15.3 mg 100 g^{-1} fw (1 mM OA) at 10 °C at the end of storage (Figure 3C). Chlorophylls are color pigments responsible for the formation of green color in horticultural crops. It has been reported that chlorophyll levels decrease significantly with ripening, and applications that accelerate ripening (such as high temperature) also accelerate chlorophyll breakdown [41]. This is supported by the fact that the chlorophyll content of rocket leaves stored at high temperatures (10 °C) was lower than those stored at cold temperatures (0 °C). The chlorophyll content of OA-treated rocket leaves was higher than the C groups at both storage temperatures. The breakdown of chlorophyll is slower, and the color remains greener in OA-treated rocket leaves (Figure 3C). Similarly, Kayashima and Katayama [42] have reported that OA, a natural antioxidant substance, plays a vital role in preventing/slowing down oxidation events, and Cefalo and Pace [19] have reported that OA delays chlorophyll breakdown. The results obtained in the yellowing part were consistent with the chlorophyll content.

Vitamin C content decreased as the storage period progressed, regardless of the application and temperature. Vitamin C decreases rapidly or slowly after harvesting depending on species, variety, and environmental factors. As shown in Figure 3D, the initial vitamin C content was 1.6 mg g^{-1} fw and was determined to be between 1.2 (0 °C, 0.5 and 1 mM OA) and 1.0 mg g^{-1} fw (0 °C: 0.25 mM OA, 10 °C: C and 0.25 mM) at the end of storage on day 10. Vitamin C content decreased at the end of storage, but no significant differences were observed between the treatments. This was associated with the preservation of rocket leaves under MAP conditions. MAP may have delayed the oxidation of vitamin C in all treatments by regulating the permeability of oxygen and carbon dioxide in it, and vitamin C can also be easily affected by many factors such as temperature, light, and the presence of oxygen, so the responses of rocket leaves to the treatments in terms of changes in vitamin C content may be different.

3.5. Indicator Statistics and Principal Component Analysis (PCA)

In the study, it was observed that many factors interact with each other to impact quality of rocket leaves during storage, so a correlation analysis study is needed. Figure 4 showed the correlation analysis results of the quality parameters of rocket leaves for the test. As expected, in terms of the analyzed parameters, a highly positive correlation was generally determined between samples stored at 0 °C and 10 °C. The highest significant negative correlation (r: −0.98; $p < 0.001$) was identified between EE (10 °C) and yellowing values (0 °C and 10 °C), while the highest significant positive correlation (r: 0.99 ; $p < 0.001$) value was identified between YY (0 °C) and YY (0 °C).

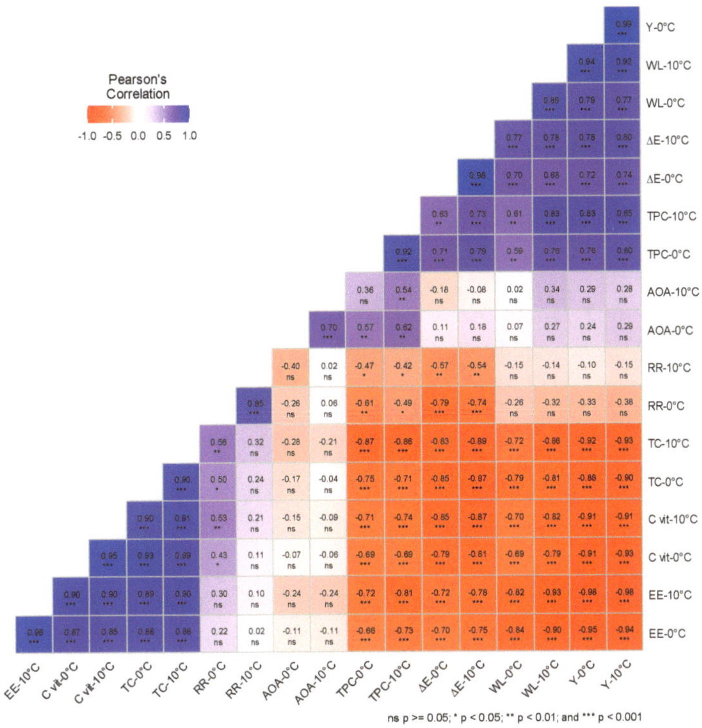

Figure 4. Correlation matrix of the parameters measured in rocket leaves. EE: External appearance, TC: Total chlorophyll content, C vit: Vitamin C, RR: Respiration rate, WL: Weight loss, Y: Yellowing, ΔE: Discoloration (color changes), TPC: Total phenolic content, AOA: Antioxidant activity.

KMO and Bartlett tests were performed on the original variables of the indicators of all samples of rocket leaves. The test results were shown in Table 2. The KMO value was 0.729 in Table 2, indicating that the data were suitable for factor analysis. The probability of

Bartlett's test statistical value was 0.000, and it was less than 0.05, which indicated that the data were correlated and could be used for factor analysis [43].

Table 2. Correlation test of KMO and Bartlett.

Test Method	KMO Measure of Sampling Adequacy	Bartlett's Test of Sphericity		
		Approx. χ^2	df	Sig.
Result	0.729	799.058	153	0.000

The squares of the coordinates (cos2) provide a measure of how well the major component expresses the relevant variable. Cos2 can be used to represent these values. The correlations between the variables and principal components are another way to express these values. The first principal component explains 68% of the variance, and the second principal component explains 12.5%. The squares of the coordinates (cos2) are an indicator of how successfully the variable of interest is represented by the principal component and is expressed as cos2. These values are also expressed as correlations between variables and principal components. When the vector directions of the properties were examined, it was seen that EE, C vit, TC, and RR exhibited negative correlations with the other properties. Among the properties that were examined, AOA had the lowest cos2 value and was considered to have the lowest effect on the principal components (Figure 5).

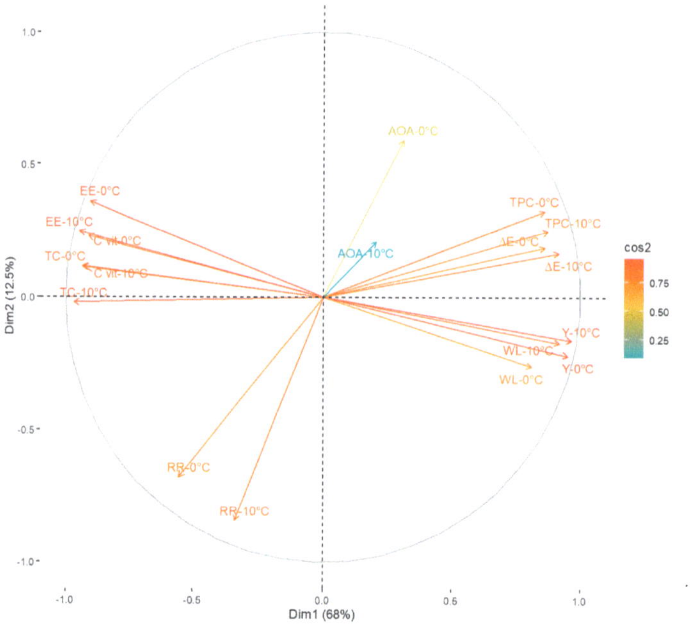

Figure 5. Principal component analysis (PCA) of the parameters measured in rocket leaves. EE: External appearance, TC: Total chlorophyll content, C vit: Vitamin C, RR: Respiration rate, WL: Weight loss, Y: Yellowing, ΔE: Discoloration (color changes), TPC: Total phenolic content, AOA: Antioxidant activity.

4. Conclusions

In this study, in which the effects of OA applied after harvesting on the quality of rocket leaves stored at different temperatures were examined, all doses of OA were effective in preserving the quality compared to the control treatment at both storage temperatures. In particular, weight loss and respiration rate were very low in OA-treated rocket leaves. The respiration rate, yellowing, and color change, an indicator of ripening, were low;

therefore, chlorophyll breakdown was also delayed in OA-treated rocket leaves. Vitamin C loss was also relatively lower in OA-treated rocket leaves compared to the untreated control group. The post-harvest quality of rocket leaves preserved by OA treatment under MAP conditions was maintained longer than the untreated group. In particular, 1 mM OA doses were the most effective treatment in terms of quality preservation at both storage temperatures. Considering the yellowing scores, which are very important for vegetables, it was determined that the rocket leaves cv. Bengi could be stored at 0 °C for 8–9 days with 1 mM OA treatment (2.67 score, 8th day) and 6 days with C treatment (2.83 score, 6th day) and at 10 °C for 6–7 days with 1 mM OA treatment (2.17 score, 6th day) and 4 days with C treatment (2.33 score, 4th day) with a minimum quality loss under MAP conditions. In conclusion, OA application is considered to be an effective application for the preservation of the post-harvest quality of rocket leaves. However, since the effects of such post-harvest applications can be affected by many factors, including variety, harvest time, application dose, application time, storage time, and storage condition, further studies with different varieties and species are needed.

Funding: This research received no external funding.

Data Availability Statement: Not applicable.

Conflicts of Interest: The author declares no conflict of interest.

References

1. Mastromatteo, M.; Conte, A.; Del Nobile, M.A. Combined used of modified atmosphere packaging and natural compounds for food preservation. *Food Eng. Rev.* **2010**, *2*, 28–38. [CrossRef]
2. Wang, C.Y. Leafy, floral and succulent vegetables. In *Postharvest Physiology and Pathology of Vegetables*; Bartzand, J.A., Brecht, J.K., Eds.; Marcel Dekker Inc.: New York, NY, USA, 2003; pp. 599–623.
3. Sharma, R.R.; Singh, D.; Singh, R. Biological control of postharvest diseases of fruits and vegetables by microbial antagonists: A review. *Biol. Control* **2009**, *50*, 205–221. [CrossRef]
4. Kasım, U.M.; Kasım, R. An alternative treatment of reduction of postharvest losses in fruits and vegetables: UV-C. *J. Agric. Sci.-Tarim Bilim. Derg.* **2007**, *13*, 413–419. [CrossRef]
5. Öz, A.T.; Süfer, Ö. Effect of edible film and coatings on the postharvest quality of fruits and vegetables. *Acad. Food J.* **2012**, *10*, 85–91.
6. Jehanno, E.; Savage, G.P. Oxalate content of commercially packed salad greens. *J. Food Agric. Environ.* **2009**, *7*, 207–208.
7. Zheng, X.; Tian, S. Effect of oxalic acid on control of postharvest browning of litchi fruit. *Food Chem.* **2006**, *96*, 519–523. [CrossRef]
8. Shafique, M.; Khan, A.S.; Malik, A.U.; Shahid, M. Exogenous application of oxalic acid delays pericarp browning and maintain fruit quality of litchi cv. Gola. *J. Food. Biochem.* **2016**, *40*, 170–179. [CrossRef]
9. Zheng, X.; Tian, S.; Meng, X.; Li, B. Physiological and biochemical responses in peach fruit to oxalic acid treatment during storage at room temperature. *Food Chem.* **2007**, *104*, 156–162. [CrossRef]
10. Zhu, Y.; Yu, J.; Brecht, J.K.; Jiang, T.; Zheng, X. Pre-harvest application of oxalic acid increases quality and resistance to Penicillium expansum in kiwifruit during postharvest storage. *Food Chem.* **2016**, *190*, 537–543. [CrossRef]
11. Huang, H.; Jing, G.; Guo, L.; Zhang, D.; Yang, B.; Duan, X.; Ashraf, M.; Jiang, Y. Effect of oxalic acid on ripening attributes of banana fruit during storage. *Postharvest. Biol. Technol.* **2013**, *84*, 22–27. [CrossRef]
12. Erbaş, D.; Koyuncu, M.A. The effects of postharvest putrescine, nitric oxide, oxalic and salicylic acid treatments on the fruit quality of plum cv. Black Diamond during storage. *J. Inst. Sci. Technol.* **2019**, *9*, 1830–1840. [CrossRef]
13. Valero, D.; Díaz-Mula, H.M.; Zapata, P.J.; Castillo, S.; Guillen, F.; Martinez-Romero, D.; Serrano, M. Postharvest treatments with salicylic acid, acetylsalicylic acid or oxalic acid delayed ripening and enhanced bioactive compounds and antioxidant capacity in sweet cherry. *J. Agric. Food. Chem.* **2011**, *59*, 5483–5489. [CrossRef] [PubMed]
14. Serna-Escolano, V.; Giménez, M.J.; Castillo, S.; Valverde, J.M.; Martínez-Romero, D.; Guillén, F.; Serrano, M.; Valero, D.; Zapata, P.J. Preharvest treatment with oxalic acid improves postharvest storage of lemon fruit by stimulation of the antioxidant system and phenolic content. *Antioxidants* **2021**, *10*, 963. [CrossRef]
15. Ali, M.; Liu, M.M.; Wang, Z.E.; Li, S.E.; Jiang, T.J.; Zheng, X.L. Pre-harvest spraying of oxalic acid improves postharvest quality associated with increase in ascorbic acid and regulation of ethanol fermentation in kiwifruit cv. Bruno during storage. *J. Integr. Agric.* **2021**, *18*, 2514–2520. [CrossRef]
16. Kant, K.; Arora, A.; Singh, V.P.; Kumar, R. Effect of exogenous application of salicylic acid and oxalic acid on postharvest shelf-life of tomato (*Solanum lycopersicon* L.). *Indian J. Plant Physiol.* **2013**, *18*, 15–21. [CrossRef]
17. Ruíz-Jiménez, J.M.; Zapata, P.J.; Serrano, M.; Valero, D.; Martínez-Romero, D.; Castillo, S.; Guillén, F. Effect of oxalic acid on quality attributes of artichokes stored at ambient temperature. *Postharvest. Biol. Technol.* **2014**, *95*, 60–63. [CrossRef]

18. Barberis, A.; Cefola, M.; Pace, B.; Azara, E.; Spissu, Y.; Serra, P.A.; Logrieco, A.F.; D'hallewin, G.; Fadda, A. Postharvest application of oxalic acid to preserve overall appearance and nutritional quality of fresh-cut green and purple asparagus during cold storage: A combined electrochemical and mass-spectrometry analysis approach. *Postharvest. Biol. Technol.* **2019**, *148*, 158–167. [CrossRef]
19. Cefola, M.; Pace, B. Application of oxalic acid to preserve the overall quality of rocket and baby spinach leaves during storage. *J. Food Process. Preserv.* **2015**, *39*, 2523–2532. [CrossRef]
20. Du, G.; Li, M.; Ma, F.; Liang, D. Antioxidant capacity and the relationship with polyphenol and vitamin C. *Food Chem.* **2009**, *113*, 557–562. [CrossRef]
21. Ergün, M.; Kösetürkmen, N. Effects of jasmonic and salicylic acids on quality of fresh-shredded carrot. *J. Fac. Agric. Harran Univ.* **2008**, *12*, 49–45.
22. TUIK. Data Portal for Statistics, Turkish Statistical Institute Web. Available online: https://data.tuik.gov.tr/Bulten/Index?p=Bitkisel-Uretim-Istatistikleri-2022-45504 (accessed on 26 February 2023).
23. Varga, J.; Apahidean, A.S.; Laczi, E.; Apahidean, A.I. Studies concerning the sowing period in the arugula (*Eruca sativa* Mill) plants development. *Acta Univ. Sapientiae Agric. Environ.* **2012**, *4*, 5–10.
24. Barazani, O.; Ziffer-Berger, J. *Eruca sativa*, a tasty salad herb with health- promoting properties. In *Medicinal and Aromatic Plants the Middle-East*; Yaniv, Z., Dudai, N., Eds.; Springer: Dordrecht, The Netherlands, 2014; Volume 2, pp. 269–280.
25. Siomos, A.S.; Koukounaras, A. Quality and postharvest physiology of rocket leaves. *Fresh Prod.* **2007**, *1*, 59–65.
26. Inestroza-Lizardo, C.; Silveira, A.C.; Escalona, V.H. Metabolic activity, microbial growth and sensory quality of arugula leaves (*Eruca vesicaria* Mill.) stored under non-conventional modified atmosphere packaging. *Sci. Hortic.* **2016**, *209*, 79–85. [CrossRef]
27. Koukounaras, A.; Siomos, A.S.; Sfakiotakis, E. 1-Methylcyclopropene prevents ethylene induced yellowing of rocket leaves. *Postharvest. Biol. Technol.* **2006**, *41*, 109–111. [CrossRef]
28. Koukounaras, A.; Siomos, A.S.; Sfakiotakis, E. Impact of heat treatment on ethylene production and yellowing of modified atmosphere packaged rocket leaves. *Postharvest. Biol. Technol.* **2009**, *54*, 172–176. [CrossRef]
29. Thaipong, K.; Boonprakob, U.; Crosby, K.; Cisneros-Zevallos, L.; Byrne, D.H. Comparison of ABTS, DPPH, FRAP, and ORAC assays for estimating antioxidant activity from guava fruit extracts. *J. Food Compost. Anal.* **2006**, *19*, 669–675. [CrossRef]
30. Lola-Luz, T.; Hennequart, F.; Gaffney, M. Effect on yield, total phenolic, total flavonoid and total isothiocyanate content of two broccoli cultivars (*Brassica oleraceae* var. Italica) following the application of a commercial brown seaweed extract (*Ascophyllum nodosum*). *Agric. Food Sci.* **2014**, *23*, 28–37. [CrossRef]
31. Gutiérrez, D.R.; Lemos, L.; Rodríguez, S.D.C. Effect of UV-C and ozone on the bioactive compounds and antioxidant capacity of minimally processed rocket (*Eruca sativa* Mill.). *Int. J. New Technol. Res.* **2018**, *4*, 23–29. [CrossRef]
32. Martínez-Sánchez, A.; Gil-Izquierdo, A.; Gil, M.I.; Ferreres, F. A comparative study of flavonoid compounds, vitamin C, and antioxidant properties of baby leaf *Brassicaceae* species. *J. Agric. Food Chem.* **2008**, *56*, 2330–2340. [CrossRef] [PubMed]
33. Arnon, D. Copper enzymes in isolated chloroplasts. Polyphenoloxidase in Beta vulgaris. *Plant Physiol.* **1949**, *24*, 1–15. [CrossRef]
34. Manolopoulou, H.; Mallidis, C. Storage and processing of apricots. *Acta Hortic.* **1999**, *488*, 567–576. [CrossRef]
35. Razzaq, K.; Khan, A.S.; Malik, A.U.; Shahid, M.; Ullah, S. Effect of oxalic acid application on Samar Bahisht Chaunsa mango during ripening and postharvest. *LWT-Food Sci. Technol.* **2015**, *63*, 152–160. [CrossRef]
36. Feliziani, E.; Lichter, A.; Smilanick, J.L.; Ippolito, A. Disinfecting agents for controlling fruit and vegetable diseases after harvest. *Postharvest. Biol. Technol.* **2016**, *122*, 53–69. [CrossRef]
37. Gutiérrez, D.R.; Char, C.; Escalona, V.H.; Chaves, A.R.; Rodríguez, S.D.C. Application of UV-C radiation in the conservation of minimally processed rocket (*Eruca sativa* Mill.). *J. Food Process. Preserv.* **2015**, *39*, 3117–3127. [CrossRef]
38. Özden, M.; Özden, A.N. Comparison of different coloured fruits in terms of total anthocyanins total phenolics and total antioxidant capacity. *Electron. J. Food Technol.* **2014**, *9*, 1–12.
39. Zheng, X.; Brecht, J.K. Oxalic acid treatments. In *Novel Postharvest Treatments of Fresh Produce*, 1st ed.; Pareek, S., Ed.; CRC Press: Boca Raton, FL, USA, 2017; pp. 35–50.
40. Pace, B.; Capotorto, I.; Palumbo, M.; Pelosi, S.; Cefola, M. Combined effect of dipping in oxalic or in citric acid and low O_2 modified atmosphere, to preserve the quality of fresh-cut lettuce during storage. *Foods* **2020**, *9*, 988. [CrossRef]
41. Çalhan, Ö.; Koyuncu, M.A. Determination of suitable criteria for optimum harvest date in cv. Eşme quince (*Cydonia oblonga* Mill.). *YYU J. Agric. Sci.* **2018**, *28*, 215–225. [CrossRef]
42. Kayashima, T.; Katayama, T. Oxalic acid is available as a natural antioxidant in some systems. *BBA Gen. Sub.* **2002**, *1573*, 1–3. [CrossRef] [PubMed]
43. Alpar, R. *Applied Multivariate Statistical Techniques*; Detay Yayıncılık: Ankara, Turkey, 2017; 43p.

horticulturae

MDPI

Review

Effects of Ethylene and 1-Methylcyclopropene on the Quality of Sweet Potato Roots during Storage: A Review

Jingjing Kou [1,2,†], Xueqian Zang [1,2,†], Maofu Li [1,2], Wenxing Li [1,2], Hongna Zhang [1,2], Yanli Chen [1,2,*] and Guopeng Zhu [1,2,*]

1 Sanya Nanfan Research Institute, Hainan University, Sanya 572025, China; jingjingkou@hainanu.edu.cn (J.K.); zangxueqian0117@126.com (X.Z.); hafu98022@126.com (M.L.); liwenxingsichuang@163.com (W.L.); 994357@hainanu.edu.cn (H.Z.)

2 Key Laboratory for Quality Regulation of Tropical Horticultural Crops of Hainan Province, School of Horticulture, Hainan University, Haikou 570228, China

* Correspondence: chenyanli@hainanu.edu.cn (Y.C.); zhuguopeng@hainanu.edu.cn (G.Z.)

† These authors contributed equally to this work.

Abstract: Sweet potato (*Ipomoea batatas* (L.) Lam.) is served as an important root crop worldwide due to its high yield, strong adaptability and nutrient richness. Sweet potato has played a significant role in ensuring food security and family income opportunities for local populations in China for years of experience. The storage roots, which provide abundant nutrition and health benefits to people, are the mainly harvested and consumed parts of sweet potato. However, after harvest, physiological disorders, such as sprouting, mechanical injury and infectious postharvest diseases, increase the magnitude of sweet potato root quality decline and nutritional compound losses. Ethylene and 1-methylcyclopropene (1-MCP) were considered to be effective commercial treatments in sweet potato postharvest. Exogenous ethylene and 1-MCP treatment could successfully inhibit root sprouts and reduce rot decay without affecting the storage quality of sweet potato. This review aims to summarize the latest available information on the effects of ethylene and 1-MCP with respect to enhancing or impairing sweet potato root quality. A better understanding of the influence of ethylene and 1-MCP on root quality parameters will be useful to further explore the role and mechanisms of action of ethylene in regulating the postharvest storage of sweet potato roots and contributions to technological development and innovation.

Keywords: *Ipomoea batatas*; postharvest; sprouting; curing; rot

Citation: Kou, J.; Zang, X.; Li, M.; Li, W.; Zhang, H.; Chen, Y.; Zhu, G. Effects of Ethylene and 1-Methylcyclopropene on the Quality of Sweet Potato Roots during Storage: A Review. *Horticulturae* **2023**, *9*, 667. https://doi.org/10.3390/horticulturae9060667

Academic Editors: Zi Teng and Zhengguo Li

Received: 19 March 2023
Revised: 28 May 2023
Accepted: 2 June 2023
Published: 5 June 2023

1. Introduction

Sweet potato (*Ipomoea batatas* (L.) Lam.) is one of the most significant root crops, used worldwide as a traditional major food source, animal feed and industrial raw material, with numerous agronomic and nutritional advantages [1,2]. As a good source of carbohydrates, carotenoids, dietary fiber, anthocyanin and other nutrients for people, sweet potato is consumed in large quantities in China [3–5]. The nutrients with various antioxidants, anti-inflammatory, anti-diabetic, anti-hypotensive properties and specific anti-cancer bioactivities play an active role in promoting health and protecting the human body from diseases [6–8]. China is the leading producer of sweet potato at the global level, accounting for about 54.74% (FAOSTAT, 2020) of the world production [9]. Sweet potato production can be a source of income for most rural and peri-urban growers in most developing countries. Although numerous benefits can be obtained from the crop, postharvest losses make its production unprofitable in most parts of the world. On average, in China, 20 to 30% percent is lost in sweet potatoes during storage after harvest. These losses affect food security and nutritional health and have negative financial impacts on both consumers and farmers.

Long-term storage of sweet potato roots allows year-round availability of the crop but induces various physicochemical changes. During storage, the shelf-life of sweet potato roots varies from a few days to months, depending on the cultivar and storage conditions. Deterioration in postharvest quality of sweet potato roots can be attributed to sprout growth, postharvest wounding, microbial attack and loss of nutritional properties (Figure 1) [10,11]. When offering the right conditions, sweet potato roots, which are dormant crops, tend to sprout after a short dormancy period. The appearance of sprouts promotes the wilting of sweet potato roots, leading to a decrease in nutritional value and causing pathogens to invade from the long germs. Freshly harvested sweet potato roots have a thin, delicate skin that is easily mechanically damaged by rough handling during harvest. The roots are particularly perishable during storage because of their high moisture content (50–80%), resulting in low mechanical strength, making damage occurrence likely during harvest and with high susceptibility to microbial decomposition. Root decay results from wound-induced fungal infections when the temperature and humidity are appropriate for fungal growth in storage. Root rot leads to a reduction in nutritional properties. The accumulation of toxins in the root caused by rot during storage should also be considered as it poses a serious threat to human health when consumed. On the other hand, a high respiratory rate produces heat and softens the textures, making them vulnerable to harm. A high rate of respiration also leads to weight loss and a reduction in nutritional quality, including sugar, phenolic, carotenoids and other nutrients [12]. Extending root dormancy, reducing root damage and decay and maintaining nutritional properties are the most important aspects for sweet potato utilization, storage, processing and marketing.

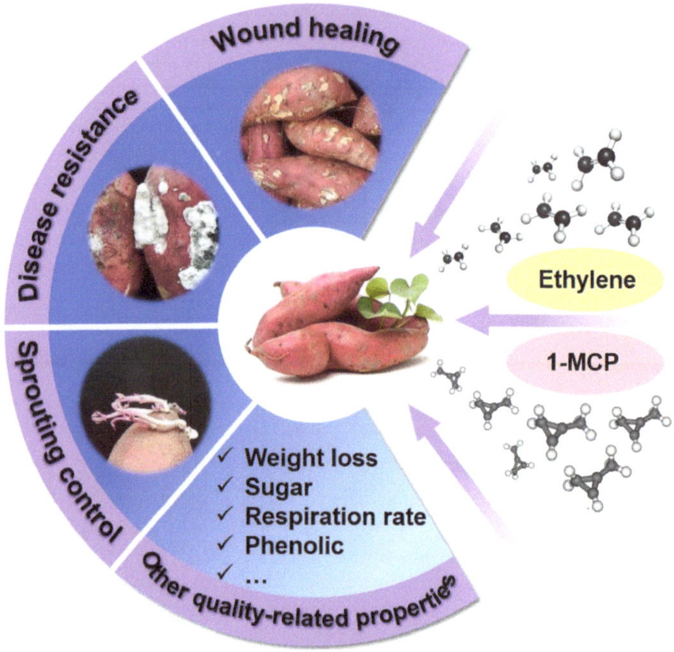

Figure 1. Main changes in the deterioration in postharvest quality of sweet potato roots under exogenous application of ethylene and 1-methylcyclopropene (1-MCP).

Maintaining the quality of sweet potato roots over a long period of storage is challenging, especially for large-volume growers. There are several options currently available for the management of postharvest in commercial sweet potato industries. Physical environmental regulation has generally been conducted through the manipulation of storage temperature and relative humidity [13]. The ideal storage conditions for sweet potato roots are at a temperature of 13–16 °C and a relative humidity of 80–95% [14,15]. However, sweet potato roots are much more sensitive to chilling injury, and cold storage facilities are difficult to promote due to high energy consumption and high financial costs for growers. On the other hand, fungicides are usually applied in industrial operation after harvest to improve the root qualities. In general, fungicide may be applied by dipping the root in a chemical suspension tank, utilizing a waterfall application or spraying the fungicide alone or combined in a wax solution as the sweet potatoes pass over a brush roller conveyor. Although synthetic sprout inhibitors and fungicides (such as CIPC) are effective in maintaining quality during storage, the use of chemicals causes environmental pollution, and the chemical residues have a harmful influence on human health [16].

Ethylene (C_2H_4) is the key endogenous plant hormone involved in plant development, maturation and rotting, which is closely related to postharvest quality maintenance in storage [17,18]. Crops produce ethylene while in storage, and pathogen attack, chilling and wounding also encourage the formation of ethylene in damaged crops. As an ethylene inhibitor, 1-MCP protects crops from both internal and external sources of ethylene and preserves the quality of fresh produce during storage [19]. In recent years, the important role of ethylene and 1-MCP has been noticed in sweet potato storage as secure, non-toxic green preservatives. Compared to the conventional technologies, ethylene and 1-MCP have the advantages of non-toxicity, good chemical stability, easy synthesis, negligible residue and low use concentration in postharvest of sweet potatoes. However, high exposures of ethylene or 1-MCP could cause root proximal disease, increased respiration rate, excessive weight loss and lower sugar content. These biological responses depend on the root sensitivity to ethylene. The responses of roots to different ethylene concentrations vary significantly with cultivars. Different treatment concentrations should be tested in advance between cultivars before use. Despite several disadvantages, since the correct application dosage is determined, using ethylene and 1-MCP is more effective, less expensive and safer than conventional technologies.

Several previous studies investigated the responses of sweet potato roots to ethylene using exogenous ethylene as well as ethylene-binding inhibitors such as 1-methylcyclopropene (1-MCP). Sweet potato has low ethylene sensitivity and produces very little ethylene during storage, which is typically considered as a non-respiratory, non-climacteric root [20,21]. However, according to another study, sweet potatoes were classified as climacteric roots due to their strong respiration after harvest [22]. The respiration type of sweet potato may vary depending on the cultivars, which needs further study and determination. Dipping in ethephon, continuous exposure to ethylene gas flow or 24 h exposure to 1-MCP effectively increased the commercial life of sweet potato roots. The effects of ethylene and 1-MCP on sweet potato postharvest are associated with respiratory rate variation, loss of water, sugar and phenolic accumulation, reactive oxygen species formation, phytoalexin protein production and interaction between phytohormones. Transcriptomic analysis has recently been employed to comprehend the mechanisms of ethylene-induced sprouting in onions [23]. The specific transcriptional responses to exogenous ethylene application in potatoes were revealed via RNA sequencing analysis [24]. Furthermore, the molecular changes associated with curing [25], carvone-treated sprouting [26], cold storage [27,28] and root development [29] in sweet potato roots were studied using RNA sequencing. The use of transcriptomic analysis may, thus, reveal molecular mechanisms involved in specific biological processes during ethylene and 1-MCP treatment.

Exogenous ethylene and 1-MCP treatment showed effective results in sprouting control, wound curing, disease resistance and maintenance of nutritional properties to maintain the postharvest storage quality of sweet potatoes (Figure 1). However, studies on the impacts of ethylene and 1-MCP on sweet potato root quality during storage are not available, and further studies are required to evaluate the parameters of sprout control, disease resistance and nutrition loss. In this review, we discuss the potential effects of ethylene and 1-MCP on the storage quality of sweet potato roots. A detailed overview of sprout growth, root decay, wound healing and other disorders, including weight loss, respiration rates and nutritional properties variation, is also provided in the review.

2. Effects of Ethylene and 1-Methylcyclopropene (1-MCP) on Sprouting Control of Sweet Potato Roots

Sprouting is nature's way of passing on the genome of the plant. Seeds begin to sprout and develop into the same plant as the seed through the process of sprouting. Under normal storage conditions, sweet potatoes are intended to sprout after a short dormancy period (about 2–4 weeks) [30]. Sprouting during storage depletes nutrients, which reduces the quality of the roots. Additionally, the longer sprouts and young leaves of sprouted sweet potato may carry bacteria or other pathogens. Although sprouted sweet potato is tender enough and generally safe to eat with no toxicity, people avoid eating them this way to prevent early sprouting.

Exogenous ethylene and 1-MCP have both been used for years in Western countries as green chemicals to suppress sprout growth of root crops, including potatoes, gingers, yams and onions, during storage. The application of ethylene and 1-MCP has been widely reported to control sprouting in stored potatoes [31,32]. Moreover, 1 $\mu L\ L^{-1}$ 1-MCP significantly suppressed the sprouting of ginger rhizomes, reduced the accumulation of reactive oxygen species (ROS) and maintained quality at room temperature storage [33]. The sprouting inhibition of onion bulbs can be achieved when treated with 10 $\mu L\ L^{-1}$ ethylene or 1 $\mu L\ L^{-1}$ 1-MCP, respectively, for as little as 24 h [34]. Just as in other root crops, 24 h exposure of sweet potato roots (cultivars 'Bushbuck' and 'Ibees') to 625 nl L^{-1} 1-MCP or continuous 10 $\mu L\ L^{-1}$ ethylene inhibited sprout growth over 4 weeks of storage at 25 °C [35]. In another study, continuous exogenous ethylene (10 $\mu L\ L^{-1}$) or 24 h 1-MCP (1 $\mu L\ L^{-1}$) treatment suppressed sprouting and the dark cooked color of 'Owairaka Red' sweet potato roots stored at 25 °C and 85% RH for 4 weeks [36]. Similarly, control of sprouting was observed in sweet potato cultivar 'Covington' stored under continuous ethylene (10 $\mu L\ L^{-1}$) during long-term storage at 25 °C [37]. Application of 1-MCP (1 $\mu L\ L^{-1}$) for 24 h also significantly inhibited sprout growth in 'Covington' roots [20]. However, sprouting of sweet potato cultivar 'Evangeline' decreased in response to 1-MCP (1–2 $\mu L\ L^{-1}$) application in three of four experiments, while 'Beauregard' sprouting decreased in two of four experiments, which provides the first evidence of the variable effect of 1-MCP on different sweet potato cultivars [38]. The sprouting control abilities of 1-MCP may depend on cultivars.

According to previous studies, 1 $\mu L\ L^{-1}$ 1-MCP applied for 24 h with continuous 10 $\mu L\ L^{-1}$ ethylene treatment is the most commonly used and most effective method to inhibit sprout growth in sweet potato roots. In general, continuous ethylene gas flow with air was applied, and 1-MCP solution was prepared using commercial tablets with fans to ensure even distribution. Sweet potato roots were exposed to the treatments in the sealed storage boxes at room temperature. Energy through sugar metabolism was used in sprouting control because the sugar concentration was reduced after 1-MCP and ethylene treatment [35]. Although ethylene increased root respiration, sprout growth was significantly suppressed. 1-MCP inhibited this increase in respiration rate and presumably counteracted the ethylene stimulation of this process. The observations that both ethylene on its own and 1-MCP inhibit sprout growth suggest that while continuous exogenous ethylene exposure inhibits sprout growth, ethylene is also essential for sprouting. Low concentrations of ethylene are necessary to break the dormancy in sweet potato roots, while high ethylene concentrations applied continuously inhibit sprouting. No sprout growth was observed when 1-MCP was applied, suggesting that ethylene production was inhibited by 1-MCP and remained dormant [35,39]. Dormancy of root crops is controlled by phytohormones, such as ethylene, ABA, cytokinins, gibberellins and auxin, in bud breaking and sprout growth. However, fewer studies were found on other phytohormones, except ethylene in sweet potato root. Reactive oxygen species (ROS) including hydrogen peroxide (H_2O_2) are also involved in breaking dormancy in root crops by affecting cellular balance or regulating the expression of genes involved in dormancy. The mechanism of ROS in the control of sprouting in sweet potato roots has been little explored.

3. Effects of Ethylene and 1-Methylcyclopropene (1-MCP) on Disease Resistance of Sweet Potato Roots

Sweet potato is susceptible to a variety of field and storage diseases. To date, postharvest diseases account for the majority of losses in stored sweet potato, severely restricting their production and commercialization. The high moisture content (50–80%) of sweet potato roots makes them highly perishable after harvest. Freshly harvested sweet potato roots generally have a thin, mechanically weak skin outside of the starchy flesh [40]. Pathogenic fungi are produced as a result of mechanical injuries and natural openings on the surface of the roots during harvest and handling. The infected roots begin to show discoloration, foul odor, textural changes or grow visible contaminations, which ultimately leads to decay [41].

Postharvest diseases of sweet potato roots can be attributed to several forms of pathogenic microorganism infections (fungi, bacteria or viruses), the majority of which are caused by fungi. The most important postharvest diseases caused by fungi mainly include soft rot (*Rhizopus stolonifer* [42], *Rhizopus oryzae* [41] and *Rhizopus nigricans* [41]), sclerotium rot (*Sclerotium rolfsii* [43]), foot rot (*Plenodomus destruens* [44]), black rot (*Ceratocystis fimbriata* [45,46]), white rot (*Globisporangium ultimum* [47]) and fusarium root rot (*Fusarium solani* [48,49]). These pathogens frequently infect sweet potato roots immediately after harvest or during storage and transport. They enter through natural plant openings and small wounds caused by insects or tools used in harvesting and transportation. Pathogens become more active, multiply rapidly and cause infections when the environment is favorable. Less air circulation, high relative humidity and warm temperature conditions combine to favor the growth of fungi [41].

Management of sweet potato postharvest diseases using synthetic fungicides is believed to increase environmental pollution and health hazards of worldwide concern. For this reason, the development of alternative methods, such as natural plant-derived products [45,46], GRAS (generally recognized as safe) compounds [48] and microbial antagonists [50], have been actively demonstrated to be most suitable to replace the synthetic fungicides. Ethylene, a naturally occurring plant hormone, increased the resistance of sweet potato against black rot in an earlier study [51]. However, in other studies, ethylene treatment promoted root proximal rot incidence in sweet potato cultivars 'Beauregard', 'Covington' and 'Chuanshanzi' during long-term storage [37,52,53]. Ethylene-induced decay predominantly initiates from the proximal region, where root splitting also occurred. End root splitting caused by ethylene treatment may increase the incidence of disease and lead to proximal rot in sweet potatoes. 1-MCP was normally evaluated in the context of its antagonistic response to ethylene. Previous work showed that 1-MCP effectively inhibited root decay in stored sweet potato roots. Application of 1 $\mu L\ L^{-1}$ 1-MCP achieved a 10% lower incidence of rot symptoms compared to the control in 'Beauregard' and 'Evangeline' roots up to 133 days [54]. 'Bushbuck' and 'Ibees' treated with 1-MCP (625 $nl\ L^{-1}$, 24 h) showed no disease when stored at 25 °C for 4 weeks [35]. Root decay was also significantly reduced in 'Covington' roots by 1 $\mu L\ L^{-1}$ 1-MCP for 24 h in 120 days [54].

Ethylene treatment usually increases disease development, simply through its acceleration of ripening or senescence. According to previous studies, ethylene increased disease in most sweet potato root cultivars. Several observations, however, indicate that when ethylene is applied before inoculation with a pathogen, it decreases or has no effect on disease development, but disease development is accelerated when plants are treated with ethylene after infection. The timing of ethylene exposure of root crops can influence whether resistance is promoted or suppressed. Ethylene can produce phytoalexins as pathogenesis-related proteins and rigidify cell walls in the induced systemic resistance of plants. Priming of defense-related genes ISR (induced systemic resistance) and pathogenesis related proteins SAR (systemic acquired resistance) requires the presence of a functional *NPR1* protein, triggered by ethylene receptor genes *etr1* and *ein2* in *Arabidopsis* [55]. 1-MCP treatment tended to maintain root quality and reduce disease development in most studies. The strongest effect of 1-MCP occurred when roots were harvested soon after treatment.

Application of 1-MCP immediately after harvest on the root crops is crucial for disease control [54,56]. The possible disease resistance mechanism after ethylene and 1-MCP treatment needs further study in sweet potato roots.

4. Effects of Ethylene and 1-Methylcyclopropene (1-MCP) on Wound Healing of Sweet Potato Roots

Some root, tuber and bulb vegetables, such as garlic, onions, potatoes and sweet potatoes, need to be cured prior to long-term storage. Following field drying, sorting and crating, within 1–2 h after harvest, sweet potato roots were cured in a controlled room in order to promote the healing of wounds acquired during harvest and handling. Curing, as a low-cost postharvest technique, allows skins to harden, wounds to heal and some of the starches to convert to sugars, which helps the tubers store for months [57]. The curing of sweet potato roots includes optimizing three conditions: temperature, relative humidity and ventilation. The optimal curing conditions are 85–95 °F (29–35 °C) and 80–90% relative humidity for 3–7 days [58].

Curing takes longer (up to 3 weeks) if conditions are less than perfect, and it can differ between sweet potato cultivars. It has been previously documented that dry matter content of sweet potato roots and wound healing efficiency are closely related. Cultivars with low dry matter content exhibited a longer shelf-life and more effective wound healing in earlier studies that investigated 34 cultivars [59]. Furthermore, the association was supported by the results of experiments conducted with 17 cultivars from different regions of the world, whose dry matter content ranged from 17.9% to 31.2% [60]. However, 'Apomuden' (19%) and 'Nane' (27%), two cultivars with low and high dry matter content, showed no significant difference in their wound healing ability [61].

Mass spectroscopic analysis of the curing process revealed that the sweet potato surface cells were desiccated, followed by the lignification of the underlying cell layers, finally covering the skin [59,62]. As a new epidermal tissue, this formative lignified layer performs as a barrier against pathogenic organisms and prevents excessive moisture loss [60,63]. Generally, no chemicals are required during the curing process. Unfortunately, although maintaining proper curing conditions is helpful for maintaining root quality, the rate of sprouting and decay during long-term storage is still very high. It is necessary to develop and combine safe and reliable chemicals in the curing process to extend the storage time of sweet potato roots. The amount of ethylene produced by sweet potato in response to wounding (2–4 days after wounding) varies among cultivars, and they respond differently to ethylene treatments [64–66]. Ethylene is involved in lignification and wound periderm formation during sweet potato wound healing [67]. Moreover, the storage protein of sweet potato sporamin, as a wound response promoter, was effectively activated by ethylene [68]. In addition, when 1-MCP was combined with curing, root sprouting was effectively inhibited during the subsequent storage and the healing of wounds [53]. Further investigation is required to evaluate the effects of ethylene and 1-MCP treatment, which are frequently used in the storage period after curing to control sprouting. Their effects on sweet potato quality also need to be examined when applied in conjunction with curing.

5. Effects of Ethylene and 1-Methylcyclopropene (1-MCP) on Other Quality-Related Properties of Sweet Potato Roots

After harvest, the sweet potato roots stay metabolically active and perform all functions of living tissues. The primary goal of storage is to maintain root quality and ensure adequate supply throughout the year by minimizing both physiological disorders and disease development [69]. According to previous research, the proper storage conditions for high-quality roots are 55 °F (13 °C) and 85–90% relative humidity, with adequate ventilation. Physiological disorders and nutritional losses are related to improper storage conditions, such as excessive light, extreme temperature, low oxygen and unsuitable moisture [70]. The following quality-related properties discussed were the other main characteristics of sweet potato roots that may be affected after harvest when ethylene and 1-MCP were applied during storage (Table 1).

Table 1. The effects of ethylene and 1-methylcyclopropene (1-MCP) on quality-related properties in different sweet potato cultivars.

Cultivars	Storage Conditions	Applications	Treatments	Quality-Related Properties	Reference
Organic 'Covington' and Portuguese-derived 'Covington'	Cold storage at 15 °C	Sprouting control and disease resistance	1-MCP (1 μL L^{-1}, 24 h)	Reduced sprouting and decay, phenolic compounds; no effect on respiration rate and sugar, maintained saleable weight	Amoah et al., 2012 [20]
'Covington'	Cured (30 °C, 90% relative humidity, 7 days) then stored at 25 °C	Sprouting control	Ethylene (10 μL L^{-1}, applied continuously)	Reduced sugar, phenolic compounds and phytohormones (abscisic acid and zeatin riboside); suppressed sprout growth, doubled root respiration; increased weight loss and incidence of proximal rots	Amoah et al., 2016 [37]
'Covington'	Cold storage at 15 °C	Disease resistance	1-MCP (1 μL L^{-1}, 24 h)	Reduced decay, weight loss; no effect on respiration rate and carbohydrates	Amoah et al., 2018 [54]
'Bushbuck'	25 °C in incubators	Sprouting control	1-MCP (625 nl L^{-1}, 24 h)	Inhibited sprouting; reduced respiration rate, weight loss, sugar content (sucrose, glucose and fructose)	Cheema et al., 2010 [71]; Cheema et al., 2013 [35],
			Ethylene (10 ppm, applied continuously)	Inhibited sprouting; increased respiration rate (3-fold), weight loss (slightly), sucrose; reduced glucose and fructose	
'Beijing 553' and 'Chuanshanzi'	Curing at 29 °C for 4 days, then stored at (13 ± 0.5) °C, 90% relative humidity	Sprouting control and disease resistance	0.045% 1-MCP cyclodextrin powder, 1.6 g/case	Improved wound healing; inhibited sprouting and decay; increased sugar content; decreased starch content, no color change	Cao et al., 2021 [53]
'BRS Rubissol'	Curing at 30 °C and 90% relative humidity for 7 days, stored in chambers at 25 °C and 90% relative humidity	Sprouting control	1-MCP 1 mg·L^{-1} in 90 L chamber for 24 h	Reduced sprouting, weight loss; increased dry matter content; processed fried chips showed less browning	Lima et al., 2021 [72]
			Ethylene 10 μL·L^{-1} in 90 L chamber for 48 h	Reduced sprouting, weight loss; increased dry matter content	
'Owairaka Red'	Curing at 30 °C and 90% relative humidity for 4 days then stored at 25 °C and 85% relative humidity	Sprouting control	1-MCP (1 μL L^{-1}, 24 h) and continuous ethylene (10 μL L^{-1})	Inhibited sprout growth; increased root respiration rates and weight loss; no color change after cook	Pankomera et al., 2016 [36]
			1-MCP (1 μL L^{-1}, 24 h)	No significantly differ from the control	
			Ethylene (10 μL L^{-1}, applied continuously)	Inhibited sprout growth; increased root respiration rates and weight loss; darken cooked flesh color	
'Belle Vue'	Curing for 4 days (25–30 °C) then stored at 20 °C	Sprouting control	Ethylene (0.001 kPa, applied continuously) with controlled atmosphere	Reduced sprouting; increased phenolics contents, sugars, weight loss and respiration rates	Sowe et al., 2018 [73]
'Beauregard'	Curing at 85 °F and 85% relative humidity for 5 days then stored at 60 °F and 75–85% relative humidity	Wound healing	Dipping 1 h in 1-MCP (1 ppm) or ethephon (2.6 mM)	Breakdown-related features on skin appeared after ethephon treatment, not detected in 1-MCP treated roots	Villordon et al., 2012 [74]

5.1. Weight Loss

During storage, high moisture content (50–80%) evaporates from the surface of sweet potato roots, resulting in weight loss and possible shriveling of the delicate skin, particularly at the root ends. Although microbial decay is the main postharvest loss in sweet potato, weight loss is considered the second most important factor for economic losses in the marketing of sweet potatoes [41,74]. Weight loss has frequently been used as a stability indicator for postharvest and shelf-life quality control [54]. There are conflicting reports regarding the role of exogenous ethylene and 1-MCP on the weight loss of sweet potato roots. After harvest, in most cases, 1-MCP treatments achieved lower weight loss in root storage [37]. Several studies indicated that ethylene treatments promote weight loss in cultivars 'Covington' [37,54], 'Owairaka Red' [36] and 'Belle Vue' [73]. Ethylene plus 1-MCP treatments increased the weight loss of 'Owairaka Red' [36], while in 'Bushbuck' and 'Ibees', weight loss was decreased by all the treatments, including 1-MCP, ethylene and ethylene plus 1-MCP, respectively [35]. Excessive weight loss was proved to be associated with moisture loss, which was accelerated by an increased respiration rate and undesirable environmental impacts during storage [37]. In general, ethylene treatment increased the respiration rate and induced weight loss, but 1-MCP inhibited ethylene release and reduced the respiration rate of the roots. This effect may differ depending on the treated cultivars' breathing patterns. Cultivars with low respiratory rates may not show significant changes in respiration rate after treatment. Weight loss varies due to different degrees of evaporation caused by different treatment methods, such as gas application or dipping. In addition, the curing process and sprouting control reduced the loss of weight due to the decreased evaporation and respiration [35,72]. Decay caused by pathogen infections leads to moisture loss and aggravated weight loss [54].

5.2. Respiration Rate

Sweet potato, as an underground storage root vegetable, typically has relatively low respiration rates. However, sweet potato continues to breathe highly in storage to release energy through the breakdown of stored carbohydrates in the roots. Exposure of sweet potato to ethylene significantly boosted the respiration rates in the roots of different cultivars, including 'Covington', 'Owairaka Red', 'Bushbuck' and 'Belle Vue' [35–37,73]. Additionally, 1-MCP showed negative impacts on changing this effect when ethylene was present in 'Owairaka Red' [36]. In general, the respiration data distinctly demonstrated the inhibitory effect of 1-MCP on ethylene. 1-MCP reduced the respiration rates of 'Bushbuck' [35] at room temperature (25 °C) stored in incubators, whereas no evident effects were observed on the respiration rates of 'Covington' and 'Owairaka Red' stored at low temperature (15 °C) and room temperature (25 °C), respectively. Genetic characteristics are the most important contributing aspect to the respiration rate of sweet potato [75]. Because of the differences in breathing patterns, various cultivars displayed low or high respiration rates after treatments. Harsh circumstances, such as mechanical damage, disease infection and sprout growth, also accelerated the rate of root respiration in storage. The differences observed above in the experiments could be attributed to the various states of the roots during storage. High respiration rates frequently coincide with active metabolic activity and lead to a short storage life that lowers dry matter content and causes weight loss.

5.3. Sugar Content

Sweet potato is known to be high in carbohydrates. Sweet potato with high sugar concentrations results in higher quality for processing. Consumer acceptability and preferences for sugar concentrations may vary regionally. Generally, people prefer low sugar concentrations when sweet potato is consumed as a staple food, whereas in the UK or US, sweet potato with high sugar concentrations is preferred. Carbohydrate macromolecules such as starch are converted into simple sugars, including sucrose, glucose and fructose, during sweet potato storage [71]. Transport and accumulation of these sugars are the sources of energy needed for root physiological metabolism and sprout growth. Conflicting data exist on the role of endogenous ethylene and 1-MCP in terms of maintaining sugar levels. In the roots of 'Covington' and 'Bushbuck', the concentration of monosaccharides was reduced by ethylene treatment [35,37]. 1-MCP decreased the amount of sucrose, glucose and fructose in 'Bushbuck' but had no effect on the sugars in 'Covington' [54,71]. Ethylene, on the other hand, increased the sugar content in 'Belle Vue' [73] and 'BRS Rubissol' [73]. 1-MCP improved the sugar content in 'Chuanshanzi' [53] and 'BRS Rubissol' [73]. In general, ethylene promotes respiration, evaporation and sugar hydrolysis, while 1-MCP inhibits ethylene production and increases sugar content. The difference in treatment results may also be related to the sugar content in various cultivars. Cultivars with high sugar content showed no significant changes following treatment. Differences in sugar determination methodologies could also potentially contribute to such variations.

5.4. Phenolic Compounds

Sweet potato roots contain high levels of phenolic compounds, especially chlorogenic acids. 1-MCP reduced phenolic levels in 'Covington' roots but increased the accumulation of phenolic compounds in the proximal ends [38]. Phenolic compounds were unevenly distributed in sweet potato roots. Higher concentrations of phenolic compounds were observed in the proximal sections of the roots, while lower concentrations were in the middle sections [38,76]. The presence of phenolic compounds was related to protection against disease and is involved in anti-resistance responses in other crops. Higher phenolic compounds in the proximal ends after 1-MCP treatment may be one of the reasons for its reduction in pathogenic attack. Since root decay and sprout growth usually tend to occur at the proximal ends of sweet potato roots, it may be possible to explore how these processes relate to the content of enriched phenolic compounds after 1-MCP treatment. Ethylene was found to decrease phenolic compounds in 'Covington' roots [37] but increase the phenolic content in 'Belle Vue' roots [73]. The conflicting results may be related to differences between cultivars and measurement methods. The phenolic content in the proximal ends needs further observation in ethylene treatment experiments, which helps to reveal the relationship of sprouting, end rotting and phenolic compounds.

6. Conclusions and Future Perspectives

Sweet potato has a short maturity period (3–5 months), drought tolerance and wide ecological adaptation with nutritional and economic value for growers and consumers in different regions of the world. Traditionally, keeping the roots in long-term storage provides a continuous food supply in the diet season when the staple crops are exhausted. The difficulties in long-term storage of sweet potato, including sprouting control, disease resistance and nutritional properties loss, pose a major challenge for extending the shelf life and improving the quality of the roots. The potential of ethylene and 1-methylcyclopropene (1-MCP) to maintain the postharvest storage quality of sweet potato has been previously observed and studied by researchers. In this review, one of the focuses was to investigate the differences between various cultivars when applying ethylene and 1-MCP. Although ethylene and 1-MCP were effective in preventing sprouting, wounding and decay during storage, their effects varied significantly depending on the cultivars and application technologies. The influence of ethylene and 1-MCP on sweet potato postharvest quality needs to be discussed separately for each cultivar to elucidate the significance of ethylene. These differing effects might be related to the ethylene sensitivity variations of cultivars by themselves. In part, investigating the trend of respiration rate and the amount of ethylene released after harvest is a useful tool to observe ethylene sensitivity in each cultivar. Furthermore, application technologies, which lead to variable ethylene reactions in the same cultivar, might play a role. Sweet potato roots are highly perishable; technologies that are simple to use, consume less and maintain root quality are highly desirable.

Additionally, attention should also be paid to the molecular mechanisms underlying the effects of ethylene and 1-MCP in the storage roots of sweet potato roots. From previous studies trying to discuss the relationships between ethylene and some of the physiological indicators, such as respiration rate changes, weight loss and variations in sugar and phenolic content, the molecular mechanism behind them is still unclear. Physiological changes are regulated by gene expressions, and using molecular techniques to reveal the important role of ethylene in sweet potato postharvest is also critical. Low levels of ethylene are necessary for sprouting but, when applied at high levels, ethylene showed an inhibition effect on sprouting. 1-MCP suppressed sprouting and produced lower levels of ethylene in the root. This suggests that ethylene and 1-MCP regulated sprouting in different ways. Further studies are needed to understand the interactions between the multiple ethylene-related genes involved in sprouting. For disease resistance, ethylene induces resistance when applied before infection, whereas increased decay when applied during infection or after symptoms. The dual action of ethylene is that it sometimes acts as a pathogenic factor and sometimes as a virulence factor for pathogens. This implies that ethylene-regulated resistance responses depend on the spatial connections of numerous signals interacting in the network type. Previous evidence suggests that study of this network is becoming available through transcriptome analyses, indicating that the interaction of various regulatory components controls numerous genes in complex ways. Future detailed analyses of the underlying molecular regulatory networks under the treatments should aid in further improving root quality in storage. As mentioned earlier, sweet potato is a good source of numerous balanced nutrients and health-promoting phytochemicals. Thus, more emphasis should be placed on studying the connections between storage characteristics and the internal nutritional compounds of the roots, which helps to illuminate the molecular mechanisms behind them. Furthermore, extensive research is also needed to develop green and ethylene-related preservatives in order to prolong the storage of sweet potato roots.

Author Contributions: Writing—original draft preparation, J.K.; conceptualization, J.K. and Y.C.; methodology, X.Z.; software, X.Z. and W.L.; investigation, M.L.; writing—review and editing, J.K. and H.Z.; supervision, Y.C. and G.Z.; funding acquisition, J.K. and Y.C. All authors have read and agreed to the published version of the manuscript.

Funding: This research was funded by Hainan Provincial Natural Science Foundation of China (323RC411), the National Natural Science Fund of China (32260443), the Earmarked Fund for CARS-10-Sweetpotato, Hainan Province Science and Technology Special Fund (ZDYF2020226) and the Scientific Research Start-up Fund Project of Hainan University (RZ2200001404).

Institutional Review Board Statement: Not applicable.

Informed Consent Statement: Not applicable.

Data Availability Statement: The data are available on request from the corresponding author.

Conflicts of Interest: The authors declare that they have no conflict of interest.

References

1. Behera, S.; Chauhan, V.B.S.; Pati, K.; Bansode, V.; Nedunchezhiyan, M.; Verma, A.K.; Monalisa, K.; Naik, P.K.; Naik, S.K. Biology and biotechnological aspect of sweet potato (*Ipomoea batatas* L.): A commercially important tuber crop. *Planta* **2022**, *256*, 40. [CrossRef]
2. Xu, X.; Wu, S.; Chen, K.; Zhang, H.; Zhou, S.; Lv, Z.; Chen, Y.; Cui, P.; Cui, Z.; Lu, G. Comprehensive Evaluation of Raw Eating Quality in 81 Sweet Potato (*Ipomoea batatas* (L.) Lam) Varieties. *Foods* **2023**, *12*, 261. [CrossRef] [PubMed]
3. Drapal, M.; Gerrish, C.; Fraser, P.D. Changes in carbon allocation and subplastidal amyloplast structures of specialised *Ipomoea batatas* (sweet potato) storage root phenotypes. *Phytochemistry* **2022**, *203*, 113409. [CrossRef]
4. Tang, C.; Lu, Y.; Jiang, B.; Chen, J.; Mo, X.; Yang, Y.; Wang, Z. Energy, Economic, and Environmental Assessment of Sweet Potato Production on Plantations of Various Sizes in South China. *Agronomy* **2022**, *12*, 1290. [CrossRef]
5. Pazos, J.; Zema, P.; Corbino, G.B.; Gabilondo, J.; Borioni, R.; Malec, L.S. Growing location and root maturity impact on the phenolic compounds, antioxidant activity and nutritional profile of different sweet potato genotypes. *Food Chem. Mol. Sci.* **2022**, *5*, 100125. [CrossRef] [PubMed]
6. Escobar-Puentes, A.A.; Palomo, I.; Rodríguez, L.; Fuentes, E.; Villegas-Ochoa, M.A.; González-Aguilar, G.A.; Olivas-Aguirre, F.J.; Wall-Medrano, A. Sweet potato (*Ipomoea batatas* L.) phenotypes: From agroindustry to health effects. *Foods* **2022**, *11*, 1058. [CrossRef] [PubMed]
7. Jiang, T.; Ye, S.; Liao, W.; Wu, M.; He, J.; Mateus, N.; Oliveira, H. The botanical profile, phytochemistry, biological activities and protected-delivery systems for purple sweet potato (*Ipomoea batatas* (L.) Lam.): An up-to-date review. *Food Res. Int.* **2022**, *161*, 111811. [CrossRef]
8. Alam, M.K. A comprehensive review of sweet potato (*Ipomoea batatas* [L.] Lam): Revisiting the associated health benefits. *Trends Food Sci. Technol.* **2021**, *115*, 512–529. [CrossRef]
9. FAOSTAT. Statistics Division of Food and Agriculture Organization of the United Nations. 2021. Available online: http://www.fao.org/faostat/zh/#data/QC (accessed on 10 January 2023).
10. Sugri, I.; Maalekuu, B.K.; Kusi, F.; Gaveh, E. Quality and shelf-life of sweet potato as influenced by storage and postharvest treatments. *Trends Hortic. Res.* **2017**, *7*, 1–10. [CrossRef]
11. Dash, S.K.; Rayaguru, K. Post-harvest processing and utilization of sweet potato: A review. *Food Rev. Int.* **2019**, *35*, 726–762.
12. Sanchez, P.D.C.; Hashim, N.; Shamsudin, R.; Nor, M.Z.M. Laser-light backscattering imaging approach in monitoring and classifying the quality changes of sweet potatoes under different storage conditions. *Postharvest Biol. Technol.* **2020**, *164*, 111163. [CrossRef]
13. Zaccari, F.; Cabrera, M.; Saadoun, A. Sweet potato and squash storage. *Encycl. Food Secur. Sustain.* **2018**, *2*, 464–472.
14. Ishiguro, K.; Yahara, S.; Yoshimoto, M. Changes in polyphenolic content and radical-scavenging activity of sweetpotato (*Ipomoea batatas* L.) during storage at optimal and low temperatures. *J. Agric. Food Chem.* **2007**, *55*, 10773–10778. [CrossRef] [PubMed]

15. De Araújo, N.O.; de Sousa Santos, M.N.; de Araujo, F.F.; Véras, M.L.M.; de Jesus Tello, J.P.; da Silva Arruda, R.; Fugate, K.K.; Finger, F.L. Balance between oxidative stress and the antioxidant system is associated with the level of cold tolerance in sweet potato roots. *Postharvest Biol. Technol.* **2021**, *172*, 111359. [CrossRef]
16. Huang, C.; Xu, M.; Lo, S.; Lai, Y. Effects of treatments of sprouting inhibitor CIPC and Co60 on sprouting and decaying rates of the storage roots of sweet potato. *J. Taiwan Agric. Res.* **2018**, *67*, 73–81.
17. Yang, S.F.; Hoffman, N.E. Ethylene biosynthesis and its regulation in higher plants. *Annu. Rev. Plant Physiol.* **1984**, *35*, 155–189. [CrossRef]
18. Schotsmans, W.C.; Prange, R.K.; Binder, B.M. 1-Methylcyclopropene: Mode of action and relevance in postharvest horticulture research. *Hortic. Res.* **2009**, *35*, 263–313.
19. Blankenship, S.M.; Dole, J.M. 1-Methylcyclopropene: A review. *Postharvest Biol. Technol.* **2003**, *28*, 1–25. [CrossRef]
20. Pankomera, P. Effects of Postharvest Treatments on Sweet Potato (*Ipomoea batatas*) Storage Quality. Ph.D. Thesis, Massey University, Auckland, New Zealand, 2015.
21. Amoah, R.; Terry, L.A. Biochemical and physiological changes in stored sweet potatoes as mediated by 1-methylcyclopropene (1-MCP). In *VII International Postharvest Symposium 1012*; International Society for Horticultural Science: Leuven, Belgium, 2012; pp. 345–351.
22. Ballesteros-Mártinez, L.; Pérez-Cervera, C.; Andrade-Pizarro, R. Effect of glycerol and sorbitol concentrations on mechanical, optical, and barrier properties of sweet potato starch film. *NFS J.* **2020**, *20*, 1–9. [CrossRef]
23. Alamar, M.C.; Anastasiadi, M.; Lopez-Cobollo, R.; Bennett, M.H.; Thompson, A.J.; Turnbull, C.G.; Mohareb, F.; Terry, L.A. Transcriptome and phytohormone changes associated with ethylene-induced onion bulb dormancy. *Postharvest Biol. Technol.* **2020**, *168*, 111267. [CrossRef]
24. Yang, X.; Chen, L.; Yang, Y.; Guo, X.; Chen, G.; Xiong, X.; Dong, D.; Li, G. Transcriptome analysis reveals that exogenous ethylene activates immune and defense responses in a high late blight resistant potato genotype. *Sci. Rep.* **2020**, *10*, 21294. [CrossRef]
25. Ji, C.Y.; Kim, Y.H.; Lee, C.J.; Park, S.U.; Lee, H.U.; Kwak, S.S.; Kim, H.S. Comparative transcriptome profiling of sweetpotato storage roots during curing-mediated wound healing. *Gene* **2022**, *833*, 146592. [CrossRef]
26. Geng, S.; Liu, Z.; Golding, J.B.; Pristijono, P.; Lv, Z.; Lu, G.; Yang, H.; Ru, L.; Li, Y. Transcriptomic analyses of carvone inhibited sprouting in sweet potato (*Ipomoea batatas* (L.) Lam cv 'Yan 25') storage roots. *Postharvest Biol. Technol.* **2023**, *195*, 112142. [CrossRef]
27. Xie, Z.; Zhou, Z.; Li, H.; Yu, J.; Jiang, J.; Tang, Z.; Ma, D.; Zhang, B.; Han, Y.; Li, Z. High throughput sequencing identifies chilling responsive genes in sweetpotato (*Ipomoea batatas* Lam.) during storage. *Genomics* **2019**, *111*, 1006–1017. [CrossRef]
28. Ji, C.Y.; Kim, H.S.; Lee, C.J.; Kim, S.E.; Lee, H.U.; Nam, S.S.; Li, Q.; Ma, D.F.; Kwak, S.S. Comparative transcriptome profiling of tuberous roots of two sweet potato lines with contrasting low temperature tolerance during storage. *Gene* **2020**, *727*, 144244. [CrossRef] [PubMed]
29. Cai, Z.; Cai, Z.; Huang, J.; Wang, A.; Ntambiyukuri, A.; Chen, B.; Zheng, G.; Li, H.; Huang, Y.; Zhan, J.; et al. Transcriptomic analysis of tuberous root in two sweet potato varieties reveals the important genes and regulatory pathways in tuberous root development. *BMC Genom.* **2022**, *23*, 473. [CrossRef] [PubMed]
30. Kim, J.S.; Jang, M.; Nie, H.; Lee, J.; Hong, E.; Kim, S.J.; Kim, S.H. Differential expression pattern of novel MADS-box genes in early root formation and differentiation of sweet potato. *Gene Expr. Patterns* **2022**, *43*, 119216. [CrossRef] [PubMed]
31. Dako, E.; Jankowski, C.K.; Gnimassou, Y.M.; Lebeau, D. Study of inhibition of germination of potato by ethylene. *Heliyon* **2021**, *7*, e06175. [CrossRef]
32. Foukaraki, S.G.; Cools, K.; Chope, G.A.; Terry, L.A. Impact of ethylene and 1-MCP on sprouting and sugar accumulation in stored potatoes. *Postharvest Biol. Technol.* **2016**, *114*, 95–103. [CrossRef]
33. Lv, J.; Bai, L.; Han, X.; Xu, D.; Ding, S.; Li, C.; Ge, Y.; Li, J. Effects of 1-MCP treatment on sprouting and preservation of ginger rhizomes during storage at room temperature. *Food Chem.* **2021**, *349*, 129004. [CrossRef] [PubMed]
34. Downes, K.; Chope, G.A.; Terry, L.A. Postharvest application of ethylene and 1-methylcyclopropene either before or after curing affects onion (*Allium cepa* L.) bulb quality during long term cold storage. *Postharvest Biol. Technol.* **2010**, *55*, 36–44. [CrossRef]
35. Cheema, M.U.A.; Rees, D.; Colgan, R.J.; Taylor, M.; Westby, A. The effects of ethylene, 1-MCP and AVG on sprouting in sweetpotato roots. *Postharvest Biol. Technol.* **2013**, *85*, 89–93. [CrossRef]
36. Pankomera, P.; Heyes, J.A.; Lewthwaite, S.L.; Roskruge, N. Effects of ethylene and 1-methylcyclopropene on sweetpotato storage root quality. *Acta Hortic.* **2016**, *1118*, 163–169. [CrossRef]
37. Amoah, R.S.; Landahl, S.; Terry, L.A. The timing of exogenous ethylene supplementation differentially affects stored sweetpotato roots. *Postharvest Biol. Technol.* **2016**, *120*, 92–102. [CrossRef]
38. Villordon, A.; Clark, C.; LaBonte, D.; Firon, N. 1-Methylcyclopropene has a variable effect on adventitious root emergence from cuttings of two sweetpotato cultivars. *HortScience* **2012**, *47*, 1764–1767. [CrossRef]
39. Cheema, M.U.A.; Rees, D.; Westby, A.; Taylor, M. Hormonal control of sprouting of sweetpotatoes in storage. In *III International Conference Postharvest Unlimited*; International Society for Horticultural Science: Leuven, Belgium, 2008; Volume 858, pp. 173–177.
40. Paul, N.C.; Park, S.; Liu, H.; Lee, J.G.; Han, G.H.; Kim, H.; Sang, H. Fungi Associated with Postharvest Diseases of Sweet Potato Storage Roots and In Vitro Antagonistic Assay of *Trichoderma harzianum* against the Diseases. *J. Fungi* **2021**, *7*, 927. [CrossRef] [PubMed]
41. Ray, R.C.; Ravi, V. Postharvest spoilage of sweetpotato in tropics and control measures. *Crit. Rev. Food Sci.* **2005**, *45*, 623–644. [CrossRef]

42. Holmes, G.J.; Stange, R.R. Influence of Wound Type and Storage Duration on Susceptibility of Sweetpotatoes to *Rhizopus* Soft Rot. *Plant Dis.* **2002**, *86*, 345–348. [CrossRef]
43. Bagam, P.H.; Kamble, S.S. Sensitivity of *Sclerotium rolfsii* sacc., causing tuber rot of sweet potato, to carbendazim. *BIOINFOLET-Q. J. Life Sci.* **2022**, *19*, 71.
44. Paul, N.C.; Nam, S.S.; Park, W.; Yang, J.W.; Kachroo, A. First report of storage tuber rot in sweetpotato (*Ipomoea batatas*) caused by *Plenodomus destruens* in Korea. *Plant Dis.* **2019**, *103*, 1020. [CrossRef]
45. Zhang, M.; Liu, M.; Pan, S.; Pan, C.; Li, Y.; Tian, J. Perillaldehyde controls postharvest black rot caused by *Ceratocystis fimbriata* in sweet potatoes. *Front. Microbiol.* **2018**, *9*, 1102. [CrossRef] [PubMed]
46. Xing, K.; Li, T.J.; Liu, Y.F.; Zhang, J.; Zhang, Y.; Shen, X.Q.; Li, X.Y.; Miao, X.M.; Feng, Z.Z.; Peng, X.; et al. Antifungal and eliciting properties of chitosan against *Ceratocystis fimbriata* in sweet potato. *Food Chem.* **2018**, *268*, 188–195. [CrossRef] [PubMed]
47. Wang, R.; Gao, B.; Ma, J.; Li, X.; Li, J.; Chen, S. Mottle necrosis caused by *Globisporangium ultimum* var. ultimum on sweet potato roots during harvest and storage in China. *J. Phytopathol.* **2022**, *170*, 693–699.
48. Pan, C.; Yang, K.; Erhunmwunsee, F.; Li, Y.X.; Liu, M.; Pan, S.; Yang, D.; Lu, G.; Ma, D.; Tian, J. Inhibitory effect of cinnamaldehyde on *Fusarium solani* and its application in postharvest preservation of sweet potato. *Food Chem.* **2023**, *408*, 135213. [CrossRef]
49. Li, Y.; Yin, Y.; Golding, J.B.; Geng, S.; Chen, G.; Yang, H. Metabolomic and Transcriptomic Analyses of Quality Deterioration in Fusarium solani-Infected Sweet Potato (*Ipomoea batatas* (L.) Lam cv Xinxiang) Storage Roots. *J. Agric. Food Chem.* **2022**, *70*, 7258–7266. [CrossRef] [PubMed]
50. Wang, C.J.; Wang, Y.Z.; Chu, Z.H.; Wang, P.S.; Liu, B.Y.; Li, B.Y.; Li, Y.X.; Luan, B.H. Endophytic *Bacillus amyloliquefaciens* YTB1407 elicits resistance against two fungal pathogens in sweet potato (*Ipomoea batatas* (L.) Lam.). *J. Plant Physiol.* **2020**, *253*, 153260. [CrossRef]
51. Stahmann, M.A.; Clare, B.G.; Woodbury, W. Increased disease resistance and enzyme activity induced by ethylene and ethylene production of black rot infected sweet potato tissue. *Plant Physiol.* **1966**, *41*, 1505–1512. [CrossRef] [PubMed]
52. Arancibia, R.A.; Main, J.L.; Clark, C.A. Sweetpotato Tip Rot Incidence Is Increased by Preharvest Applications of Ethephon and Reduced by Curing. *Horttechnology* **2013**, *23*, 288–293. [CrossRef]
53. Cao, J.; Liu, P.; Wang, X.; Wang, Q.; Shi, J. Combination of wound healing with 1-methylcyclopropene and wound detection by iodine solution to maintain the quality of sweet potato during long-term storage. *Int. J. Agric. Biol. Eng.* **2021**, *14*, 241–246. [CrossRef]
54. Amoah, R.S.; Terry, L.A. 1-Methylcyclopropene (1-MCP) effects on natural disease resistance in stored sweet potato. *J. Sci. Food Agric.* **2018**, *98*, 4597–4605. [CrossRef]
55. van Loon, L.C.; Geraats, B.P.; Linthorst, H.J. Ethylene as a modulator of disease resistance in plants. *Trends Plant Sci.* **2006**, *11*, 184–191. [CrossRef] [PubMed]
56. Harper, G. *1-Methylcyclopropene: A Review of Its Use on Potato Tubers*; Sutton Bridge Crop Storage Research, Potato Council: Warwickshire, UK, 2015; p. 8.
57. Morris, L.; Mann, L. Wound healing, keeping quality, and compositional changes during curing and storage of sweet potatoes. *Hilgardia* **1955**, *24*, 143–183. [CrossRef]
58. Sugri, I.; Maalekuu, B.K.; Gaveh, E.; Kusi, F. Compositional and shelf-life indices of sweet potato are significantly improved by pre-harvest dehaulming. *Ann. Agric. Sci.* **2019**, *64*, 113–120. [CrossRef]
59. Van Oirschot, Q.E.A.; Rees, D.; Aked, J.; Kihurani, A. Sweetpotato cultivars differ in efficiency of wound healing. *Postharvest Biol. Technol.* **2006**, *42*, 65–74. [CrossRef]
60. Rees, D.; van Oirschot, Q.E.; Aked, J. The role of carbohydrates in wound-healing of sweetpotato roots at low humidity. *Postharvest Biol. Technol.* **2008**, *50*, 79–86. [CrossRef]
61. Atuna, R.A.; Carey, E.E.; Low, J.W.; Amagloh, F.K. Wound healing and dry matter content of orange-fleshed sweetpotato cultivars as influenced by curing methods. *Open Agric.* **2017**, *2*, 274–279. [CrossRef]
62. Artschwager, E.; Starrett, R.C. Suberization and wound-periderm formation in sweetpotato and gladiolus as affected by temperature and relative humidity. *J. Agric. Res.* **1931**, *43*, 353–364.
63. Walter, W.M.; Schadel, W.E. A Rapid Method for Evaluating Curing Progress in Sweet Potatoes1. *J. Am. Soc. Hortic. Sci.* **1982**, *107*, 1129–1133. [CrossRef]
64. Randle, W.M.; Woodson, W.R. The effect of storage and wounding on ethylene production by sweet potato. *HortScience* **1986**, *21*, 1018–1019. [CrossRef]
65. Saltveit, M.E.; Locy, R.D. Cultivar Differences in Ethylene Production by Wounded Sweet Potato Roots1. *J. Am. Soc. Hortic. Sci.* **1982**, *107*, 1114–1117. [CrossRef]
66. Cabrera, R.M.; Saltveit, M.E. Survey of wound-induced ethylene production by excised root segments. *Physiol. Plant.* **2003**, *119*, 203–210. [CrossRef]
67. Amand, P.S.; Randle, W.M. Ethylene production and wound healing in sweet potato roots. *HortScience* **1989**, *24*, 805–807. [CrossRef]
68. Wang, S.J.; Lan, Y.C.; Chen, S.F.; Chen, Y.M.; Yeh, K.W. Wound-response regulation of the sweet potato sporamin gene promoter region. *Plant Mol. Biol.* **2002**, *48*, 223–231. [CrossRef] [PubMed]

69. Grace, M.H.; Yousef, G.G.; Gustafson, S.J.; Truong, V.D.; Yencho, G.C.; Lila, M.A. Phytochemical changes in phenolics, anthocyanins, ascorbic acid, and carotenoids associated with sweetpotato storage and impacts on bioactive properties. *Food Chem.* **2014**, *145*, 717–724. [CrossRef] [PubMed]
70. Dandago, M.A.; Gungula, D.T. Effects of various storage methods on the quality and nutritional composition of sweet potato (*Ipomea batatas* L.) in Yola Nigeria. *Int. Food Res. J.* **2011**, *18*, 271–278.
71. Cheema, M.U.A. Dormancy and Sprout Control in Root and Tuber Crops. Ph.D. Thesis, University of Greenwich, London, UK, 2010.
72. Lima, P.C.C.; Santos, M.N.D.S.; Guimaraes, M.E.D.S.; Araujo, N.O.D.; Krause, M.R.; Finger, F.L. Ethylene and its inhibitors affect the quality of processed sweet potatoes. *Food Sci. Technol.* **2021**, *41*, 825–832. [CrossRef]
73. Sowe, S. The Effects of Controlled Atmosphere and Ethylene on the Postharvest Quality of Sweet Potato during Storage. Ph.D. Thesis, Cranfield University, Cranfield, UK, 2018.
74. Villordon, A.Q. Differential response of 'Beauregard' sweetpotato storage roots to ethephon, 1-methylcyclopropene and water submergence treatments. *HortScience* **2012**, *47*, S324.
75. Han, S.H.; Jang, H.D.; Lee, S.J. Modeling respiration rates of *Ipomoea batatas* (sweet potato) under hermetic storage system. *Food Sci. Biotechnol.* **2020**, *29*, 227–234. [CrossRef]
76. Jung, J.K.; Lee, S.U.; Kozukue, N.; Levin, C.E.; Friedman, M. Distribution of phenolic compounds and antioxidative activities in parts of sweet potato (*Ipomoea batata* L.) plants and in home processed roots. *J. Food Compos. Anal.* **2011**, *24*, 29–37. [CrossRef]

horticulturae

Article

Effect of Alginate-Based Edible Coating Containing Thyme Essential Oil on Quality and Microbial Safety of Fresh-Cut Potatoes

Sarengaowa [1], Ke Feng [2,*], Yuanzheng Li [3], Ya Long [4] and Wenzhong Hu [1,*]

[1] School of Pharmacy and Food Science, Zhuhai College of Science and Technology, Zhuhai 519041, China; kuailexiaosa@sina.com
[2] Faculty of Medicine, Macau University of Science and Technology, Macau 999078, China
[3] Taiyuan Distillery Co., Ltd., Taiyuan 030000, China
[4] Pharmaron Beijing Co., Ltd., Beijing 100000, China
* Correspondence: fengkesky@163.com (K.F.); hwz@dlnu.edu.cn (W.H.)

Abstract: Fresh-cut potatoes (*Solanum tuberosum* L.) are a favorite product on account of their freshness, convenience, and health benefits. However, cutting causes potatoes to lose their protective tissue and suffer mechanical damage, which greatly increases the quality deterioration and safety risk of potatoes. The background microorganism and foodborne pathogens on fresh-cut potatoes might rapidly grow during transportation, processing, and marketing, and cause high health risks for consumers. In this study, the quality and safety of fresh-cut potatoes coated with an alginate-based edible coating containing thyme essential oil (AEC-TEO) was evaluated during a storage period of 16 days at 4 °C. Samples were coated with AEC-TEO at different concentrations (0, 0.05, 0.35, and 0.65%, v/v). The quality characteristics of fresh-cut potatoes including color, weight loss, firmness, and sensory attributes were evaluated over 4 days. The viability of the background microorganism of fresh-cut potatoes and artificially inoculated bacteria involving *Listeria monocytogenes* (LM) was measured every 4 days. The research showed that treatment with AEC-TEO at a 0.05% concentration was the most beneficial for maintaining quality and inhibiting the microorganism of fresh-cut potatoes. The increase in L and firmness was 10.55 and 8.24 N, respectively, and the decrease in browning was 4.19 compared to that in the control. Sensory attributes represent an assessment between "indifferent" and "like a little". The reductions in total plate counts, total coliform counts, yeast and mold counts, and *Lactobacillus* counts were 2.41 log cfu/g, 1.37 log cfu/g, 1.21 log cfu/g, and 2 log cfu/g, and *Listeria monocytogenes* decreased by 3.63 log cfu/g on fresh-cut potatoes after 16 days. Therefore, AEC-TEO effectively improved the quality of fresh-cut potatoes and, to a certain extent, prolonged their shelf life. This represents a potential application prospect for the preservation of fresh-cut potatoes.

Keywords: background microorganisms; coating; *Listeria monocytogenes*; preservation; vegetables

Citation: Sarengaowa; Feng, K.; Li, Y.; Long, Y.; Hu, W. Effect of Alginate-Based Edible Coating Containing Thyme Essential Oil on Quality and Microbial Safety of Fresh-Cut Potatoes. *Horticulturae* **2023**, *9*, 543. https://doi.org/10.3390/horticulturae9050543

Academic Editor: Othmane Merah

Received: 24 February 2023
Revised: 20 March 2023
Accepted: 27 April 2023
Published: 29 April 2023

1. Introduction

Potatoes (*Solanum tuberosum* L.) are economically important staple crop plants that produce high yields of nutritionally valuable food and are important in our daily lives [1]. Potatoes, as a main crop plant, play an important role in food provision all over the world. In China and across the world, potatoes rank as the fourth most common agricultural crop after rice, wheat, and corn. Potatoes are cultured in 79% of the world's countries, making this the largest vegetable crop [2]. Potato has been identified by the Food and Agriculture Organization of the United Nations (FAO) as a staple and sustainable food for the growing world population [3]. However, a large amount of soil attached to the surface of fresh potatoes needs to be removed, and potatoes need to be washed before cutting or cooking. The cleaning and processing of potatoes take more time before the end consumer is reached. Fresh-cut fruits and vegetables (FFVs) are peeled, cubed, trimmed, and sliced prior to sale and have the characteristics of freshness, nutrition, and convenience [4]. Such products are increasingly becoming popular with consumers. Potato is a kind of preprocessed vegetable that is suitable for cutting. However, the mechanical damage caused to potatoes by cutting ruptures the cells and tissues [5], causing texture softening, flavor loss, microbial growth, and other undesirable phenomena, making storage more difficult and impacting the commercial value [6]. Hence, it is important to develop preservation technology that can enhance the quality attributes of fresh-cut potatoes.

Edible coatings (ECs) can maintain freshness as a wrapping on FFVs by improving the tactile and visual properties of product surfaces [7]. FFVs have been preserved with various types of ECs, including alginates, pectin, starch, chitosan, gums, and carrageenan [8,9]. Among these, the alginate-based edible coating (AEC) has attracted widespread interest. Alginate is an anionic polysaccharide found in the outer cell wall of brown algae. In terms of its physical properties, sodium alginate itself is nontoxic and stable in the environment. It is able to gel, form films, and bind with numerous molecules [10]. Sodium alginate is classified as GRAS (generally regarded as safe) by the Food and Drug Administration (FDA) and can be used as an emulsifier, stabilizer, thickener, or gel [11]. Alginic acid and its salts are food additives approved by the European Commission [12]. The most common salt of alginate is sodium alginate [13]. Alginate can be used as an edible, biodegradable preservative for coating fruits and vegetables [14]. AEC affects fruits and vegetables by controlling the gas exchange, reducing the moisture transfer, and delaying the ripening process. Some reports have shown that AEC also has beneficial effects on the preservation of FFVs [15–17].

Essential oils (EOs), also known as volatile oils, can be obtained from all plant organs, i.e., flowers, seeds, leaves, roots, wood, twigs, fruits, and bark; they are stored in secretory cells, cavities, canals, epidermal cells, or glandular trichomes [18]. There is widespread recognition that essential oils have antibacterial, antifungal, antiparasitic, and antiviral properties [19]. In one of our previous reports, using in vitro experiments, we demonstrated that thyme essential oil (TEO) exhibited the strongest antibacterial activity against foodborne pathogens among many well-known EOs (thyme oil, cinnamon oil, oregano oil, lemongrass oil, mint oil, rosemary oil, clove oil, eucalyptus oil, lavender oil, tea tree oil, blumea oil, valerian oil, atractylodes oil, and zingiber oil) [20]. TEO is among the world's top ten essential oils and is noted for its antimicrobial, antimycotic, antioxidative, food preservative, antifungal, and mammalian-age-delaying properties [21]. Antimicrobial properties are attributed to the alcohols, phenols, terpenes, and ketones in EOs [22]. Researchers have demonstrated that the main component of TEO is thymol [23]. TEO can readily enhance antibacterial activity by volatilizing and inhibiting microbial growth on fresh collard greens and sweet basil leaves during storage [24,25]. However, the volatile odor of TEO can affect the quality, flavor, and sensory attributes of FFVs. Therefore, TEO might be combined with EC to reduce these negative impacts. The preservation effects of alginate EC with oregano essential oils, carboxymethyl chitosan-pullulan EC with galangal essential oil, EC with cinnamon essential oil, and alginate-based EC with thyme essential oil have all been evaluated on fruits and vegetables such as tomatoes, mangos, strawberries, and cantaloupes [26–29]. Moreover, the quality and safety of fresh-cut potatoes treated with alginate-based EC containing thyme essential oil have been not reported. The development of the preservative technology can provide a potential antibacterial agent for fresh-cut potatoes.

The objective of this study is to further evaluate the preservation effects of AEC-TEO on the quality and sensory attributes of fresh-cut potatoes, as well as its impact on background microorganisms and *Listeria monocytogenes*.

2. Materials and Methods

2.1. Bacterial Inoculum

Listeria monocytogenes (LM, CICC 21633) was purchased from the China Center of Industrial Culture Collection (CICC, Beijing, China). A tryptic soy broth containing yeast extract (TSB-YE) was used to culture LM at 37 °C for 12 h. The bacterium suspension was centrifuged at 5000 r/min for 5 min. Peptone water at 0.1% (*w*/*v*) was added to mix the LM for washing. Dilution was carried out using 0.1% (*w*/*v*) peptone water, in a ratio of 1:10, in order to reach the proper inoculum. The population of LM was expressed as log cfu/mL [30].

2.2. Fresh-Cut Potatoes

Fresh potatoes were obtained from a New-Mart in Dalian City (China). They were uniform in size, color, and the absence of defects. The experiment site is located at 38.9° N and 121.6° E. Fresh potatoes were stored at a low temperature (approximately 4 °C) before being processed. They were cleaned with fresh water to remove surface dirt and washed again with distilled water. The surface of the samples was sterilized using alcohol (75% *v*/*v*). A biosafety cabinet was used to air-dry samples for 10 min at 25 °C. The potatoes were then cut into cubes (1 cm × 1 cm × 1 cm) using a sterile knife.

2.3. Preparation of AEC-TEO

The preparation of AEC in this study was carried out in line with previous research [29]. A mixture of sodium alginate at 1.29% (*w*/*v*) and glycerol at 1.5% (*w*/*v*) was stirred in ultrapure water at 70 °C until the solution became transparent [31]. TEO at different concentrations (0.05%, 0.35%, and 0.65%, *v*/*v*) was added to the AEC and stirred for 3 min at 12,500 rpm using Ultra Turrax T25 mixer (IKA® WERKE, Staufen, Germany). To produce a cross-linking reaction necessary for gel formation, a 2% (*w*/*v*) calcium chloride solution (food grade) containing 1% (*w*/*v*) ascorbic acid (food grade) and 1% (*w*/*v*) citric acid (food grade) was prepared. To the calcium chloride solution, ascorbic acid and citric acid were added as antioxidants and color fixatives, respectively [32].

2.4. Fresh-Cut Potato Coating

Fresh-cut potatoes were soaked in the solution of sodium alginate containing TEO for 2 min. Samples were then placed in the calcium chloride solution for 2 min. Fresh-cut potatoes treated with AEC, AEC-TEO (0.05%), AEC-TEO (0.35%), and AEC-TEO (0.65%) were evaluated. Uncoated samples were used as controls. Fresh-cut potatoes (10 cubes) were placed on polystyrene trays as a group and wrapped with PVC films for the evaluation of quality and sensory attributes every 4 days, for a 16-day period, at 4 °C.

2.5. Evaluation of Fresh-Cut Potato Color, Weight Loss, and Firmness

Five group experiments were planned to measure the color of fresh-cut potatoes over 16 days (one test every 4 days). The surfaces of three fresh-cut potato cubes were randomly selected from one group (10 cubes) for each test. The color of the cut surfaces of the potato cubes was measured using a CR400/CR410 colorimeter (Minolta, Tokyo, Japan). The color parameters L^* (lightness), a^* (green chromaticity), and b^* (yellow chromaticity) were measured. Each measurement was carried out on potato cubes. The browning index (BI) was calculated using Equation (1), as follows:

$$\mathrm{BI} = [100(\mathrm{x} - 0.31)]/0.172 \quad (1)$$

where

$$\mathrm{x} = (a^* + 1.75\,L^*)/(5.645\,L^* + a^* - 3.012\,b^*) \quad (2)$$

Three group experiments were planned to measure the weight loss of fresh-cut potatoes over 16 days. Polystyrene trays containing the potato cubes were measured using a digital balance (PL-2002, METTLER TOLEDO, Greifensee, Switzerland) over a period of 16 days. The weight loss rate was calculated as follows:

$$\text{Weight loss rate (\%)} = [(\mathrm{m}_1 - \mathrm{m}_2)/\mathrm{m}_1] \times 100 \quad (3)$$

where m_1 is the initial weight (g) and m_2 is the weight at the specified time point (g).

Five group experiments were planned to measure the firmness of fresh-cut potatoes over 16 days (one test every 4 days). Three fresh-cut potato cubes were randomly selected from one group (10 cubes) for each test. The firmness of the fresh-cut potatoes was determined using a TA.XT texture analyzer (Stable Micro Systems Ltd., Godalming, UK). We measured the firmness of the fresh-cut potato cubes based on the force (N) exerted on the cubes using the P5 compression probe. Each experiment was carried out three times.

2.6. Sensory Attribute Analysis

The sensory characteristics of the potato cubes were evaluated after 8 days by regular consumers of potatoes. A total of twenty individuals, including students and staff, were recruited from the food science and technology faculty. Twenty group experiments (10 cubes in each group) were planned to evaluate the sensory attributes. Fresh-cut potatoes were randomly presented to assessors in different treatment groups. Testing was carried out in individual rooms for each candidate. The odor, color, texture, appearance, and acceptability of the fresh-cut potatoes were evaluated according to a 9-point hedonic scale test. The hedonic evaluation scale recorded degrees of appreciation using the following scoring system: 9 = like very much; 8 = like a lot; 7 = like moderately; 6 = like slightly; 5 = indifferent; 4 = dislike slightly; 3 = dislike moderately; 2 = dislike a lot; 1 = dislike very much [33,34]. The assessors recorded their responses on paper scorecards. Panelists were advised to sip water between the evaluations of two different samples.

2.7. Analysis of Background Microorganisms

Fresh-cut potatoes coated with AEC and AEC-TEO of different concentrations (0.05%, 0.35%, and 0.65%, v/v) were stored at 4 °C for 16 days. Samples without the coating served as the control. Samples were taken to measure the background microorganism at an interval of 4 days during storage time. Samples were homogenized with 0.1% peptone water (1:10) in a sterile blender. Suspensions of 0.1 mL were taken from the homogenate and cultured on plate count agar (PCA) at 37 °C for 24 h to measure the total plate counts; on potato dextrose agar (PDA) at 28 °C for 48 h to measure yeast and mold counts; on violet red bile dextrose agar (VRBDA) at 37 °C for 24 h to measure total coliform counts; and on Lactobacilli MRS agar at 37 °C for 24 h to measure *Lactobacillus* counts [35,36].

2.8. Analysis of Listeria monocytogenes (LM)

Fresh-cut potato cubes (10 g) were placed in sterile Petri dishes. For the challenges, the potato cube surfaces were inoculated with LM suspensions (8–9 log CFU/mL, 500 μL). Samples were air-dried in a biosafety cabinet at 25 °C for 1 h. As described previously, samples were treated with AEC and AEC-TEO at different concentrations. Untreated fresh-cut potatoes were used as a control. During the 16-day storage period, fresh-cut potato cubes were put in a sterile blender bag and stored at 4 °C. The population of LM on the fresh-cut potatoes was measured at 4-day intervals during the 16-day storage period. Each experiment was conducted three times. The suspension of LM was taken and cultured on an Oxford agar base at 37 °C to determine the bacterial count, which was expressed as log cfu/g.

2.9. Statistical Analysis

Experiments were conducted in triplicate and the mean + standard deviation values were obtained for each experiment. The data were analyzed using SPSS software (Version 14.0; SPSS, Chicago, IL, USA). The significance of differences between variables was tested using a one-way ANOVA (between groups) and a repeated-measures ANOVA (within groups). Duncan's multiple range test was used to compare the means. The statistical significance was determined at $p < 0.05$.

3. Results

3.1. Effects of AEC-TEO on the Quality of Fresh-Cut Potatoes

3.1.1. Color

Color is a critical factor in consumer acceptance of fruit and vegetables. Enzymatic browning is an important process that compromises the color of fresh-cut potatoes. In this study, L^* (lightness) indicates the brightening or darkening of fresh-cut potatoes and BI indicates enzymatic browning during the storage period.

Figure 1a shows changes in the L^* of fresh-cut potatoes treated with AEC with or without TEO. The L^* of fresh-cut potatoes decreased significantly in the control and AEC treatment during the storage period ($p < 0.05$). The L^* in AEC-TEO (0.05%) was higher than that in other groups after 12 days. Decreases were more dramatic in fresh-cut potatoes treated with AEC-TEO (0.35% and 0.65%) than in other treatment groups after 12 days.

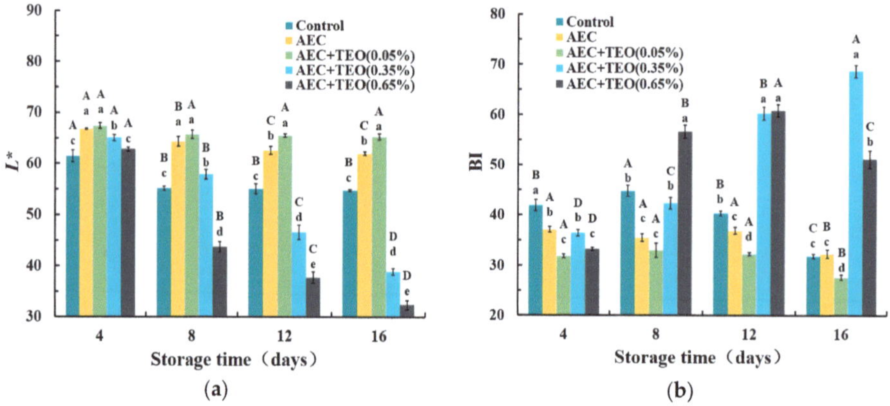

Figure 1. The effect of alginate-based EC with thyme essential oil on color of fresh-cut potatoes at 4 °C. (**a**) L^* (lightness); (**b**) BI (Browning index). Control: Uncoated; AEC: Alginate-based edible coatings; TEO: Thyme essential oil. The different lowercase letters indicate significant differences across different treatments. The different uppercase letters indicate significant differences across different storage times ($p < 0.05$).

Figure 1b shows changes in the BI of fresh-cut potatoes treated with AEC and AEC-TEO. The BI of fresh-cut potatoes in the control was higher than that of the other groups on the fourth day and then decreased more significantly during the storage period. The BI in AEC was lower than that of the control group after 12 days. The BI of fresh-cut potatoes treated with AEC-TEO (0.05%) also decreased significantly during the storage period, reaching the lowest value among all groups on the 16th day. The BI of fresh-cut potatoes treated with AEC-TEO (0.35% and 0.65%) increased significantly during the storage period, more dramatically than in other groups after 4 days.

3.1.2. Weight Loss

It was important to assess the quality of the fresh-cut potatoes in terms of weight loss during the 16-day storage period (Figure 2). The measured weight loss values of the fresh-cut potatoes increased significantly during storage times ($p < 0.05$). There were no significant differences in the control, AEC, and AEC-TEO (0.05%) during storage time ($p \geq 0.05$). The weight loss in AEC-TEO (0.35% and 0.65%) groups was significantly higher than that in the other treatment groups during storage times ($p < 0.05$).

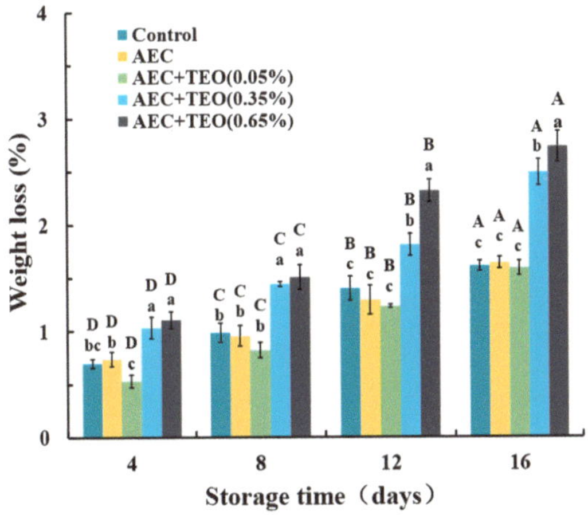

Figure 2. The effect of alginate-based EC with thyme essential oil on weight loss of fresh-cut potatoes at 4 °C. Control: Uncoated; AEC: Alginate-based edible coatings; TEO: Thyme essential oil. The different lowercase letters indicate significant differences across different treatments. The different uppercase letters indicate significant differences across different storage times ($p < 0.05$).

3.1.3. Firmness

The firmness of fruits and vegetables depends on the composition of the cells and the structure of the cell wall. The softening of fruit occurs during ripening and postharvest storage and affects the quality, shelf life, and commercial value of the fruit. The process of softening involves the hydrolysis of starch into sugars and pectin degradation in the cell wall of the fruit. Figure 3 shows changes in the firmness of fresh-cut potatoes treated with AEC with or without TEO. The firmness in the control group showed the lowest value among all treatment groups on the 16th day ($p < 0.05$). The firmness in the AEC and AEC-TEO (0.05%, 0.35%, and 0.65%) was higher than that in the control on the 16th day ($p < 0.05$). There was no significant difference between AEC and AEC-TEO (0.05%, 0.35%, and 0.65%) on the 16th day ($p \geq 0.05$).

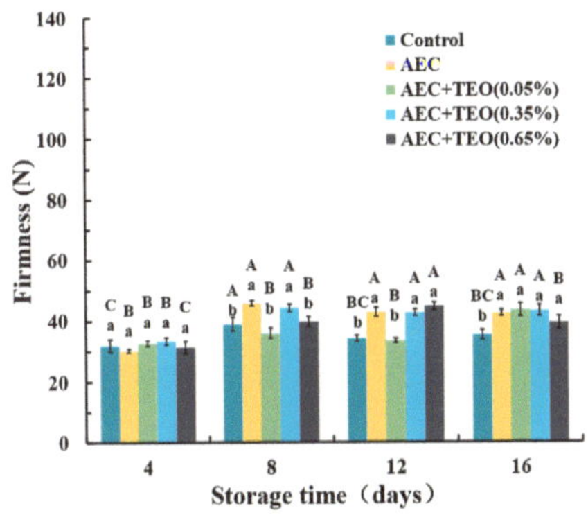

Figure 3. The effect of alginate-based EC with thyme essential oil on firmness of fresh-cut potatoes at 4 °C. Control: Uncoated; AEC: Alginate-based edible coatings; TEO: Thyme essential oil. The different lowercase letters indicate significant differences across different treatments. The different uppercase letters indicate significant differences across different storage times ($p < 0.05$).

3.1.4. Sensory Analysis of Fresh-Cut Potatoes

EOs are known to have strong odors and flavors that might affect the organoleptic characteristics of fruits. The sensory evaluation of fresh-cut potatoes in the AEC and AEC-

TEO (0.05%, 0.35%, and 0.65%) treatments after 8 days is presented in Figure 4. Hedonic data analysis revealed score ranges of 1.55–5.05 for color, 1.5–5 for appearance, 3.3–5.1 for odor, 2.65–5.6 for texture, and 1.75–5.15 for acceptability. In terms of acceptability, the score of the fresh-cut potatoes in the AEC-TEO (0.05%) treatment represented an assessment between "indifferent" and "like a little". The score for fresh-cut potatoes in the control, AEC, and AEC-TEO (0.35%) groups represented an assessment between "dislike moderately" and "dislike a little". For samples in the AEC-TEO (0.65%) treatment, the acceptability score represented an assessment between "dislike very much" and "dislike a lot." Therefore, fresh-cut potatoes in AEC-TEO (0.05%) obtained significantly higher scores than those in the control and the AEC-TEO (0.35% and 0.65%) treatments ($p < 0.05$), and their sensory attributes might be acceptable for further commercialization. A visualization diagram of the fresh-cut potatoes during storage times is shown in Figure 5. This shows that fresh-cut potatoes in the AEC-TEO (0.05%) treatment group were more acceptable visually than those in other treatment groups during the storage period.

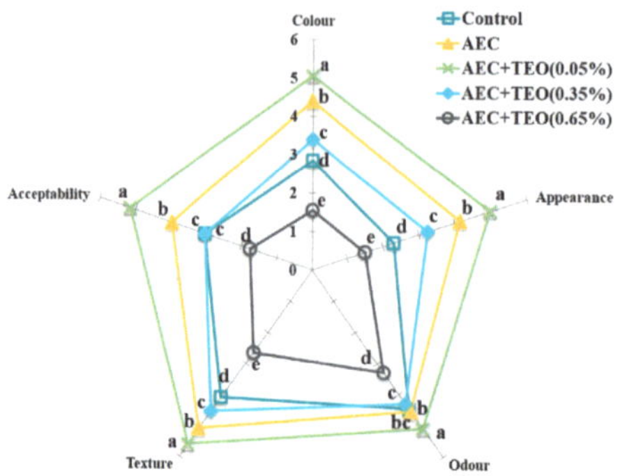

Figure 4. Effect of alginate-based EC with thyme essential oil on sensory attribute of fresh-cut potatoes after 8 days of storage at 4 °C. Control: Uncoated; AEC: Alginate-based edible coatings; TEO: Thyme essential oil. The hedonic evaluation scale used a 9-point scale as follows: 9 represented like very much and 1 represented dislike very much. For each sample, the means designated by different letters are significantly different ($p < 0.05$).

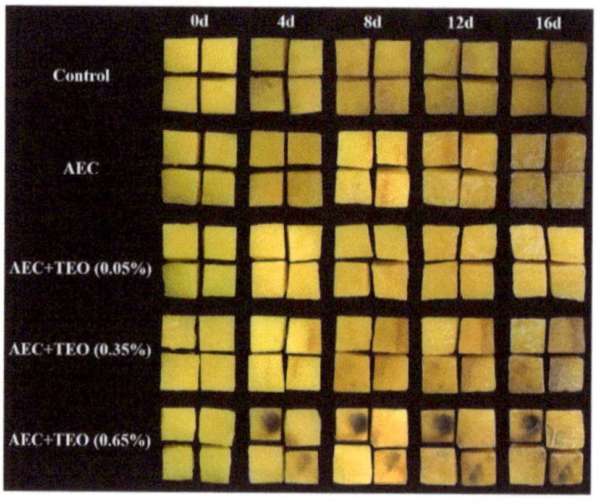

Figure 5. The effect of alginate-based EC with thyme essential oil on appearance of fresh-cut potatoes at 4 °C. Control: Uncoated; AEC: Alginate-based edible coatings; TEO: Thyme essential oil.

3.2. Background Microorganisms

Populations of background microorganisms on fresh-cut potatoes in the AEC treatment with or without TEO were measured (Figure 6). The results showed considerable increases for fresh-cut potatoes in the control, AEC, and AEC-TEO groups ($p < 0.05$). Total plate counts, total coliform counts, yeast and mold counts, and lactic acid bacteria counts in the AEC and AEC-TEO treatments were all lower than those of fresh-cut potatoes in the control over 16 days ($p < 0.05$). The total plate counts, total coliform counts, yeast and mold counts, and lactic acid bacteria counts were much lower in the AEC-TEO treatment than in the AEC treatment ($p < 0.05$). The addition of TEO to AEC caused the population of the background microorganisms to decrease compared with that of fresh-cut potatoes treated with AEC alone. The total plate counts and total coliform counts of fresh-cut potatoes in the AEC-TEO (0.05%) treatment group were lower than those in the AEC treatment group after 12 days. The yeast and mold counts and *Lactobacillus* counts of fresh-cut potatoes in the AEC-TEO (0.05%) treatment were lower than those of potatoes treated with AEC alone after 16 days. The total plate counts in the AEC-TEO (0.35% and 0.65%) treatments were higher than those in the AEC-TEO (0.05%) treatment after 12 days. The yeast and mold counts and total coliform counts in the AEC-TEO (0.35% and 0.65%) treatment were higher than those in the AEC-TEO (0.05%) treatment after 16 days. There were no significant differences in the *Lactobacillus* counts among fresh-cut potatoes in the AEC-TEO (0.05%) and AEC-TEO (0.35%) at 16 days. Interestingly, the values for total plate counts and total coliform counts recorded in the AEC-TEO (0.05%) treatment increased significantly after 12 days. The counts for yeast and mold and for *Lactobacillus* bacteria were also obtained for fresh-cut potatoes treated with AEC-TEO (0.05%) after 16 days.

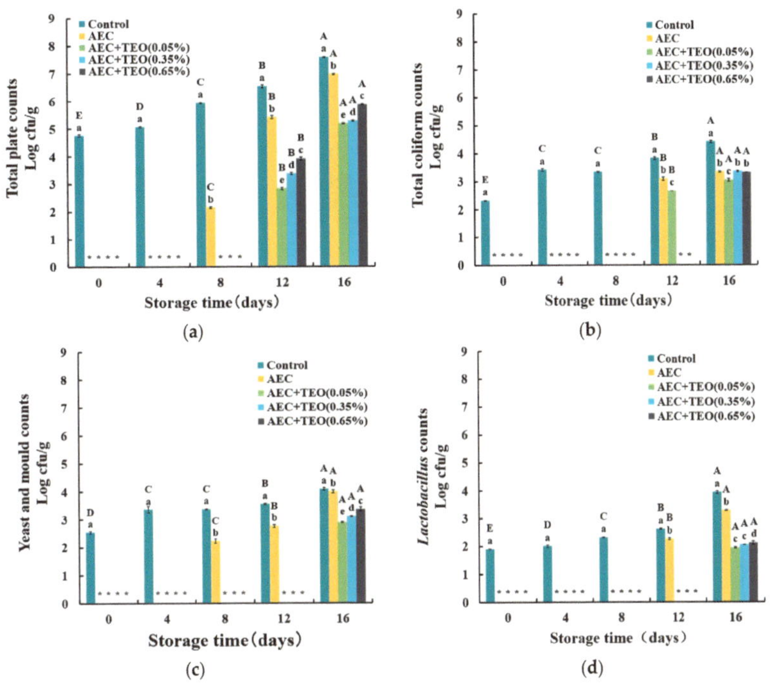

Figure 6. The effect of alginate-based EC with thyme essential oil on background microorganisms on fresh-cut potatoes at 4 °C. (**a**) Total plate counts; (**b**) total coliform counts; (**c**) yeast and mold counts; (**d**) lactobacillus counts. Control: Uncoated; AEC: Alginate-based edible coatings; TEO: Thyme essential oil. Asterisk indicate microorganisms were not detected. The different lowercase letters indicate significant differences across different treatments. The different uppercase letters indicate significant differences across different storage times ($p < 0.05$).

3.3. *Listeria monocytogenes* Analysis

Changes in the populations of *L. monocytogenes* inoculated on the fresh-cut potatoes in the AEC treatment with and without TEO were also measured (Figure 7). The population of *L. monocytogenes* on fresh-cut potatoes in the AEC, AEC-TEO (0.05%, 0.35%, and 0.65%) treatments decreased significantly during the storage period. The populations of *L. monocytogenes* in the AEC were lower than those in the control during the storage period, and the populations of *L. monocytogenes* in the AEC-TEO were lower than those treated with AEC alone. The populations of *L. monocytogenes* decreased as the concentration of the essential oil increased. After 16 days, the *L. monocytogenes* population was lowest on fresh-cut potatoes with the highest tested concentration of TEO (0.65%).

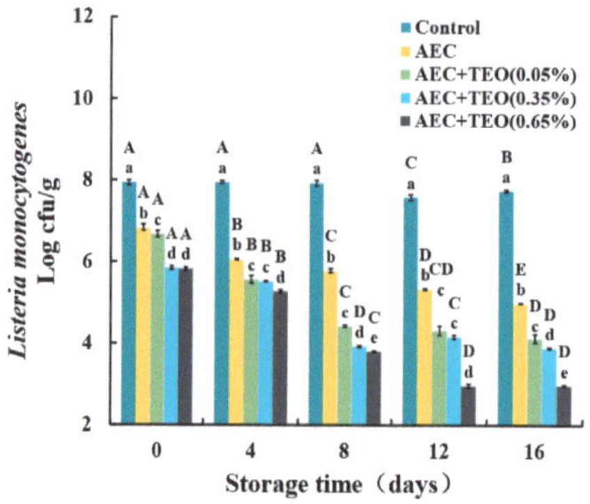

Figure 7. The effect of alginate-based EC with thyme essential oil on *Listeria monocytogenes* on fresh-cut potatoes at 4 °C. Control: Uncoated; AEC: Alginate-based edible coatings; TEO: Thyme essential oil. The different lowercase letters indicate significant differences across different treatments. The different uppercase letters indicate significant differences across different storage times ($p < 0.05$).

4. Discussion

Potatoes are suitable for processing into fresh-cut vegetables as a type of important food material. Fresh-cut potatoes were one of the earliest FFV products in the catering and retail industries. However, mechanical cutting damages the integrity of the potato tissue structure and causes physiological and biochemical changes such as enzymatic browning, water loss, and tissue softening. In particular, the nutrient composition of potatoes provides ideal growth and reproduction conditions for microorganisms [37]. The combination of EC and essential oils as preservative agents can inhibit the growth of microorganisms while maintaining the quality of fresh-cut potatoes. Essential oils are natural and safe and exhibit antibacterial and antioxidant properties. Other research indicated that perillaldehyde, anethole, carvacrol, cinnamaldehyde, eugenol, thymol, and anethole essential oils have a positive effect on enhancing the antioxidant capacities of raspberries, strawberries, and blueberries [38,39]. However, the volatility of essential oils affects their antibacterial activity. EC is an excellent preservative for FFVs. It can reduce weight loss, delay tissue softening, inhibit enzymatic browning, and prevent microbial infection [40]. The volatility of essential oils might be delayed by mixing with EC and maintaining the flavor and texture of fruit and vegetables. In addition, EC can also be used as a carrier of active ingredients such as antioxidants (ascorbic acid, citric acid, and oxalic acid) and nutrients (Vitamin E), which can increase the nutritional value of FFVs [41].

Color has a significant impact on the quality and appearance of FFVs. In this study, L^* and BI were used to evaluate the color of fresh-cut potatoes. L^* (lightness) indicates the brightening or darkening of fresh-cut potatoes, and BI indicates enzymatic browning during the storage period. Lower L^* and higher BI values indicate more serious browning. L^* significantly decreased and BI significantly increased during the storage period ($p < 0.05$). This may be due to the cutting and peeling stimulating signal molecular transmission in fruits and vegetables, activating the phenylalanine ammonia lyase, and synthesizing phenols. The substrate for enzymatic browning is phenols, which are oxidized to quinones under catalysis [42]. During the 4–12-day period, L^* values in the AEC and AEC-TEO (0.05%) treatments were higher, while BI values were lower, compared with the control group. These results indicate that AEC inhibited the browning of fresh-cut potatoes by reducing the reaction between oxygen and PPO. Another study demonstrated that the combination of an edible coating and cinnamon essential oil reduced phenolic content and the browning index compared to uncoated fresh-cut apples after 25 days [43]. The addition of ascorbic acid and citric acid to AEC can also contribute to anti-browning and color protection [44]. Several studies have reported similar results. For example, fresh-cut potatoes soaked in ascorbic acid (1%) maintained good coloration for 6 days [45]. A combination of ascorbic acid (1%) and aloe vera gel (50%) significantly reduced the browning of fresh-cut lotus root [46]. The combination of lginate-CaCl2 and an acetylate monoglyceride coating reduced BI compared with the control fresh-cut apples [47]. However, in the current study, the L^* values in the AEC-TEO (0.35% and 0.65%) treatments significantly decreased as concentrations of TEO increased, while the BI values significantly increased, indicating that TEO in a high concentration seriously affected the appearance of fresh-cut potatoes. This may be because a high concentration of TEO damages the potato tissues. The polyphenol oxidase and substrates in potato tissues lose their separation, and this accelerates enzymatic browning [48]. In another study, fresh-cut potatoes experienced serious browning when exposed to high concentrations of cinnamon essential oil [49]. In this study, we demonstrated that adding a higher concentration of TEO affected the sensory quality of fresh-cut potatoes. This observation is in line with the findings of other studies. For example, the study found that adding thyme essential oil reduced appearance scores to acceptable levels compared with chitosan films for fresh collard greens after 8 days [24]. Fresh-cut vegetables treated with coriander, marjoram, and origanum essential oil show lower odor and taste scores, which reduced the acceptability of the samples [50]. The addition of 0.5% (v/v) lemongrass essential oil significantly affected the sensory properties of fresh-cut pineapple, leading to flavor scores of less than five, and high concentrations of essential oil accelerated the softening of fresh-cut samples [31].

Weight loss is a crucial index to evaluate the quality of FFVs. In this study, weight loss in fresh-cut potatoes in the control, AEC, and AEC-TEO groups increased significantly during the storage period. Because fresh-cut products lose their integrity when peeled, cut, sliced, or shredded, they are more likely to lose water than uncut fruit and vegetable products. Fresh-cut potatoes in the AEC and AEC-TEO (0.05%) treatment showed lower weight loss among all treatment groups during the storage period. Related research has demonstrated that the edible coating can affect weight loss in fresh-cut papaya. The authors of [51] found that a combination of alginate and pectin in an edible coating significantly reduced weight loss in fresh-cut papaya and also reduced juice loss at the end of storage times. However, in the current study, when the TEO concentration increased to 0.35% and 0.65%, the weight loss of the fresh-cut potato samples increased significantly. Indeed, high concentrations of essential oil are toxic to FFVs, causing the destruction of the tissue structure and increased weight loss [52], as demonstrated by previous authors who found that higher concentrations of TEO damaged the tissue of fresh-cut apple, resulting in juice loss, increased respiration, and weight loss [20].

Softening is a serious problem that affects FFVs during storage periods. Starch and pectin function to hydrolyze during fruit softening [53]. This limits the shelf life of FFVs and is an important factor affecting the acceptability of consumers. In this study, the AEC and AEC-TEO maintained the firmness of the fresh-cut potatoes compared with those in the control. The components used in coating solutions might help fresh-cut potatoes to maintain their firmness, and the presence of $CaCl_2$ in the EC may be important in this regard. Ca^{2+} can bind with pectin to stabilize the cell wall, reduce the activity of cell wall hydrolase, and slow down the softening rate [54].

This study reveals that background microorganisms and *Listeria monocytogenes* are able to survive and grow on fresh-cut potatoes during storage periods. Fresh-cut potatoes may provide ideal survival conditions for microorganisms due to the nutrients released after cutting. Pathogens and spoilage microorganisms have also been reported to grow on different types of potato products [55,56]. In the current study, background microorganisms of fresh-cut potatoes in the AEC-TEO treatments were lower than those in the control, and it is reasonable to suggest that TEO exhibits antibacterial activity and that the coating process itself might wash away some microorganisms. Other research showed that an edible coating containing cinnamon oil inhibited yeast and mold on strawberries [28]. An alginate edible coating with thyme oil reduced mold and yeast growth compared to the control and alginate without thyme oil [57]. Bacterial counts were significantly reduced in fruits that were coated with edible coatings with ethanolic essential oil [58]. This study also demonstrated that the populations of background microorganisms and LM on fresh-cut potatoes in the AEC-TEO treatments significantly decreased during the storage period, and antibacterial activity was enhanced with increased concentrations of essential oil. In a previous study, TEO was found to be effective in inhibiting *Listeria monocytogenes*, *Salmonella typhimurium*, *Staphylococcus aureus*, and *E. coli* O157:H7 [20]. TEO exhibits strong antibacterial activity due to its own volatile components. It is well-known that the chemical composition of EOs plays an important role in inhibiting microorganisms [59]. TEO is primarily composed of thymol and carvacrol, and these have biological properties including anti-inflammatory and liver-protecting effects [60]. Other studies have found that thymol can effectively inhibit Gram-positive bacteria by interfering with cell membrane lipids, and proteins, causing an increase in membrane permeability [61]. However, in the current study, high concentrations of TEO did not significantly inhibit the decline of background microorganisms on fresh-cut potatoes after 16 days. The chemical structure of fresh-cut potatoes may be damaged by TEO at concentrations of 0.35% to 0.65%. The juice and nutrition composition provides rich growth conditions for microorganism growth [62]. Therefore, fresh-cut potatoes in the AEC-TEO (0.35% and 0.65%) treatments were prone to microorganism growth. In addition, background microorganisms of 6 log cfu/g are considered an acceptable limit for the shelf life of a fruit product according to the standard of the Institute of Food Science and Technology (IFST) [63]. In this study, therefore, the shelf life of fresh-cut potatoes in the AEC-TEO (0.05%) treatment was extended to 16 days.

5. Conclusions

In this study, the AEC-TEO (0.05%) treatment effectively enhanced the *L*, firmness, and sensory attributes and reduced the weight loss and browning index regarding the quality of fresh-cut potatoes after 16 days. In addition, AEC-TEO exhibited increasing antibacterial activity against background microorganisms and LM on fresh-cut potatoes with increasing concentrations of TEO. Therefore, the AEC-TEO preservation agent can effectively control the infection of LM and maintain the sensory quality of fresh-cut potatoes during the process of packaging, storage, logistics, and sales. The development of AEC-TEO can ensure the wholesomeness and safety of consumers to meet the present-day growing consumer demands for fresh-cut potatoes.

Author Contributions: Conceptualization, K.F. and W.H.; methodology, S. and K.F.; software, S. and K.F.; validation, W.H.; investigation, Y.L. (Yuanzheng Li), resources, W.H.; data curation, Y.L. (Yuanzheng Li) and Y.L. (Ya Long); writing—original draft preparation S.; writing—review and editing, K.F. and W.H.; project administration, K.F. and W.H.; funding acquisition, S. All authors have read and agreed to the published version of the manuscript.

Funding: This study was supported by the Guangdong Basic and Applied Basic Research Foundation (Grant No. 2023A1515011473) and the Doctor Promotion Program of Zhuhai College of Science and Technology.

Data Availability Statement: The data used to support the findings of this study are available from the corresponding author upon request.

Conflicts of Interest: The authors declare no conflict of interest.

References

1. Li, Y.; Zhao, D. Transcriptome analysis of scions grafted to potato rootstock for improving late blight resistance. *BMC Plant Biol.* **2021**, *21*, 272. [CrossRef]
2. Batool, T.; Ali, S.; Seleiman, M.F.; Naveed, N.H.; Ali, A.; Ahmed, K.; Abid, M.; Rizwan, M.; Shahid, M.R.; Alotaibi, M.; et al. Plant growth promoting rhizobacteria alleviates drought stress in potato in response to suppressive oxidative stress and antioxidant enzymes activities. *Sci. Rep.* **2020**, *10*, 16975. [CrossRef] [PubMed]
3. Ciccone, M.; Chambers, D.; Iv, E.C.; Talavera, M. Determining which cooking method provides the best sensory differentiation of potatoes. *Foods* **2020**, *9*, 451. [CrossRef] [PubMed]
4. Van Haute, S.; Sampers, I.; Holvoet, K.; Uyttendaele, M. Physicochemical quality and chemical safety of chlorine as a reconditioning agent and wash water disinfectant for fresh-cut lettuce washing. *Appl. Environ. Microbiol.* **2013**, *79*, 2850–2861. [CrossRef]
5. Kang, W.; Robitaille, M.C.; Merrill, M.; Teferra, K.; Kim, C.; Raphael, M.P. Mechanisms of cell damage due to mechanical impact: An in vitro investigation. *Sci. Rep.* **2020**, *10*, 12009. [CrossRef] [PubMed]
6. Agriopoulou, S.; Stamatelopoulou, E.; Sachadyn-Król, M.; Varzakas, T. Lactic acid bacteria as antibacterial agents to extend the shelf life of fresh and minimally processed fruits and vegetables: Quality and safety aspects. *Microorganisms* **2020**, *8*, 952. [CrossRef]
7. Valencia-Chamorro, S.A.; Palou, L.; Del Río, M.A.; Pérez-Gago, M.B. Antimicrobial edible films and coatings for fresh and minimally processed fruits and vegetables: A review. *Crit. Rev. Food Sci.* **2011**, *51*, 872–900. [CrossRef]
8. Chen, G.; Zhang, B.; Zhao, J. Dispersion process and effect of oleic acid on properties of cellulose sulfate-oleic acid composite film. *Materials* **2015**, *8*, 2346–2360. [CrossRef]
9. Duan, C.; Meng, X.; Meng, J.; Khan, M.I.H.; Dai, L.; Khan, A.; An, X.; Zhang, J.; Huq, T.; Ni, Y. Chitosan as a preservative for fruits and vegetables: A review on chemistry and antimicrobial properties. *J. Bioresour. Bioprod.* **2019**, *4*, 11–21. [CrossRef]
10. Hu, Q.; Nie, Y.; Xiang, J.; Xie, J.; Si, H.; Li, D.; Zhang, S.; Li, M.; Huang, S. Injectable sodium alginate hydrogel loaded with plant polyphenol-functionalized silver nanoparticles for bacteria-infected wound healing. *Int. J. Biol. Macromol.* **2023**, *234*, 123691. [CrossRef]
11. U.S. Food & Drug Administration. Code for Federal Regulations Title 21 Part 184—Direct Food Substances Affirmed as Generally Recognized as Safe. Available online: https://www.accessdata.fda.gov/scripts/cdrh/cfdocs/cfcfr/CFRSearch.cfm?fr=184.1724 (accessed on 5 April 2018).
12. Younes, M.; Aggett, P.; Aguilar, F.; Crebelli, R.; Filipic, M.; Jose Frutos, M.; Galtier, P.; Gott, D.; Gundert-Remy, U.; Georg Kuhnle, G.; et al. Re-evaluation of alginic acid and its sodium, potassium, ammonium and calcium salts (e 400–e 404) as food additives. *EFSA J.* **2017**, *15*, 5049.
13. Yoo, S.; Krochta, J.M. Whey protein–polysaccharide blended edible film formation and barrier, tensile, thermal and transparency properties. *J. Sci. Food Agric.* **2011**, *91*, 2628–2636. [CrossRef] [PubMed]
14. Chit, C.S.; Olawuyi, I.F.; Park, J.J.; Lee, W.Y. Effect of composite chitosan/sodium alginate gel coatings on the quality of fresh-cut purple-flesh sweet potato. *Gels* **2022**, *8*, 747. [CrossRef]
15. Malvano, F.; Corona, O.; Pham, P.L.; Cinquanta, L.; Pollon, M.; Bambina, P.; Farina, V.; Albanese, D. Effect of alginate-based coating charged with hydroxyapatite and quercetin on colour, firmness, sugars and volatile compounds of fresh cut papaya during cold storage. *Eur. Food Res. Technol.* **2022**, *248*, 2833–2842. [CrossRef]
16. Glicerina, V.; Siroli, L.; Betoret, E.; Canali, G.; Dalla, R.M.; Lanciotti, R.; Romani, S. Characterization and evaluation of the influence of an alginate, cocoa and a bilayer alginate-cocoa coating on the quality of fresh-cut oranges during storage. *J. Sci. Food Agric.* **2022**, *102*, 4454–4461. [CrossRef] [PubMed]
17. Huang, D.; Wang, C.; Zhu, S.; Anwar, R.; Hussain, Z.; Khadija, F. Combination of sodium alginate and nitric oxide reduces browning and retains the quality of fresh-cut peach during cold storage. *Food Sci. Technol. Int.* **2022**, *28*, 735–743. [CrossRef]
18. Patrignani, F.; Siroli, L.; Serrazanetti, D.I.; Gardini, F.; Lanciotti, R. Innovative strategies based on the use of essential oils and their components to improve safety, shelf-life and quality of minimally processed fruits and vegetables. *Trends Food Sci. Technol.* **2015**, *46*, 311–319. [CrossRef]
19. Zhang, Y.; Ma, Q.; Critzer, F.; Davidson, P.M.; Zhong, Q. Organic thyme oil emulsion as an alternative washing solution to enhance the microbial safety of organic cantaloupes. *Food Control* **2016**, *67*, 31–38. [CrossRef]
20. Sarengaowa; Hu, W.Z.; Jiang, A.L.; Xiu, Z.L.; Feng, K. Effect of thyme oil–alginate-based coating on quality and microbial safety of fresh-cut apples. *J. Sci. Food Agric.* **2018**, *98*, 2302–2311. [CrossRef]
21. Fatma, G.; Mouna, B.F.; Mondher, M.; Ahmed, L. In-vitro assessment of antioxidant and antimicrobial activities of methanol extracts and essential oil of *Thymus hirtus* sp. *algeriensis*. *Lipids Health Dis.* **2014**, *13*, 114. [CrossRef]
22. Abers, M.; Schroeder, S.; Goelz, L.; Sulser, A.; St. Rose, T.; Puchalski, K.; Langland, J. Antimicrobial activity of the volatile substances from essential oils. *BMC Complement. Med. Ther.* **2021**, *21*, 124. [CrossRef]
23. Ibrahim, D.; Abdelfattah-Hassan, A.; Badawi, M.; Ismail, T.A.; Bendary, M.M.; Abdelaziz, A.M.; Mosbah, R.A.; Mohamed, D.I.; Arisha, A.H.; El-Hamid, M.I.A. Thymol nanoemulsion promoted broiler chicken's growth, gastrointestinal barrier and bacterial community and conferred protection against Salmonella Typhimurium. *Sci. Rep.* **2021**, *11*, 7742. [CrossRef] [PubMed]

24. Zehra, A.; Wani, S.M.; Jan, N.; Bhat, T.A.; Rather, S.A.; Malik, A.R.; Hussain, S.Z. Development of chitosan-based biodegradable films enriched with thyme essential oil and additives for potential applications in packaging of fresh collard greens. *Sci. Rep.* **2022**, *12*, 16923. [CrossRef] [PubMed]
25. Hassan, F.A.S.; Ali, E.F.; Mostafa, N.Y.; Mazrou, R. Shelf-life extension of sweet basil leaves by edible coating with thyme volatile oil encapsulated chitosan nanoparticles. *Int. J. Biol. Macromol.* **2021**, *177*, 517–525. [CrossRef]
26. Pirozzi, A.; Del Grosso, V.; Ferrari, G.; Donsì, F. Edible coatings containing oregano essential oil nanoemulsion for improving postharvest quality and shelf life of tomatoes. *Foods* **2020**, *9*, 1605. [CrossRef]
27. Zhou, W.; He, Y.; Liu, F.; Liao, L.; Huang, X.; Li, R.; Zou, Y.; Zhou, L.; Zou, L.; Liu, Y.; et al. Carboxymethyl chitosan-pullulan edible films enriched with galangal essential oil: Characterization and application in mango preservation. *Carbohydr. Polym.* **2021**, *256*, 117579. [CrossRef]
28. Piechowiak, T.; Skóra, B. Edible coating enriched with cinnamon oil reduces the oxidative stress and improves the quality of strawberry fruit stored at room temperature. *J. Sci. Food Agric.* **2023**, *103*, 2389–2400. [CrossRef] [PubMed]
29. Sarengaowa; Hu, W.; Feng, K.; Xiu, Z.; Jiang, A.; Lao, Y. Thyme oil alginate-based edible coatings inhibit growth of pathogenic microorganisms spoiling fresh-cut cantaloupe. *Food Biosci.* **2019**, *32*, 100467. [CrossRef]
30. Joshua, P.V.; Di, L.; Linda, J.H.; Donald, W.S.; Michelle, D.D. Fate of *Escherichia coli* O157:H7, *Listeria monocytogenes*, and *Salmonella* on fresh-cut celery. *Food Microbiol.* **2013**, *34*, 151–157.
31. Azarakhsh, N.; Osman, A.; Ghazali, H.M.; Tan, C.P.; Adzahan, N.M. Lemongrass essential oil incorporated into alginate-based edible coating for shelf-life extension and quality retention of fresh-cut pineapple. *Postharvest Biol. Technol.* **2014**, *88*, 1–7. [CrossRef]
32. Bhat, T.A.; Rather, A.H.; Hussain, S.Z.; Naseer, B.; Qadri, T.; Nazir, N. Efficacy of ascorbic acid, citric acid, ethylenediaminetetraacetic acid, and 4-hexylresorcinol as inhibitors of enzymatic browning in osmo-dehydrated fresh cut kiwis. *J. Food Meas. Charact.* **2021**, *15*, 4354–4370. [CrossRef]
33. Peryam, D.R.; Pilgrim, F.J. Hedonic scale method of measuring food preferences. *Food Technol.* **1957**, *11*, 9–14.
34. Harich, M.; Maherani, B.; Salmieri, S.; Lacroix, M. Evaluation of antibacterial activity of two natural bio-preservatives formulations on freshness and sensory quality of ready to eat (RTE) foods. *Food Control* **2018**, *85*, 29–41. [CrossRef]
35. Gómez, P.L.; Salvatori, D.M. Pulsed light treatment of cut apple: Dose effect on color, structure, and microbiological stability. *Food Bioprocess Technol.* **2012**, *5*, 2311–2322. [CrossRef]
36. Siroli, L.; Patrignani, F.; Serrazanetti, D.I.; Tabanelli, G.; Montanari, C.; Gardini, F.; Lanciotti, R. Lactic acid bacteria and natural antimicrobials to improve the safety and shelf-life of minimally processed sliced apples and lamb's lettuce. *Food Microbiol.* **2015**, *47*, 74–84. [CrossRef]
37. Chen, C.; Hu, W.; He, Y.; Jiang, A.L.; Zhang, R.D. Effect of citric acid combined with UV-C on the quality of fresh-cut apples. *Postharvest Biol. Technol.* **2016**, *111*, 126–131. [CrossRef]
38. Jin, P.; Wang, S.Y.; Gao, H.; Chen, H.; Zheng, Y.; Wang, C.Y. Effect of cultural system and essential oil treatment on antioxidant capacity in raspberries. *Food Chem.* **2012**, *132*, 399–405. [CrossRef]
39. Wang, S.Y.; Chen, C.T.; Sciarappa, W.; Wang, C.Y.; Camp, M.J. Fruit quality, antioxidant capacity, and flavonoid content of organically and conventionally grown blueberries. *J. Agric. Food Chem.* **2008**, *56*, 5788–5794. [CrossRef]
40. Guerreiro, A.C.; Gago, C.M.L.; Faleiro, M.L.; Miguel, M.G.C.; Antunes, M.D.C. The effect of edible coatings on the nutritional quality of 'Bravo de Esmolfe' fresh-cut apple through shelf-life. *LWT-Food Sci. Technol.* **2017**, *75*, 210–219. [CrossRef]
41. Arnon-Rips, H.; Porat, R.; Poverenov, E. Enhancement of agricultural produce quality and storability using citral-based edible coatings; the valuable effect of nano-emulsification in a solid-state delivery on fresh-cut melons model. *Food Chem.* **2019**, *277*, 205–212. [CrossRef]
42. Cofelice, M.; Lopez, F.; Cuomo, F. Quality control of fresh-cut apples after coating application. *Foods* **2019**, *8*, 189. [CrossRef] [PubMed]
43. Solís-Contreras, G.A.; Rodríguez-Guillermo, M.C.; de la Luz Reyes-Vega, M.; Aguilar, C.N.; Rebolloso-Padilla, O.N.; Corona-Flores, J.; de Abril Alexandra Soriano-Melgar, L.; Ruelas-Chacon, X. Extending shelf-life and quality of minimally processed golden delicious apples with three bioactive coatings combined with cinnamon essential oil. *Foods* **2021**, *10*, 597. [CrossRef] [PubMed]
44. Rizzo, V.; Lombardo, S.; Pandino, G.; Barbagallo, R.N.; Mazzaglia, A.; Restuccia, C.; Mauromicale, G.; Muratore, G. Shelf-life study of ready-to-cook slices of globe artichoke 'Spinoso sardo': Effects of anti-browning solutions and edible coating enriched with Foeniculum vulgare essential oil. *J. Sci. Food Agric.* **2019**, *99*, 5219–5228. [CrossRef] [PubMed]
45. Gong, W.; Shi, B.; Zeng, F.K.; Dong, N.; Lei, Z.; Liu, J. Evaluation of cooking, nutritional, and quality characteristics of fresh-cut potato slice pretreated with acetic acid. *J. Food Sci.* **2022**, *87*, 427–437. [CrossRef]
46. Ali, S.; Anjum, M.A.; Nawaz, A.; Naz, S.; Hussain, S.; Ejaz, S.; Sardar, H. Effect of pre-storage ascorbic acid and Aloe vera gel coating application on enzymatic browning and quality of lotus root slices. *J. Food Biochem.* **2020**, *44*, e13136. [CrossRef]
47. Salvia-Trujillo, L.; Rojas-Graü, M.; Soliva-Fortuny, R.; Martín-Belloso, O. Use of antimicrobial nanoemulsions as edible coatings: Impact on safety and quality attributes of fresh-cut Fuji apples. *Postharvest Biol. Technol.* **2015**, *105*, 8–16. [CrossRef]
48. Bøjer Rasmussen, C.; Enghild, J.J.; Scavenius, C. Identification of polyphenol oxidases in potato tuber (*Solanum tuberosum*) and purification and characterization of the major polyphenol oxidases. *Food Chem.* **2021**, *365*, 130454. [CrossRef]

49. Sarengaowa; Wang, L.; Liu, Y.; Yang, C.; Feng, K.; Hu, W. Screening of Essential Oils and Effect of a chitosan-based edible coating containing cinnamon oil on the quality and microbial safety of fresh-cut potatoes. *Coatings* **2022**, *12*, 1492. [CrossRef]
50. Kraśniewska, K.; Kosakowska, O.; Pobiega, K.; Gniewosz, M. The influence of two-component mixtures from spanish origanum oil with spanish marjoram oil or coriander oil on antilisterial activity and sensory quality of a fresh cut vegetable mixture. *Foods* **2020**, *9*, 1740. [CrossRef]
51. Brasil, I.M.; Gomes, C.; Puerta-Gomez, A.; Castell-Perez, M.E.; Moreira, R.G. Polysaccharide-based multilayered anti-microbial edible coating enhances quality of fresh-cut papaya. *LWT-Food Sci. Technol.* **2012**, *47*, 39–45. [CrossRef]
52. Sánche ez-González, L.; Vargas, M.; González-Martínez, C.; Chiralt, A.; Cháfer, M. Use of essential oils in bioactive edible coatings: A review. *Food Eng. Rev.* **2011**, *3*, 1–16. [CrossRef]
53. Fagundes, C.; Moraes, K.; Pérez-Gago, M.B.; Palou, L.; Maraschin, M.; Monteiro, A.R. Effect of active modified atmosphere and cold storage on the postharvest quality of cherry tomatoes. *Postharvest Biol. Technol.* **2015**, *109*, 73–81. [CrossRef]
54. Chong, J.; Lai, S.; Yang, H. Chitosan combined with calcium chloride impacts fresh-cut honeydew melon by stabilising nanostructures of sodium-carbonate-soluble pectin. *Food Control* **2015**, *53*, 195–205. [CrossRef]
55. Tamminga, S.K.; Beumer, R.R.; Keijbets, M.J.H.; Kampelmacher, E.M. Microbial spoilage and development of food poisoning bacteria in peeled, completely or partly cooked vacuum-packed potatoes. *Arch. Lemensmittelhyg.* **1976**, *29*, 215–219.
56. Rashid, M.H.; Khan, M.R.; Roobab, U.; Rajoka, M.S.R.; Inam-ur-Raheem, M.; Anwar, R.; Ahmed, W.; Jahan, M.; Ijaz, M.R.A.; Asghar, M.M.; et al. Enhancing the shelf stability of fresh-cut potatoes via chemical and nonthermal treatments. *J. Food Process Pres.* **2021**, *45*, e15582. [CrossRef]
57. Hashemi, M.; Dastjerdi, A.M.; Shakerardekani, A.; Mirdehghan, S.H. Effect of alginate coating enriched with Shirazi thyme essential oil on quality of the fresh pistachio (*Pistacia vera* L.). *J. Food Sci. Technol.* **2021**, *58*, 34–43. [CrossRef]
58. Naeem, A.; Abbas, T.; Ali, T.M.; Hasnain, A. Effect of guar gum coatings containing essential oils on shelf life and nutritional quality of green-unripe mangoes during low temperature storage. *Int. J. Biol. Macromol.* **2018**, *113.*, 403–410. [CrossRef] [PubMed]
59. Inouye, S.; Takizawa, T.; Yamaguchi, H. Antibacterial activity of essential oils and their major constituents against respiratory tract pathogens by gaseous contact. *J. Antimicrol. Chemoth.* **2001**, *47*, 565–573. [CrossRef]
60. Vaclav, V.; Jana, V. Essential oils from thyme (*Thymus vulgaris*): Chemical composition and biological effects in mouse model. *J. Med. Food* **2016**, *19*, 1180–1187.
61. Burt, S. Essential oils: Their antibacterial properties and potential applications in foods-a review. *Int. J. Food Microb.* **2004**, *94*, 223–253. [CrossRef]
62. Feng, K.; Hu, W.Z.; Jiang, A.L.; Sarengaowa; Xu, Y.P.; Ji, Y.R.; Shao, W.J. Growth of *Salmonella* spp. and *Escherichia coli* O157:H7 on fresh-cut fruits stored at different temperatures. *Foodborne Pathog. Dis.* **2017**, *14*, 510–517. [CrossRef] [PubMed]
63. IFST. *Development and Use of Microbiological Criteria for Foods*; Institute of Food Science and Technology: London, UK, 1999; p. 76.

horticulturae

MDPI

Article

Effect of Wounding Intensity on Edible Quality by Regulating Physiological and ROS Metabolism in Fresh-Cut Pumpkins

Wenzhong Hu [1,*,†], Yuge Guan [2,†], Yi Wang [3] and Ning Yuan [3]

1 School of Pharmacy and Food Science, Zhuhai College of Science and Technology, Zhuhai 519041, China
2 School of Food and Health, Zhejiang Agricultural and Forestry University, Hangzhou 311300, China
3 College of Life Science, Dalian Minzu University, Dalian 116600, China
* Correspondence: wenzhongh@sina.com
† These authors contributed equally to this work.

Abstract: Fresh-cut pumpkin is favored by consumers for its environmental protection, safety, and convenience at home and abroad. To investigate the effect of different wounding intensities (piece, strip and slice, corresponding to 1.90, 3.53 and 6.29 $m^2\ kg^{-1}$) on the quality of fresh-cut pumpkin, the critical indexes involved in reactive oxygen species (ROS) metabolism, vitamin C-glutathione cycle, phenylpropanoid metabolism and membrane lipid peroxidation were monitored for pumpkin during storage at 4 °C for 6 d. The results showed that with the increase in cutting injury strength, the lightness, whiteness index, respiration rate, ethylene content, lipoxygenase activity and malondialdehyde content of fresh-cut pumpkin increased, while the hardness, sensory quality, appearance and total soluble solid content continuously decreased. The quality deterioration was the most severe in the slice group, while a higher sensory quality was maintained in the piece after 6 d of storage. However, the activity of phenylalanine ammonia-lyase increased and then contributed to the synthesis of the phenolic compound, which resulted in enhancements of 79.13%, 29.47% and 16.14% in piece, strip and slice, respectively. Meanwhile, cutting enhanced the activity of antioxidant enzymes including ascorbate peroxidase, glutathione reductase, superoxide dismutase and catalase, resulting in the enhancement of antioxidant activity in fresh-cut pumpkin. The collected results showed that the wounding intensities have an obvious influence on the quality by regulating physiological and ROS metabolism.

Keywords: fresh-cut pumpkins; wounding intensities; reactive oxygen species; edible quality

Citation: Hu, W.; Guan, Y.; Wang, Y.; Yuan, N. Effect of Wounding Intensity on Edible Quality by Regulating Physiological and ROS Metabolism in Fresh-Cut Pumpkins. *Horticulturae* **2023**, *9*, 512. https://doi.org/10.3390/horticulturae9040512

Academic Editor: Isabel Lara

Received: 19 March 2023
Revised: 12 April 2023
Accepted: 13 April 2023
Published: 20 April 2023

1. Introduction

Pumpkin (*Cucurbita moschata* Duch.) is one of the most common vegetables in the life of consumers, and it is favored by worldwide consumers due to its prominent yellow color and nutritional and functional compounds, such as anthocyanins, phenolic acid derivatives, flavonoids and vitamins [1]. Fresh pumpkin has a large shape and is not convenient for consumers to eat, so the market sale of pumpkin is greatly limited, and the pumpkin product subjected to fresh-cutting treatment can solve the problem. Fresh-cut pumpkins have the characteristics of convenience, safety, nutrition and freshness [2]. Simultaneously, the skin residue, core and other waste material produced via fresh-cut processing can also be recycled and reused for animal feed, which has the characteristics of reducing urban domestic waste and environmental protection [3,4]. In recent years, fresh-cut pumpkins have been increasingly favored by consumers due to their convenient characteristics of ready-to-eat, ready-to-use and ready-to-cook [5,6]. However, cutting causes pumpkins to suffer irreparable mechanical damage and lose their protective tissue, which may induce quality deterioration reactions [7].

In general, when fruits and vegetables are in fresh-cutting processing, a systematic responsive mechanism would be activated. Previous studies have reported that wounding stress could stimulate the production of reactive oxygen species (ROS) [2]. A study on fresh-cut carrots suggested that ROS may increase the resistance of fresh-cut products by inducing the synthesis and accumulation of secondary metabolites, including polyphenols and flavonoids [8,9]. Chai et al. [10] found that excessive ROS accumulation will break the balance of ROS metabolism in plants, increase the burden of scavenging free radicals and lead to browning and membrane lipid peroxidation, which further reduces the edible quality of fresh-cut products. Therefore, it is essential to regulate ROS metabolism aimed to maintain the quality of fresh-cut products.

It has been reported that cutting styles corresponding to different wounding intensity have an impact on the production of ROS signal in fresh-cut products, and then contribute to the difference in enzymatic reaction and the phenylpropanoid metabolic pathway [11–13]. Generally, during the processing of fresh-cut pumpkins, different cutting methods subjected to piece, strip and slice are necessary for the food industry or at home. However, there is still lacking scientific certification and systematic research on whether the different wounding intensities affect the quality of fresh-cut pumpkins. Moreover, whether ROSs are involved in wound-induced quality change in fresh-cut pumpkins is still unknown. Therefore, three different cutting methods (piece, strip and slice) on the physiological metabolism and quality of fresh-cut pumpkins were evaluated in this study to select the optimal wounding intensities of production and consumption. Meanwhile, the regulation mechanism of ROS on edible quality change of fresh-cut pumpkins was explored, which further provides a certain theoretical basis for fresh-cut pumpkin during processing and storage.

2. Materials and Methods

2.1. Sample Preparation and Treatment

Our study was conducted from 10 March 2022 to 16 June 2022 in the greenhouse at the Dalian Shengyuan Farmer, PR China (39°15′58.72″ N, 121°55′41.41″ E). Sixty seeds for the Luli pumpkin were sowed on 10 March 2022 in a greenhouse in plastic seedling trays containing potting mix and starter fertilizer in Shengyuan farm, located in Jinzhou District, Dalian, China. At two weeks old, the seedlings of each germplasm line were transplanted on 26 March 2022 into a single plot in the field. The mature tropical pumpkins were harvested at mature green stage (65 days after flowering).

The pumpkin (*Cucurbita moschata,* Duch.) variety named "Luli" was harvested in July 2022. Fresh pumpkins were sorted to ensure uniformity of size, color and ripeness and were maintained at 4 °C before processing. Whole samples were thoroughly washed with tap water and then soaked in sodium hypochlorite (0.2 mL L^{-1}) solution for 2 min for sterilization. After being rinsed twice with water, the selected pumpkins were cut into piece, strip and slice. The wounded surface diameter of piece, strip and slice was manually measured, and then wounding intensities were calculated and found to be 1.90, 3.53 and 6.29 m^2 kg^{-1}, respectively. Fresh-cut pumpkins (120 g) were loaded into a polypropylene container and packaged with polyethylene films (Miuge Chemical Commodities Science and Technology Co., Ltd., Hangzhou, China), then stored at 4 °C. The whole pumpkin was used as the control in this experiment; all samples were collected at 0, 1, 2, 3, 4, 5 and 6 d. In this experiment, 6 of the samples were taken for analysis of lightness, whiteness index, respiration rate and ethylene content, and the other samples were frozen with liquid nitrogen and stored at −80 °C for further analysis.

2.2. Lightness, Whiteness Index and Appearance Assay

Tristimulus color values including lightness (L^*), redness (a^*) and yellowness (b^*) of fresh-cut and control samples were determined by using a HunterLab MiniScan@XE plus colorimeter (Hunter Associates Laboratory, Reston, VA, USA); the white and black plank was used as the calibrate standard in this work. For fresh-cut pumpkins, three samples from each group were chosen to measure surface color, taking an average of three surface measurements per sample.

The whiteness index (WI) value was estimated using the formula as shown below, $WI = 100 - [(100 - L^*)^2 + (a^*)^2 + (b^*)^2]^{1/2}$ [14], where L^*, a^* and b^* are the color values of lightness, redness and yellowness of samples. The change in the appearance of fresh-cut pumpkins during storage was recorded by obtaining pictures.

2.3. Respiration Rate and Ethylene Content Assay

The respiration rate and ethylene content were determined via a gas analyzer (F-940 STORE II, Felix, USA) according to the method reported by Zhou et al. [15]. The respiration rate of fresh-cut pumpkin was calculated as follows:

$$\text{Respiration rate} = \frac{(C_1 - C_2)/100 * V}{m * t} * 1.96$$

where C_1 represents the volume fraction of CO_2 in the container (%); C_2 represents the volume fraction of CO_2 in the air (%); V is the volume of the container (mL); m represents the weight of the sample (kg); t represents time (h); 1.96 is the density of CO_2 (g L^{-1}).

The ethylene content (μL kg^{-1} h^{-1}) of the sample was calculated as follows:

$$\text{Ethylene content} = \frac{(C_1 - C_2) * V}{m * t * 1000}$$

where C_1 represents the volume fraction of ethylene in the container (%); C_2 represents the volume fraction of ethylene in the air (%); V is the volume of the container (mL); m represents the weight of the sample (kg); t represents time (h).

2.4. $O_2^{-\bullet}$ and H_2O_2 Content Assay

The $O_2^{-\bullet}$ and H_2O_2 content were measured by using commercially available kits (Shanghai Yuanye Biotechnology Co., Ltd., Shanghai, China). The absorbance of $O_2^{-\bullet}$ and H_2O_2 reaction system was monitored at 530 nm and 412 nm, respectively, and then the result was expressed as mmol kg^{-1}.

2.5. Malondialdehyde (MDA) Content and Lipoxygenase (LOX) Activity Assay

The MDA content was determined by using the color reaction between MDA and thiobarbituric acid, the measurement method was in accordance with the plant kit (www.cominbio.com, accessed on 15 September 2022) and the result was expressed in mmol kg^{-1}.

The MDA content was determined by using the color reaction between MDA and thiobarbituric acid by using the plant kit (Coming Biotechnology Co., Ltd., Suzhou, China). The MDA content was determined according to the instructions of the manufacturer [16], and the result was expressed mmol kg^{-1}.

LOX catalyzes the oxidation of linoleic acid, the oxidation product has a characteristic absorption peak at 280 nm, and the LOX activity was calculated by measuring the increase rate at 280 nm absorbance. The determination method was in accordance with the plant kit (Coming Biotechnology Co., Ltd., Suzhou, China) [17]; the result was expressed in U kg^{-1}, where $U_{LOX} = 0.01 \times \Delta A_{280}$ nm per min based on fresh tissue weight.

2.6. Lignin Content and Phenylalanine Ammonia-Lyase (PAL) Activity Assay

The samples were dried to constant weight at 80 °C, and then crushed and sieved through a 40 mesh sieve, which was prepared for the extraction of lignin. The lignin content was measured with plant kits (Coming Biotechnology Co., Ltd., Suzhou, China), the method was in accordance with the instructions of the manufacturer [18] and the result was expressed as g kg^{-1} based on fresh weight.

The PAL activity was assayed as described by Guan et al. [6]. Frozen samples (1.0 g) were extracted in 5 mL of ice-cold sodium borate buffer consisting of 5 mmol L^{-1} β-mercaptoethanol. The 2.5 mL reaction system comprised borate buffer (2 mL), L-phenylalanine (1 mL; 20 mmol L^{-1}) and extracted supernatant (0.5 mL). The absorbance change was measured at 290 nm and one U of PAL was described as the amount of enzyme that causes a 0.01 absorbance change per minute.

2.7. Antioxidant Activity Assay

The antioxidant activity was assessed by analyzing the hydroxyl radical ($OH^•$), 2,2′-azinobis-3-ethylbenzthiazoline-6-sulphonate (ABTS) scavenging ability and ferric reducing antioxidant power (FRAP) by using plant kits (Nanjing Jiancheng Bioengineering Institute Co., Ltd., Nanjing, China). The $OH^•$, ABTS and FRAP scavenging ability was determined according to the instructions of the manufacturer; the specific steps were shown at the company home page [19–21]. The result of $OH^•$ and ABTS scavenging ability was expressed as the formula: scavenging ability (%) = $100 \times (A_{control} - A_{sample})/A_{control}$, where $A_{control}$ and A_{sample} are the absorbances of the control and of the sample, respectively. The FRAP result was expressed as mmol kg^{-1} based on a Trolox calibration curve.

2.8. Antioxidant Enzyme Activity Assay

The pumpkin samples were extracted using ice-cold sodium phosphate buffer (0.2 mol L^{-1}, pH 6.4) and then centrifuged at 12,000× *g* for 30 min at 4 °C, the collected supernatant was prepared to measure polyphenol oxidase (PPO), peroxidase (POD) and catalase (CAT) activity. The PPO, POD and CAT activity was determined as reported by Tian et al. [22] and Chen et al. [23], and the results were expressed as U kg^{-1}.

The activity of superoxide dismutase (SOD), ascorbate peroxidase (APX) and glutathione reductase (GR) was measured using kits (Comin Biotechnology Co., Ltd., Suzhou, China). The SOD activity was determined according to the instructions of the manufacturer [24–26]. The results were expressed as U kg^{-1}, where $U_{SOD} = 0.01 \times \Delta_{450}$ nm per min, $U_{APX} = 0.01 \times \Delta_{290}$ nm per min and $U_{GR} = 0.01 \times \Delta_{340}$ nm per min.

2.9. Antioxidant Compounds and Total Soluble Solid (TSS) Content Assay

The antioxidant compounds of pumpkin including phenols, flavonoids, anthocyanin, Vitamin C (Vc) and glutathione (GSH) content affected by wounding intensities were evaluated in this study. For total phenol content evaluation, the frozen pumpkin sample (5 g) was mixed with 20 mL 80% ethanol and extracted in the dark at 40 °C for 40 min and the obtained supernatant was prepared to determine the value, expressed as mg kg^{-1} based on the gallic acid standard curve [6].

The method of flavonoid content measurement was in accordance with Santana-Galvez [7]. Fresh tissues (0.5 g) were mixed with 5 mL of 99% HCl-methanol for 20 min at 4 °C. The absorbance changes of the reaction system were measured at 325 nm to determine total flavonoid content based on the rutin equivalent standard curve (g kg^{-1}).

Then, the content of anthocyanin, Vc and GSH was determined by using plant kits (Solarbio Science & Technology Co, Ltd., Beijing, China) in accordance with the manufacturer's instructions. The absorbance of anthocyanin, Vc and GSH was measured at 500 nm, 420 nm and 412 nm, respectively. Anthocyanin, Vc and GSH content are expressed as g kg^{-1}, mg kg^{-1} and mmol kg^{-1}, respectively.

TSS content was assayed according to the process of Li et al. [8] by using a hand-held digital refractometer (PAL-1, ATAGO, Tokyo, Japan), expressed as %.

2.10. Statistical Analysis

All the experiments were performed in triplicates, and data were presented as the mean ± SD (standard deviation). Orignpro 2021 software (OriginLab., Northampton, MA, USA) was used for the drawing of figures. One-way analysis of variance (ANOVA) and least significant difference (LSD) tests via SPSS software, version 20 (IBM Corp., Armonk, NY, USA) were used to differentiate mean values at the $p < 0.05$ level.

3. Results and Discussion

3.1. Colorimetric Values and Appearance

The colorimetric value evaluation involved the important quality parameters of food [27]. In order to quickly and intuitively illuminate the visual color change of fresh-cut pumpkin during storage at 4 °C, we measured the lightness (L^*) as shown in Figure 1a. The result revealed that the L^* of cutting treatment groups increased, and the rising trend of color difference L^* value in a whole pumpkin was significantly lower than that in other fresh-cutting treatment groups. On storage for 6 d, the L^* of whole, piece, strip and slice was 74.9, 78.6, 79.2 and 81.07, respectively, which indicated that the degree of L^* level was strip > piece > slice > whole (Figure 1a); the above results show that with the increase in the wounding intensities, the color difference L^* value of the fresh-cut pumpkin has an increasing trend. Furthermore, the whiteness index (WI) was used to indicate the degree of blanching of fresh-cut pumpkins during storage. According to Figure 1b, the WI was increased during the whole storage period. On storage for 6 d, the WI of the slice was reached at the maximum value (56.79), which was 1.69 times higher than the control group. The result of the sensory evaluation (Figure 1c) also demonstrated that the score of appearance, color, taste, texture and overall appearance in slice was lower than strip and piece, which indicated the piece pumpkin had the best visible quality.

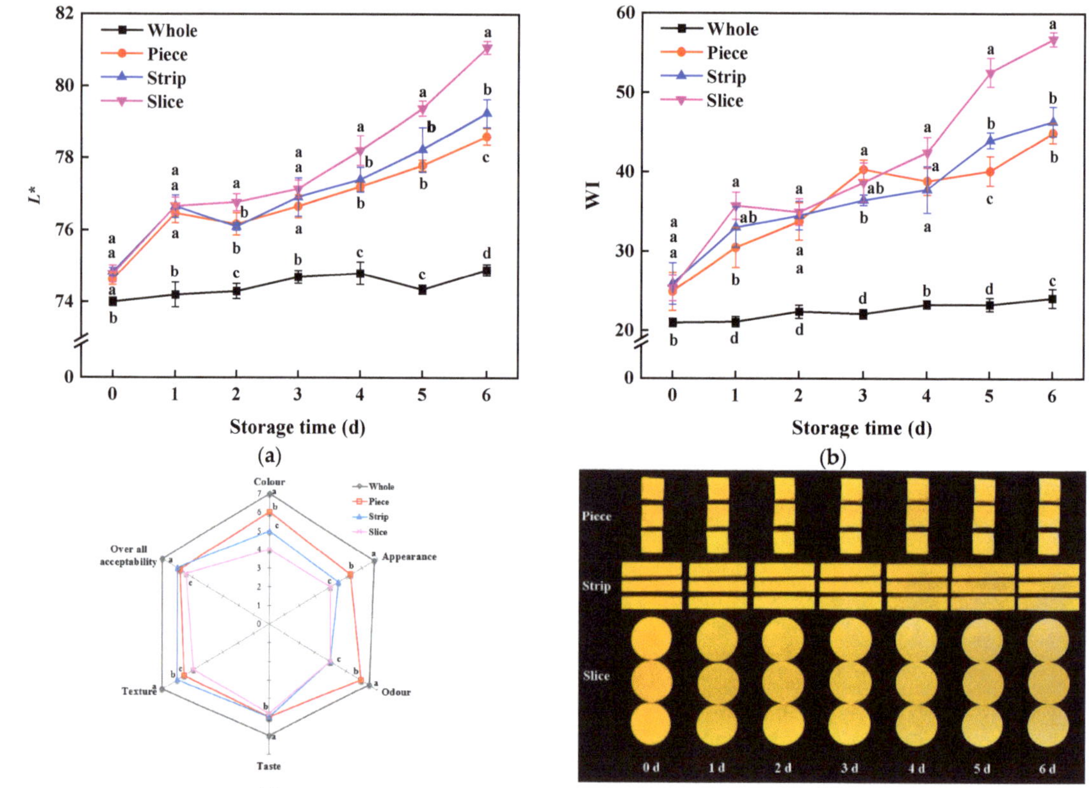

Figure 1. Effect of wounding intensities on lightness (L^*) (**a**), whiteness index (WI) (**b**), sensory evaluation (**c**) and appearance (**d**) of fresh-cut pumpkin during storage at 4 °C for 6 days. Vertical bars represent the standard deviation of the mean (n = 3). Different letters indicate significant differences between the groups ($p < 0.05$).

The phenomenon of whitening on the cut surface of pumpkin was easy to observe based on the picture of the appearance (Figure 1d); it was also found in fresh-cut carrots that the surface is apt to white blush [28]. The reason for this may be that the color compounds of fresh-cutting processed fruits and vegetables changed during storage due to incomplete inactivation of enzymes, which would result in the white blush on the surface [29]. However, this result was opposite to the quality change of fresh-cut potatoes [11] and apples [30] where the degree of browning phenomenon was deep. The above results also showed that the color changes were significantly different in fresh fruits and vegetables with different biological characteristics after being sliced.

3.2. Weight Loss and Total Soluble Solid (TSS) Content

Weight loss was increased significantly for fresh-cut pumpkin (Figure 2a). Among these groups, the weight loss of sliced pumpkin was highest (1.49%), followed by strip and piece. As for TSS content in fresh-cut pumpkin, it was affected by wounding intensities, significantly decreasing during the whole storage period (Table 1), yet it surprisingly remained almost constant in whole pumpkin during storage. Contrary to the weight loss, the TSS content of the slice group was lowest; it decreased by 32.49% at 7 d compared with the original values. A significant negative correlation (R = −0.885) was observed between weight loss and TSS content, which indicated that fresh-cutting operations easily cause water loss and further resulted in TSS content change in fresh-cut pumpkin [31,32].

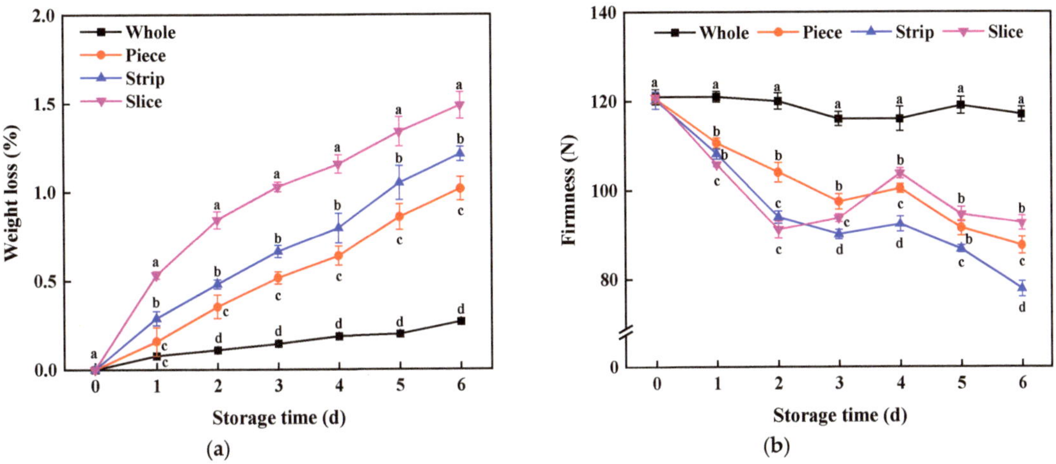

Figure 2. Effect of wounding intensities on weight loss (**a**) and firmness (**b**) of fresh-cut pumpkin during storage at 4 °C for 6 days. Vertical bars represent the standard deviation of the mean (n = 3). Different letters indicate significant differences between the groups ($p < 0.05$).

Firmness is an important apparent property of fresh-cut products [33]; just after fresh-cutting, the firmness measured on pumpkin was 120.16 N, then it decreased sharply (Figure 2b). Notably, the firmness of piece, strip and slice showed a slight significant increase at 3–4 d. Probably, the cutting operation with wounding stress applied during the experiment resulted in wound healing and then provided an essential barrier for a compact superficial surface layer, which contributed to surface hardening [1].

Table 1. Influence of wounding intensities on the quality of fresh-cut pumpkin.

Index	Treatment	0 d	1 d	2 d	3 d	4 d	5 d	6 d
TSS (%)	Whole	16.00 ± 0.54 ^{a}	15.70 ± 0.20 ^{a}	15.30 ± 0.50 ^{a}	15.61 ± 0.06 ^{a}	15.13 ± 0.44 ^{a}	15.52 ± 0.11 ^{a}	16.03 ± 0.07 ^{a}
	Piece	15.65 ± 0.39 ^{a}	13.81 ± 0.11 ^{c}	12.45 ± 0.39 ^{b}	14.57 ± 0.34 ^{b}	14.23 ± 0.37 ^{b}	12.97 ± 0.11 ^{b}	12.54 ± 0.29 ^{b}
	Strip	15.87 ± 0.13 ^{a}	14.62 ± 0.40 ^{b}	11.67 ± 0.19 ^{c}	12.32 ± 0.33 ^{c}	13.12 ± 0.29 ^{c}	11.97 ± 0.13 ^{c}	11.87 ± 0.15 ^{c}
	Slice	15.45 ± 0.21 ^{a}	13.52 ± 0.16 ^{d}	11.55 ± 0.37 ^{c}	10.75 ± 0.05 ^{d}	12.22 ± 0.22 ^{d}	11.07 ± 0.07 ^{d}	10.43 ± 0.11 ^{d}
Total phenols content (g kg^{-1})	Whole	10.74 ± 0.42 ^{c}	10.19 ± 0.25 ^{c}	10.04 ± 0.11 ^{c}	9.92 ± 0.23 ^{d}	10.19 ± 0.44 ^{d}	10.35 ± 0.22 ^{d}	10.53 ± 0.22 ^{c}
	Piece	11.16 ± 0.26 ^{b}	12.01 ± 0.19 ^{b}	13.47 ± 0.21 ^{a}	12.74 ± 0.26 ^{c}	12.70 ± 0.38 ^{c}	12.02 ± 0.29 ^{c}	11.07 ± 0.32 ^{b}
	Strip	11.32 ± 0.35 ^{b}	12.32 ± 0.20 ^{ab}	12.41 ± 0.23 ^{b}	13.52 ± 0.21 ^{b}	14.58 ± 0.33 ^{b}	13.40 ± 0.22 ^{b}	11.61 ± 0.23 ^{b}
	Slice	12.07 ± 0.28 ^{a}	12.71 ± 0.31 ^{a}	13.92 ± 0.29 ^{a}	15.74 ± 0.04 ^{a}	16.30 ± 0.44 ^{a}	18.54 ± 0.24 ^{a}	12.49 ± 0.34 ^{a}
Flavonoids content (g kg^{-1})	Whole	5.95 ± 0.17 ^{a}	5.46 ± 0.21 ^{c}	5.57 ± 0.35 ^{c}	5.89 ± 0.16 ^{d}	5.45 ± 0.11 ^{d}	5.49 ± 0.31 ^{c}	5.58 ± 0.15 ^{b}
	Piece	5.83 ± 0.19 ^{a}	7.60 ± 0.25 ^{a}	7.58 ± 0.38 ^{b}	8.40 ± 0.19 ^{b}	8.97 ± 0.28 ^{b}	7.89 ± 0.18 ^{b}	7.08 ± 0.19 ^{a}
	Strip	5.63 ± 0.44 ^{a}	6.66 ± 0.12 ^{b}	7.37 ± 0.16 ^{b}	7.69 ± 0.14 ^{c}	7.79 ± 0.16 ^{c}	8.21 ± 0.11 ^{b}	7.20 ± 0.16 ^{a}
	Slice	5.98 ± 0.55 ^{a}	7.65 ± 0.24 ^{a}	8.46 ± 0.18 ^{a}	9.10 ± 0.41 ^{a}	11.04 ± 0.28 ^{a}	8.57 ± 0.07 ^{a}	7.44 ± 0.26 ^{a}
Anthocyanin content (g kg^{-1})	Whole	0.11 ± 0.04 ^{a}	0.13 ± 0.01 ^{c}	0.13 ± 0.03 ^{c}	0.11 ± 0.01 ^{c}	0.16 ± 0.01 ^{d}	0.13 ± 0.04 ^{c}	0.15 ± 0.01 ^{c}
	Piece	0.17 ± 0.03 ^{a}	0.19 ± 0.02 ^{ab}	0.21 ± 0.02 ^{b}	0.24 ± 0.01 ^{b}	0.55 ± 0.02 ^{b}	0.24 ± 0.03 ^{b}	0.22 ± 0.02 ^{b}
	Strip	0.15 ± 0.01 ^{a}	0.18 ± 0.02 ^{b}	0.20 ± 0.03 ^{ab}	0.28 ± 0.02 ^{a}	0.35 ± 0.01 ^{c}	0.45 ± 0.03 ^{a}	0.21 ± 0.02 ^{b}
	Slice	0.14 ± 0.02 ^{a}	0.24 ± 0.03 ^{a}	0.26 ± 0.03 ^{a}	0.31 ± 0.01 ^{a}	0.72 ± 0.01 ^{a}	0.50 ± 0.02 ^{a}	0.28 ± 0.03 ^{a}
Vitamin C (mg kg^{-1})	Whole	347.94 ± 1.55 ^{a}	350.30 ± 11.28 ^{a}	348.57 ± 5.77 ^{a}	333.85 ± 1.46 ^{a}	337.62 ± 11.41 ^{a}	334.68 ± 2.08 ^{a}	340.39 ± 1.42 ^{a}
	Piece	342.21 ± 10.40 ^{a}	345.30 ± 10.00 ^{a}	315.64 ± 5.13 ^{b}	315.76 ± 8.62 ^{b}	299.77 ± 7.64 ^{b}	287.70 ± 3.54 ^{b}	245.30 ± 2.03 ^{b}
	Strip	344.06 ± 1.85 ^{a}	275.20 ± 7.61 ^{c}	230.15 ± 2.76 ^{c}	218.55 ± 7.36 ^{c}	222.85 ± 7.80 ^{c}	231.59 ± 1.92 ^{c}	228.97 ± 5.30 ^{c}
	Slice	325.56 ± 3.11 ^{b}	322.27 ± 2.96 ^{b}	222.69 ± 5.04 ^{c}	165.27 ± 7.23 ^{d}	159.38 ± 13.80 ^{d}	179.79 ± 3.88 ^{d}	193.57 ± 7.68 ^{d}
Glutathione (mmol kg^{-1})	Whole	0.23 ± 0.004 ^{a}	0.20 ± 0.015 ^{c}	0.25 ± 0.011 ^{d}	0.27 ± 0.007 ^{d}	0.24 ± 0.012 ^{d}	0.25 ± 0.015 ^{d}	0.26 ± 0.015 ^{d}
	Piece	0.20 ± 0.006 ^{c}	0.15 ± 0.007 ^{d}	0.29 ± 0.005 ^{c}	0.52 ± 0.015 ^{c}	0.52 ± 0.006 ^{c}	0.35 ± 0.007 ^{c}	0.43 ± 0.014 ^{c}
	Strip	0.22 ± 0.004 ^{b}	0.24 ± 0.011 ^{b}	0.41 ± 0.004 ^{b}	0.68 ± 0.002 ^{b}	0.75 ± 0.016 ^{b}	0.58 ± 0.009 ^{b}	0.58 ± 0.005 ^{b}
	Slice	0.23 ± 0.017 ^{ab}	0.33 ± 0.003 ^{a}	0.54 ± 0.004 ^{a}	0.88 ± 0.006 ^{a}	0.80 ± 0.016 ^{a}	0.77 ± 0.006 ^{a}	0.87 ± 0.014 ^{a}

Note: Influence of wounding intensities on TSS, total phenols, flavonoids, anthocyanin, vitamin C and glutathione content of fresh-cut pumpkin during 6 d of storage at 4 °C. Data are expressed as the mean ± SD. Values with different letters were significantly different at $p < 0.05$. Lowercase letters represented significant difference among treatment factors.

3.3. *Effect of Wounding Intensities on Physiological Metabolism of Fresh-Cut Pumpkins*

3.3.1. Respiration Rate and Ethylene Content

Respiratory rate is an important condition for maintaining the freshness and shelf life of produce [34]. In this study, the trend for respiration rate of slice, strip and piece was increased first and reached the peak value on the 3rd day, and increased 1.83, 2.26, and 2.55 times compared with the whole sample (Figure 3a), revealing that the cutting operation treatment caused wounding stress in fresh tissues and further resulted in triggering respiratory metabolism of fresh-cut pumpkin [35]. Similar results were found in fresh-cut potatoes [15], apples [23] and cabbage [27]; all these results indicate that cutting causes fruit and vegetables to suffer irreparable mechanical damage and lose their protective tissue.

As for ethylene, an important plant signal molecule, it was inspired by wounding stress in pumpkins. The highest ethylene content in slice, strip and piece was 14.32, 5.72 and 1.29-fold higher than the control, respectively. During the whole storage period, the enhancement of ethylene content was as follows: slice > strip > piece ($p < 0.05$, Figure 3b). The result of a significant positive correlation (R = 0.636) between ethylene and respiration rate confirmed that ethylene plays a crucial role in regulating the respiratory process during the processing of fresh-cut pumpkin, and this may be an important reason that directly affected the pumpkin tissue senescence [36,37].

3.3.2. Phenylpropane Metabolism-Related Parameters

When plants are subjected to mechanical damage and stress, the comprehensive defense system would be induced [2,38]. Among these, the phenylpropanoid pathway plays a critical role in the process of resisting wounding stress, by activating the production of beneficial secondary metabolites including phenolics, flavonoids and lignin [39]. The previous study of fresh-cut cabbage [27], potato [11], carrot [40] and pitaya [10] illustrated that wounding stress enhances phenolic and flavonoid content, whereas the effects of different wounding intensities on these secondary metabolites in pumpkins are scarcely reported. In this study, we found that the wounding stress increased the total phenol contents in pumpkin (Table 1). Total phenol content of slice, strip and piece increased by 79.13%, 29.47% and 16.14%, respectively, at 5 d compared with the whole pumpkin (Table 1). As one of the important biological compounds in plants, flavonoids have been reported to contain important antioxidant substances [41,42]. In this study, we found that the flavonoid content increased first, at 4 d; the total flavonoid content of slice, strip and piece pumpkins increased 1.03-fold, 0.43-fold and 0.65-fold, respectively, compared with the control group (Table 1). It can be clearly seen from the above results that total phenol and flavonoid content was higher with the increasing wounding intensity, similar results were found in fresh-cut broccoli [6] and potato [11].

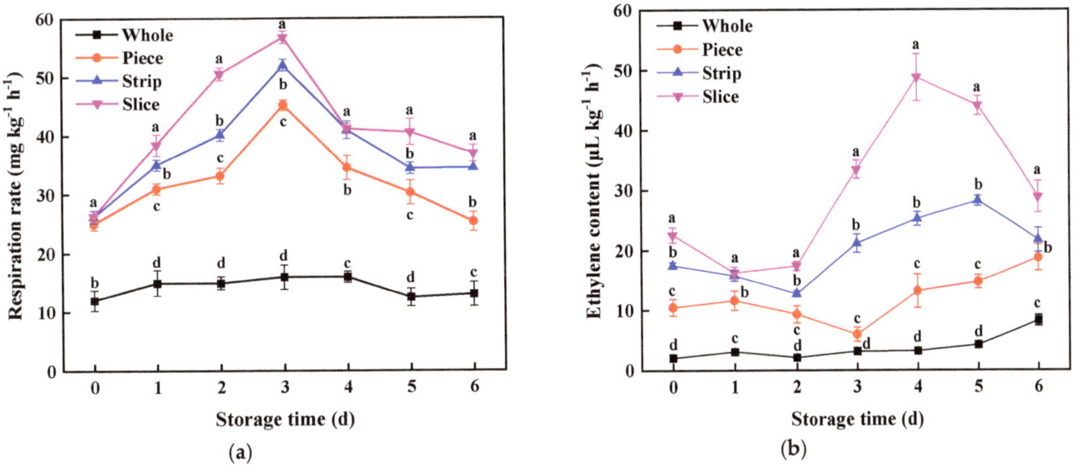

Figure 3. Effect of wounding intensities on respiration rate (**a**) and ethylene content (**b**) of fresh-cut pumpkin during storage at 4 °C for 6 days. Vertical bars represent the standard deviation of the mean (n = 3). Different letters indicate significant differences between the groups ($p < 0.05$).

Lignin is widely regarded as an essential secondary metabolite in agricultural products and it plays crucial role in repairing wounding and maintaining a solid shape as well as alleviating oxidative damage and preventing the invasion of pathogenic bacteria [2,43]. The present work revealed that fresh-cut pumpkins accumulated higher lignin content (Figure 4a), and the lignin levels of fresh-cut pumpkin were slice > strip > piece, and the total lignin content in slice, strip and piece increased by 164.18%, 134.32% and 68.76% compared with whole pumpkins (Figure 4a), respectively. In the phenylpropanoid pathway, PAL is the critical enzyme in the biosynthesis of phenols [13]. According to Figure 4b, PAL activity of piece, strip and slice pumpkins was all induced by the cutting process; they arrived at their peaks at 2 d, 3 d and 4 d, respectively, and then declined overall afterward. The significant positive correlation results among PAL activity, phenols and lignin content clarified that the fresh-cutting operation increased PAL activity to promote the synthesis of phenolic substances, which further provide sufficient precursor substances to generate lignin.

3.3.3. Membrane Lipid Peroxidation-Related Parameters

Generally, MDA is the end product of membrane lipid peroxidation, and it is considered a biomarker in plants [44,45]. As shown in Figure 5a, the MDA content of the fresh-cut pumpkin group fluctuated throughout, and the overall change was significant. Throughout the storage period, the MDA content of the slice, strip and piece was obviously higher than that of the whole group. At the end of the storage time (6 d), the MDA content in slice, strip and piece increased by 88.13%, 80.48% and 56.06% compared with whole pumpkins, respectively. The enhancement in MDA content indicated that wounding stress accelerated the membrane lipid peroxidation in fresh-cut pumpkins.

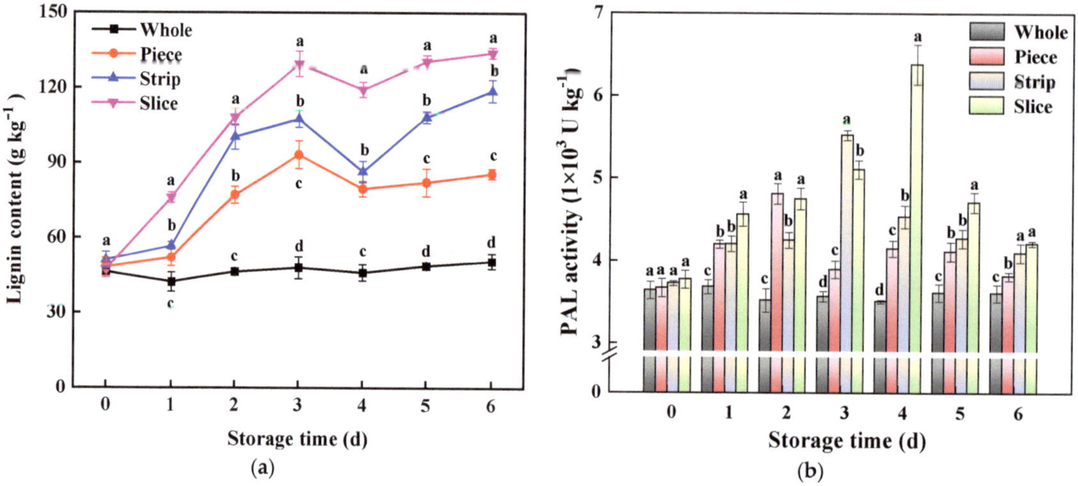

Figure 4. Effect of wounding intensities on lignin content (**a**) and phenylalanine ammonia-lyase (PAL) activity (**b**) of fresh-cut pumpkin during storage at 4 °C for 6 days. Vertical bars represent the standard deviation of the mean (n = 3). Different letters indicate significant differences between the groups ($p < 0.05$).

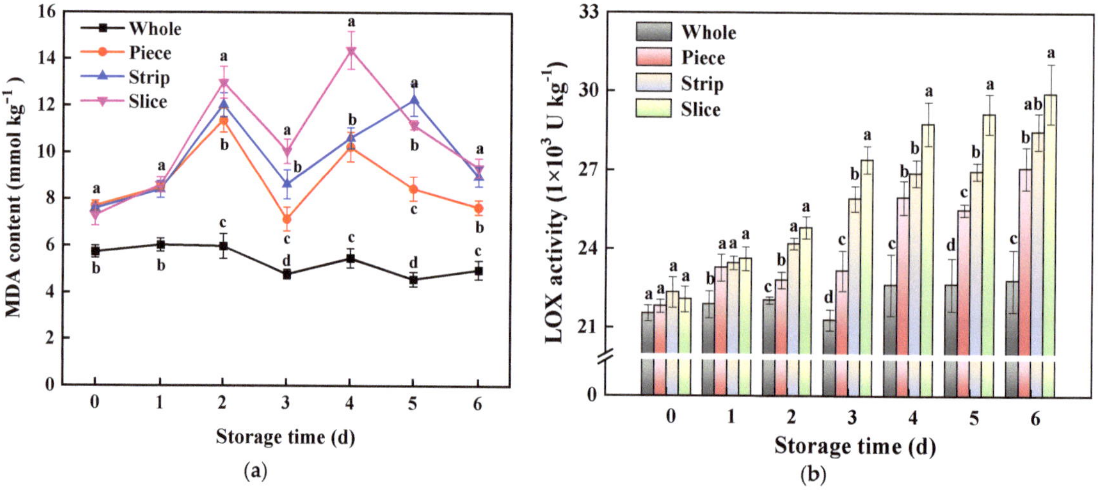

Figure 5. Effect of wounding intensities on MDA content (**a**) and LOX activity (**b**) of fresh-cut pumpkin during storage at 4 °C for 6 days. Vertical bars represent the standard deviation of the mean (n = 3). Different letters indicate significant differences between the groups ($p < 0.05$).

LOX is a vital enzyme in membrane lipid metabolism by catalyzing polyunsaturated fatty acids to produce hydrogen peroxide with conjugated double bonds [46]. According to Figure 5b, LOX activity was continuous in the upward trend. At 6 d, the LOX activity in slice, strip and piece increased by 31.22%, 24.91% and 18.82% compared with the whole pumpkin, respectively. In this study, we found that there was a significant positive correlation (R = 0.649) between LOX and MDA; similar results were found in fresh-cut potatoes [11], lotus roots [47], pears [46] and Chinese water chestnuts [48] where LOX could enhanced MDA accumulation through oxidative damage. However, significant negative correlations were found among LOX and TSS content (−0.822) and firmness (−0.772), indicating that fresh-cutting reduced TSS content and firmness of pumpkins by stimulating LOX enzyme activity. This may be because the enhancement of LOX led to the membrane lipid peroxidation, and then promoted leakage of cellular substances and further contributed to the reduction in TSS and firmness.

3.4. Effect of Wounding Intensities on ROS Metabolism of Fresh-Cut Pumpkins

3.4.1. ROS Content

As the key signal molecules in plants, ROS regulate the integration of signal transmission and activation of the wounding stress response network [49]. The major forms of ROS in fruits and vegetables, which vary greatly in their functions and characteristics, mainly include superoxide ($O_2^{-\bullet}$) and hydrogen peroxide (H_2O_2) [50]. According to Figure 6a, the $O_2^{-\bullet}$ content in fresh-cut pumpkin was immediately induced by the cutting process; similar results were found in various products, such as apple, which suggests that fresh-cut treatment caused excessive oxidative stress. Moreover, this study revealed that $O_2^{-\bullet}$ content increases with the increase in wounding intensities. This may be because with the wounding intensity increasing, the larger wounded surface is in contact with O_2, which speeds up the process of peroxidation and leads to the accumulation of $O_2^{-\bullet}$ [7].

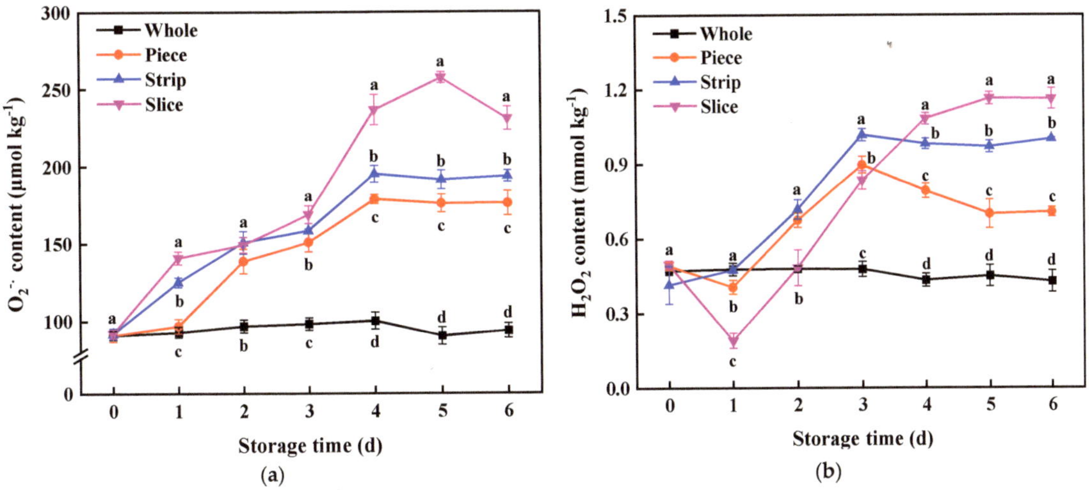

Figure 6. Effect of wounding intensities on $O_2^{-\bullet}$ (**a**) and H_2O_2 content (**b**) of fresh-cut pumpkin during storage at 4 °C for 6 days. Vertical bars represent the standard deviation of the mean (n = 3). Different letters indicate significant differences between the groups ($p < 0.05$).

Different from the $O_2^{-\bullet}$ change trend, H_2O_2 content in slice pumpkin with the highest wounding intensity was lower than strip and piece during the first 3 d storage time (Figure 6b). Whereas H_2O_2 content also increased with the enhancement of wounding intensity after 3 d storage, the H_2O_2 content in slice, strip and piece groups increased 1.70 times, 1.33 times and 0.65 times, respectively, compared with that of whole pumpkins. There was a significant positive correlation between ROS ($O_2^{-\bullet}$, H_2O) and membrane lipid peroxidation-related parameters (LOX, MDA); similar results were obtained from the obvious study that cutting induced the ROS to burst and further caused higher membrane integrity and oxidative stress to accelerate quality deterioration [51,52]. Additionally, a significant positive correlation between ROS ($O_2^{-\bullet}$, H_2O) and phenylpropane metabolism-related indexes (PAL, phenols and lignin content) was also found in this study, revealing the fact that ROS involves signaling molecules in fresh-cut pumpkins.

3.4.2. Antioxidant Compound Content

In plants, anthocyanin is a natural pigment existing and is given the color of vegetables [53]. As shown in Table 1, anthocyanin content increased in all treatments during the early stage of storage, and then decreased after 5 d of storage time. At 4 d of storage, the anthocyanin content of slice and piece reached the maximum values, which increased 3.50 times and 2.44 times in comparison with that of whole pumpkins, respectively. The anthocyanin content of the strip reached the maximum values at 5 d of storage, which was an increase from 0.15 g kg^{-1} to 0.45 g kg^{-1}. This finding was similar to the previous study on red cabbage that the cutting operation resulted in a remarkable enhancement of anthocyanin [27]. The anthocyanin content decreased at the later storage period, which could be explained by the color fading appearance by the increase in WI values infresh-cut pumpkins.

Vc is derived from fresh fruits and vegetables and is an essential nutrient for the human body [49]. The effects of wounding intensities on the Vc of pumpkin are shown in Table 1. Vc content decreased continuously, and the Vc content of fresh-cut groups was lower than the whole group throughout the storage. At the end of the storage time, the Vc content of slice, strip and piece was decreased by 43.24%, 32.65% and 27.94%, respectively, compared with the whole pumpkin. Clearly, remarkable differences were observed in the Vc content in the three wounding intensities. However, a previous study on fresh-cut broccoli found that the cutting types had no significant effect on Vc content [6]. This may be because of a unique property of pumpkin in which Vc decomposition and oxidation process was relatively sensitive to wounding stress. Furthermore, different from the previous reports that the temperature was 20 °C during the storage of fresh-cut broccoli, 4 °C was applied in this study [6].

GSH was the most important antioxidant sulfhydryl substance involved in the Vc-GSH metabolism; it interacted with Vc to scavenge free radicals in plants [54]. In this study, GSH content increased sharply in fresh-cut pumpkins, which is associated with wounding intensity (Table 1). At 6 d of storage time, the GSH contents of slice, strip and piece increased 2.35, 1.23 and 0.65 times compared with that of the whole group, respectively. A similar phenomenon was also observed in fresh-cut red cabbage [55] and potato [56].

3.4.3. ROS Metabolism-Related Enzyme Activity

It is well known that cutting causes fruit and vegetables to suffer irreparable mechanical damage and lose their protective tissue, which may induce ROS accumulation [57]. At the same time, the plant machinery will start its own enzyme system to regulate the balance of ROS to reduce oxidative damage. In general, there were two types of resistance enzymes, oxidase and antioxidant enzymes. Among these, polyphenol oxidase (PPO) and peroxidase (POD) were the most important oxidase enzyme that catalyzes phenolic substrates to quinone with the presence of ROS [58]. According to Figure 7a, the PPO activity of slice, strip and piece increased sharply and reached the maximum at 1 d, which was an of increase 3.91, 2.66, and 1.50 times, respectively, compared with that of the whole pumpkin. A similar result was found in POD activity; it was induced sharply by wounding stress, and POD activity during storage was ranked as follows: slice > strip > piece (Figure 7b). Furthermore, a significant positive correlation appeared between oxidase enzyme (PPO, POD) activity and phenol content, and was also found in PPO, POD activity and MDA content, which confirmed that PPO and POD accelerated the oxidation process of phenols and promoted the oxidation process of membrane lipids in fresh-cut pumpkin. The previous study on fresh-cut apple [23] also addressed that the PPO and POD play a catalytic role in the process of oxidative damage.

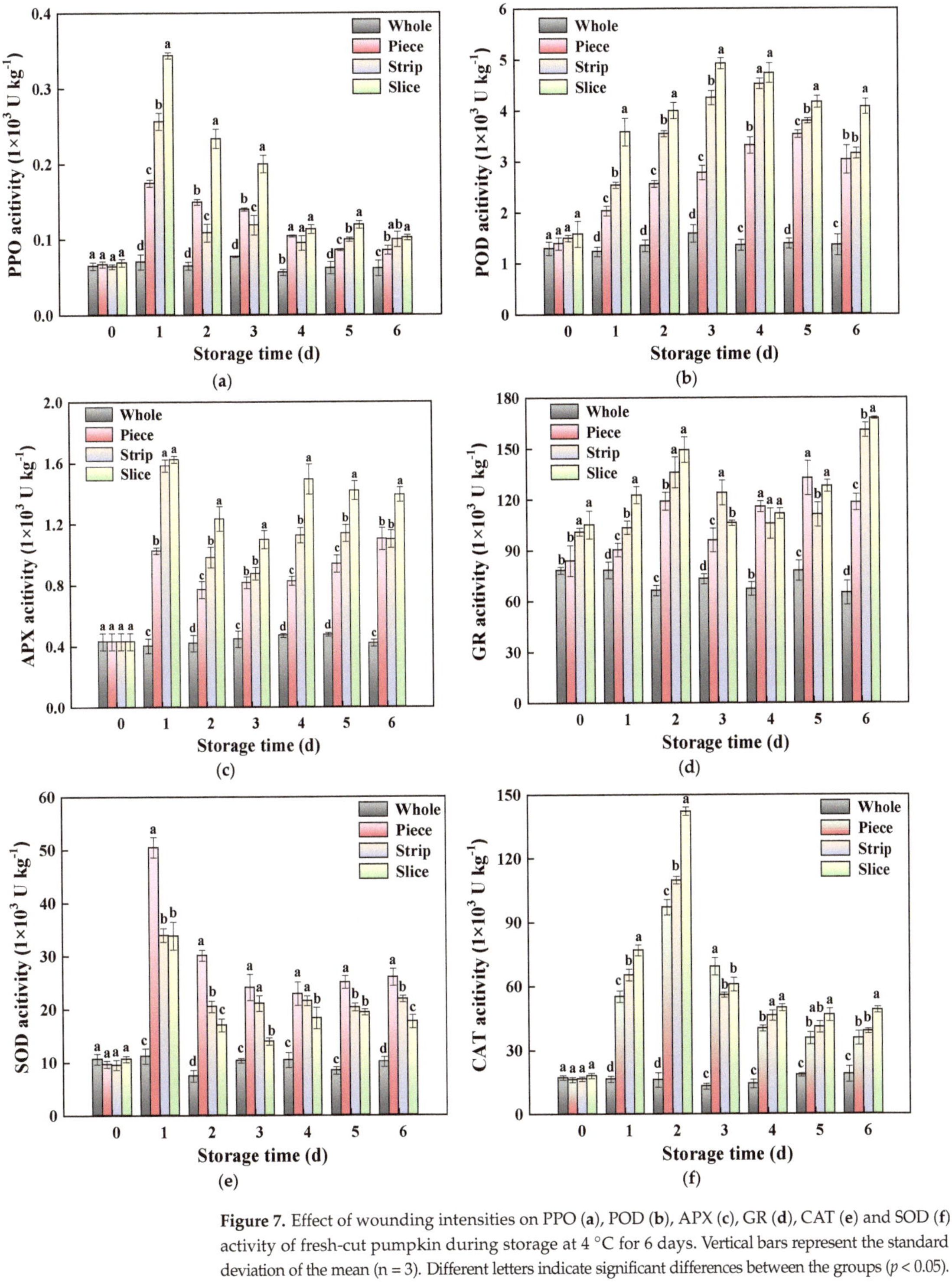

Figure 7. Effect of wounding intensities on PPO (**a**), POD (**b**), APX (**c**), GR (**d**), CAT (**e**) and SOD (**f**) activity of fresh-cut pumpkin during storage at 4 °C for 6 days. Vertical bars represent the standard deviation of the mean (n = 3). Different letters indicate significant differences between the groups ($p < 0.05$).

It has been reported that the Vc-GSH cycle was one of the antioxidant enzyme systems, in which ascorbate peroxidase (APX) and glutathione reductase (GR) interacted with antioxidant substances (Vc and GSH) to scavenge excess ROS and alleviate oxidative damage in fresh-cut products [59,60]. In the present work, the fresh-cutting operation induced the APX and GR activity, and the APX and GR activity of cutting groups was higher than whole pumpkin during the whole storage time (Figure 7c,d). It is worth mentioning that APX activity was significantly negatively correlated with Vc content (−0.727), whereas GR activity was significantly positively correlated with GSH content (0.599). Meanwhile, the APX and GR activity was significantly positively correlated with ROS content. The above results revealed that the existence of ROS promoted the process of Vc-GSH cycle by regulating APX and GR activity.

In plants, SOD could catalyze $O_2^{-\bullet}$ into H_2O_2, which is then decomposed by CAT [50,61]. As depicted in Figure 7e, the SOD activity of fresh-cut kiwifruits increased substantially and reached the maximum on the first day of storage, then followed by a slight decrease until the end of storage. Different from the other antioxidant enzymes, the order of SOD activity was: slice < strip < piece. Similarly, CAT activity was induced by wounding stress in fresh-cut pumpkin (Figure 7f), and it was enhanced with the higher wounding intensity. At 2 d, the CAT activity in slice, strip and piece was increased by 7.94, 5.90 and 5.11 times compared with whole pumpkin, respectively.

3.4.4. Antioxidant Activity

Studies on lettuce [62], cauliflower [63] and carrot [64] have demonstrated that ABTS and FRAP are very important assays for the evaluation of antioxidant activity in vegetables. In this study, the antioxidant activity including $OH^{\bullet}$, ABTS radical scavenging capacity and FRAP of fresh-cut pumpkin during storage was analyzed. As shown in Figure 8a, the $OH^{\bullet}$ radical scavenging capacity increased on day 1 of storage, followed by a gradual decrease, and then increased again. At the end of storage time (6 d), the $OH^{\bullet}$ radical scavenging capacity of slice, strip and piece was 31.65%, 16.19% and 18.15% higher than in whole pumpkin, respectively. Meanwhile, this study found that the cutting operation also promotes the ABTS radical scavenging capacity (Figure 8b) and FRAP (Figure 8c), and wounding intensities have an obvious influence on antioxidant activity in fresh-cut pumpkins. As for ABTS radical scavenging capacity, the levels were ranked as follows: slice > strip > piece. Unlike the ABTS change trend, the order of FRAP levels in fresh-cut pumpkins was slice < strip < piece. In order to explain the mechanism of the enhancement of antioxidant activity in fresh-cut pumpkins induced by wounding stress, the relationship among antioxidant-related indicators was performed.

According to Figure 9, the antioxidant activity (FRAP, $OH^{\bullet}$, ABTS) was significantly positively correlated with antioxidant enzyme (APX, GR) activity, which indicated that the Vc-GSH cycle facilitated the promotion of antioxidant activity in wounded pumpkin. Studies on fresh-cut broccoli [60] and celery [65] also demonstrated this result. Moreover, there was a positive correlation among anthocyanin, GSH, phenol content and antioxidant activity (ABTS, FRAP), whereas there was no correlation among Vc, flavonoid content and ABTS, FRAP. Our findings contradict the previous study on red cabbage [27], in which Vc contributed to improving the antioxidant activity. This was likely due to Vc being decomposed for providing precursors for GSH synthesis in this study. Various studies on fresh-cut carrot [66], celery [67], lettuce [68], broccoli [60], mushroom [69], onion [70], mango [71], pitaya [8], potato [11,72] and red cabbage [28] were consistent with our work which found that the phenol accumulation would further contribute to the increase in antioxidant activity. These results confirmed that the antioxidant compounds play different roles in the process of resisting oxidative damage in fresh-cut products.

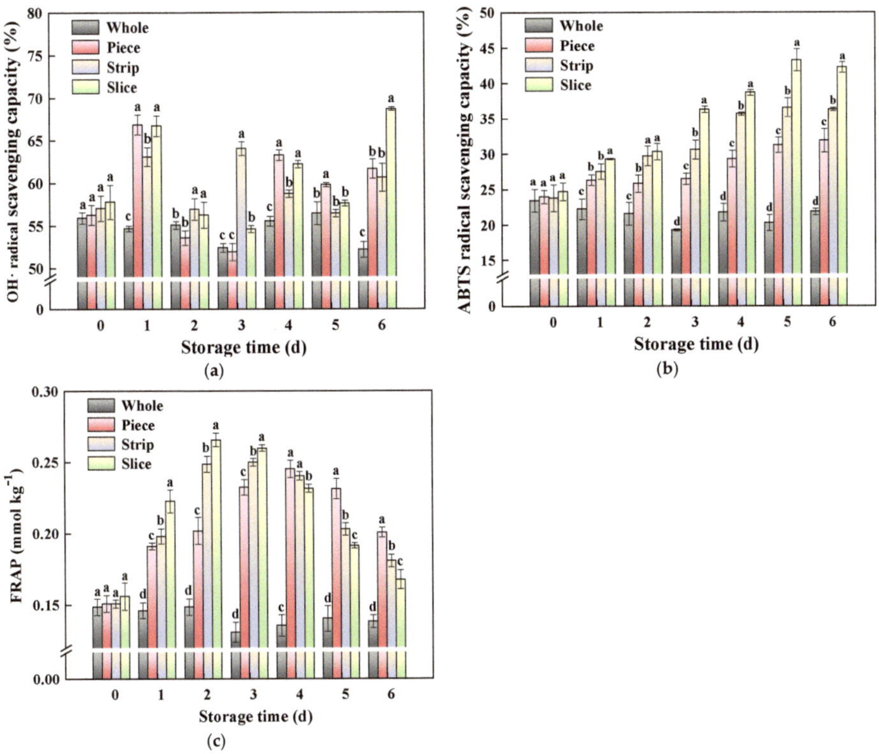

Figure 8. Effect of wounding intensities on OH$^{\bullet}$ (**a**), ABTS (**b**) radical scavenging capacity and FRAP (**c**) of fresh-cut pumpkin during storage at 4 °C for 6 days. Vertical bars represent the standard deviation of the mean (n = 3). Different letters indicate significant differences between the groups ($p < 0.05$).

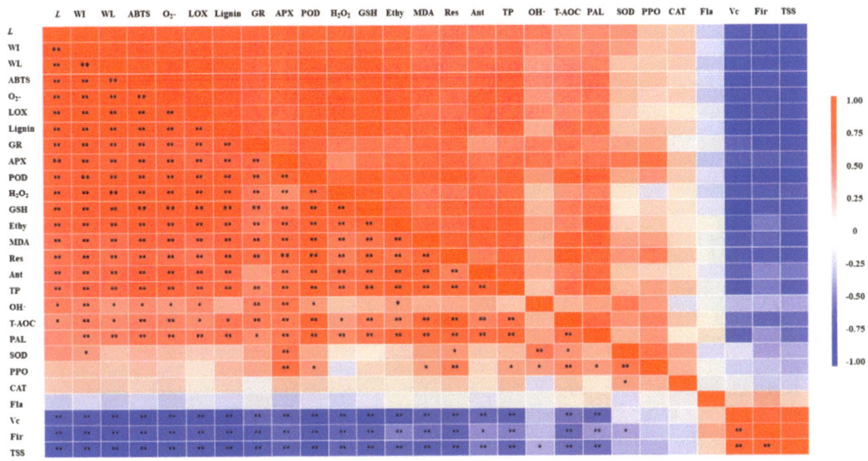

Figure 9. Pearson correlation matrix of indexes. * indicates significance at the $p < 0.05$ probability level. ** indicates significance at the $p < 0.01$ probability level. WL, Ethy, Res, Ant, TP, Fla and Fir represent weight loss, ethylene, respiration, anthocyanin, total phenols, flavonoids and firmness, respectively.

4. Conclusions

Fresh-cutting operation induced obvious increases in lightness, whiteness index, respiration rate and ethylene content in fresh-cut pumpkin, and there was a significant difference in wounding intensities in these parameters. Meanwhile, wounding stress stimulated the production of ROS, and then improved the process of membrane lipid peroxidation, which resulted in a decrease in appearance, firmness, sensory quality and TSS content. Furthermore, fresh-cutting treatment induced the activity of the critical enzyme involved in phenylpropanoid and ROS metabolism, including PAL, SOD, CAT, APX and GR, resulting in a synthesis of phenols, flavonoids and GSH and further improved the ability to clear ROS, which contributed to the increase in antioxidant activity (OH·, ABTS radical scavenging capacity and FRAP) in wounded pumpkin. These phenomena were more obvious in the cutting method with the enhancement of wounding intensity degree. The collected results revealed that different wounding intensities have an obvious influence on ROS metabolism, Vc-GSH cycle, phenylpropanoid metabolism and membrane lipid peroxidation, which further affect the quality of fresh-cut pumpkin.

Author Contributions: Conceptualization, W.H. and Y.G.; Data curation, W.H., Y.G. and Y.W.; Formal analysis, W.H. and Y.G.; Funding acquisition, W.H. and Y.G.; Investigation, W.H. and Y.G.; Methodology, W.H. and Y.G.; Project administration, W.H.; Resources, W.H.; Software, W.H. and Y.G.; Supervision, Y.W. and N.Y.; Visualization, Y.G., Y.W. and N.Y.; Writing—original draft, W.H. and Y.G.; Writing—review and editing, Y.G., Y.W. and N.Y. All authors have read and agreed to the published version of the manuscript.

Funding: This research was supported by the "Thirteenth Five-Year Plan" for National Key Research and Development Program (No. 2016YFD0400903), National Natural Science Foundation of China (No. 31471923), National Natural Science Foundation of China (No. 32202120), Zhejiang Agricultural and Forestry University Scientific Research Development Fund Project (Grant No. 2022LFR043).

Data Availability Statement: The datasets used and/or analyzed during the current study are available from the corresponding author on reasonable request.

Conflicts of Interest: The authors declare no conflict of interest.

References

1. Liang, J.Y.; Guo, F.; Cao, S.F.; Zhao, K.; Zhao, K.X.; Wang, H.F.; Shao, X.F.; Wei, Y.Y.; Zhang, C.D.; Zheng, Y.H.; et al. γ-aminobutyric acid (GABA) alleviated oxidative damage and programmed cell death in fresh-cut pumpkins. *Plant Physiol. Biochem.* **2022**, *180*, 9–16. [CrossRef] [PubMed]
2. Cisneros-Zevallos, L.; Jacobo-Velazquez, D.A. Controlled abiotic stresses revisited: From homeostasis through hormesis to extreme stresses and the impact on nutraceuticals and quality during pre- and postharvest applications in horticultural crops. *J. Agric. Food Chem.* **2020**, *68*, 11877–11879. [CrossRef] [PubMed]
3. Gade, S.R.; Meghwal, M.; Prabhakar, P.K.; Giuffre, A.M. A comparative study on the nutritional, antioxidant, thermal, morphological and diffraction properties of selected cucurbit seeds. *Agronomy* **2022**, *12*, 2242. [CrossRef]
4. Valdez-Arjona, L.P.; Ramírez-Mella, M. Pumpkin waste as livestock feed: Impact on nutrition and animal health and on quality of meat, milk, and egg. *Animals* **2019**, *9*, 769. [CrossRef] [PubMed]
5. Hu, W.Z.; Sarengaowa; Guan, Y.G.; Feng, K. Biosynthesis of phenolic compounds and antioxidant activity in fresh-cut fruits and vegetables. *Front. Microbiol.* **2022**, *13*, 906069. [CrossRef] [PubMed]
6. Guan, Y.G.; Hu, W.Z.; Jiang, A.L.; Xu, Y.P.; Yu, J.X.; Zhao, M.R.; Ji, Y.R.; Feng, K.; Sa, R.G.W.; Yang, X.Z. Influence of cut type on quality, antioxidant substances and antioxidant activity of fresh-cut broccoli. *Int. J. Food Sci. Technol.* **2020**, *55*, 3019–3030. [CrossRef]
7. Han, C.; Ji, Y.; Li, M.L.; Li, X.A.; Jin, P.; Zheng, Y.H. Influence of wounding intensity and storage temperature on quality and antioxidant activity of fresh-cut Welsh onions. *Sci. Hortic.* **2016**, *212*, 203–209. [CrossRef]
8. Li, X.A.; Long, Q.H.; Gao, F.; Han, C.; Jin, P.; Zheng, Y.H. Effect of cutting styles on quality and antioxidant activity in fresh-cut pitaya fruit. *Postharvest Biol. Technol.* **2017**, *124*, 1–7. [CrossRef]
9. Han, C.; Jin, P.; Li, M.L.; Wang, L.; Zheng, Y.H. Physiological and transcriptomic analysis validates previous findings of changes in primary metabolism for the production of phenolic antioxidants in wounded carrots. *J. Agric. Food Chem.* **2017**, *65*, 7159–7167. [CrossRef]
10. Chai, Z.Y.; Zhang, F.; Liu, B.J.; Chen, X.F.; Meng, X.H. Antibacterial mechanism and preservation effect of curcumin-based photodynamic extends the shelf life of fresh-cut pears. *LWT-Food Sci. Technol.* **2021**, *142*, 110941. [CrossRef]

11. Hu, W.Z.; Guan, Y.G.; Ji, Y.R.; Yang, X.Z. Effect of cutting styles on quality, antioxidant activity, membrane lipid peroxidation, and browning in fresh-cut potatoes. *Food Biosci.* **2021**, *44*, 101435.
12. Han, C.; Li, J.; Jin, P.; Li, X.; Wang, L.; Zheng, Y. The effect of temperature on phenolic content in wounded carrots. *Food Chem.* **2017**, *215*, 116–123. [CrossRef] [PubMed]
13. Guan, Y.G.; Hu, W.Z.; Jiang, A.L.; Xu, Y.P.; Zhao, M.R.; Yu, J.X.; Ji, Y.R.; Sa, R.G.W.; Yang, X.Z.; Feng, K. The effect of cutting style on the biosynthesis of phenolics and cellular antioxidant capacity in wounded broccoli. *Food Res. Int.* **2020**, *137*, 109565. [CrossRef]
14. Sarengaowa; Hu, W.Z.; Jiang, A.L.; Xiu, Z.L.; Feng, K. Effect of thyme oil-alginate based coating on quality and microbial safety of fresh-cut apples. *J. Sci. Food Agric.* **2018**, *98*, 2302–2311. [CrossRef]
15. Zhou, F.H.; Xu, D.Y.; Liu, C.H.; Chen, C.; Tian, M.X.; Jiang, A.L. Ascorbic acid treatment inhibits wound healing of fresh-cut potato strips by controlling phenylpropanoid metabolism. *Postharvest Biol. Technol.* **2021**, *181*, 111644. [CrossRef]
16. Coming Biotechnology Company Home Page of MDA Content Kit. Available online: http://www.cominbio.com/a/shijihe/shenghuashiji/yanghuayukangyanghuaxilie/2018/0708/109.html (accessed on 15 November 2022).
17. Coming Biotechnology Company Home Page of LOX Activity Kit. Available online: http://www.cominbio.com/a/shijihe/shenghuashiji/zhifangsuandaixiexilie/2014/1112/509.html (accessed on 25 November 2022).
18. Coming Biotechnology Company Home Page of Lignin Content Kit. Available online: http://www.cominbio.com/a/shijihe/shenghuashiji/qitaxilie/2016/0502/969.html (accessed on 3 November 2022).
19. Jiancheng Bioengineering Institute Company Home Page of OH. Scavenging Ability Kit. Available online: http://www.njjcbio.com/products.asp?id=318 (accessed on 20 November 2022).
20. Jiancheng Bioengineering Institute Company Home Page of ABTS Scavenging Ability Kit. Available online: http://www.njjcbio.com/product.asp?commend=1&q=ABTS&sort=hitsorder (accessed on 20 November 2022).
21. Jiancheng Bioengineering Institute Company Home Page of DPPH Scavenging Ability Kit. Available online: http://www.njjcbio.com/product.asp?commend=1&q=FRAP&sort=hitsorder (accessed on 20 November 2022).
22. Tian, S.P.; Xu, Y.; Jiang, A.; Gong, Q.Q. Physiological and quality responses of longan fruit to high O_2 or high CO_2 atmospheres in storage. *Postharvest Biol. Technol.* **2002**, *24*, 335–340. [CrossRef]
23. Chen, C.; Hu, W.; He, Y.; Jiang, A.; Zhang, R. Effect of citric acid combined with UV-C on the quality of fresh-cut apples. *Postharvest Biol. Technol.* **2016**, *111*, 126–131. [CrossRef]
24. Coming Biotechnology Company Home Page of SOD Activity Kit. Available online: http://www.cominbio.com/a/shijihe/shenghuashiji/yanghuayukangyanghuaxilie/2019/0411/1795.html) (accessed on 18 December 2022).
25. Coming Biotechnology Company Home Page of APX Activity Kit. Available online: http://www.cominbio.com/a/shijihe/shenghuashiji/weishengsuCdaixiexilie/2014/1111/439.html (accessed on 28 December 2022).
26. Coming Biotechnology Company Home Page of GR Activity Kit. Available online: http://www.cominbio.com/a/shijihe/shenghuashiji/guguanggataixilie/2014/1111/432.html (accessed on 28 December 2022).
27. Guan, Y.G.; Ji, Y.R.; Yang, X.Z.; Pang, L.J.; Cheng, J.Y.; Lu, X.H.; Zheng, J.; Yin, L.Q.; Hu, W.Z. Antioxidant activity and microbial safety of fresh-cut red cabbage stored in different packaging films. *LWT-Food Sci. Technol.* **2023**, *17*, 114478. [CrossRef]
28. Song, Z.; Feng, L.; Hui, G.; Xu, Y.; Li, D. Combination of nisin and ε-polylysine with chitosan coating inhibits the white blush of fresh-cut carrots. *Food Control.* **2017**, *74*, 34–44. [CrossRef]
29. Chen, C.C.; Xu, L.; Wang, S.W.; Chen, X.N.; Shen, Y.Y. Characterization of abscisic acid and ethylene in regulating the white blush in fresh-cut carrots. *Int. J. Mol. Sci.* **2022**, *23*, 12788. [CrossRef]
30. Zhao, P.; Li, W.; Zhen, C.; Wang, K.; Qin, Z.; Gao, H. Transcriptomic analysis of the effect of γ-aminobutyric acid treatment on browning and induced disease resistance in fresh-cut apples. *Postharvest Biol. Technol.* **2021**, *181*, 111686. [CrossRef]
31. Wang, L.; Guo, Y.Y.; Wang, X.Y.; Zhang, X.H. Short-term O_2/CO_2 controlled atmosphere altered the water status and thus promoted phenolic biosynthesis during wound healing of fresh-cut white mushroom (*Agaricus bisporus*). *Postharvest Biol. Technol.* **2022**, *188*, 111879. [CrossRef]
32. Zhou, Z.C.; Han, P.; Bai, S.Q.; Ma, N.; Fang, D.L.; Yang, W.J.; Hu, Q.H.; Pei, F. Caffeic acid-grafted-chitosan/polylactic acid film packaging enhances the postharvest quality of *Agaricus bisporus* by regulating membrane lipid metabolism. *Food Res. Int.* **2022**, *158*, 111557. [CrossRef]
33. Renna, M.; Pace, B.; Cefola, M.; Santamaria, P.; Serio, F.; Gonnella, M. Comparison of two jam-making methods to preserve the quality of colored carrots. *LWT-Food Sci. Technol.* **2013**, *53*, 547–554. [CrossRef]
34. Lin, X.; Yang, R.; Dou, Y.; Zhang, W.; Du, H.; Zhu, L.; Chen, J. Transcriptome analysis reveals delaying of the ripening and cell-wall degradation of kiwifruit by hydrogen sulfide. *J. Sci. Food Agric.* **2020**, *100*, 2280–2287. [CrossRef]
35. Baccio, D.D.; Ederli, L.; Marabottini, R.; Badiani, M.; Francini, A.; Nali, C.; Antonelli, M.; Santangelo, E.; Sebastiani, L.; Pasqualini, S. Similar foliar lesions but opposite hormonal patterns in a tomato mutant impaired in ethylene perception and its near isogenic wild type challenged with ozone. *Environ. Exp. Bot.* **2012**, *75*, 286–297. [CrossRef]
36. Wang, S.; Zhou, Q.; Zhou, X.; Zhang, F.; Ji, S. Ethylene plays an important role in the softening and sucrose metabolism of blueberries postharvest. *Food Chem.* **2020**, *310*, 125965. [CrossRef]
37. Zhao, T.; Nakano, A.; Iwasaki, Y. Differences between ethylene emission characteristics of tomato cultivars in tomato production at plant factory. *J. Agric. Food Res.* **2021**, *5*, 100181. [CrossRef]

38. Fan, X.; Du, Z.; Cui, X.; Ji, W.; Gong, H. Preharvest methyl salicylate treatment enhance the chilling tolerance and improve the postharvest quality of apricot during low temperature storage. *Postharvest Biol. Technol.* **2021**, *177*, 111535. [CrossRef]
39. Ferrer, J.L., Austin, M.B., Stewart, C.; Noel, J.P. Structure and function of enzymes involved in the biosynthesis of phenylpropanoids. *Plant Physiol. Biochem.* **2008**, *46*, 356–370. [CrossRef]
40. Jacobo-Velazquez, D.A.; Martinez-Hernandez, G.B.; Rodriguez, S.D.C.; Cao, C.M.; Cisneros-Zevallos, L. Plants as biofactories: Physiological role of reactive oxygen species on the accumulation of phenolic antioxidants in carrot tissue under wounding and hyperoxia stress. *J. Agric. Food Chem.* **2011**, *59*, 6583–6593. [CrossRef]
41. Giuffre, A.M.; Zappia, C.; Capocasale, M. Physico-chemical stability of blood orange juice during frozen storage. *Int. J. Food Prop.* **2017**, *20*, 1930–1943.
42. Mele, M.A. Bioactive compounds and biological activity of ginger. *J. Multidiscip. Sci.* **2019**, *1*, 1–7. [CrossRef]
43. Fan, K.; Zhang, M.; Jiang, F.J. Ultrasound treatment to modified atmospheric packaged fresh-cut cucumber: Influence on microbial inhibition and storage quality. *Ultrason. Sonochem.* **2019**, *54*, 162–170. [CrossRef]
44. Huang, H.; Guo, L.F.; Wang, L.; Wang, H.; Ma, S.M.; Jiang, Y.M.; Qu, H.X. 1-Methylcyclopropene (1-MCP) slows ripening of kiwifruit and affects energy status, membrane fatty acid contents and cell membrane integrity. *Postharvest Biol. Technol.* **2019**, *156*, 110941. [CrossRef]
45. Mao, L.C.; Pang, H.Q.; Wang, G.; Zhu, C.G. Phospholipase D and lipoxygenase activity of cucumber fruit in response to chilling stress. *Postharvest Biol. Technol.* **2007**, *44*, 42–47. [CrossRef]
46. Hu, K.D.; Wang, Q.; Hu, L.Y.; Gao, S.P.; Wu, J.; Li, Y.H. Hydrogen sulfide prolongs postharvest storage of fresh-cut pears (*Pyrus pyrifolia*) by alleviation of oxidation of oxidative damage and inhibition of fungal growth. *PLoS ONE* **2014**, *9*, e85524.
47. Gao, H.; Chai, H.K.; Cheng, N.; Cao, W. Effects of 24-epibrassinolide on enzymatic browning and antioxidant activity of fresh-cut lotus root slices. *Food Chem.* **2017**, *217*, 45–51. [CrossRef]
48. You, Y.L.; Jiang, Y.M.; Sun, J.; Liu, H.; Song, L.L.; Duan, X.W. Effects of short-term anoxia treatment on browning of fresh-cut Chinese water chestnut in relation to antioxidant activity. *Food Chem.* **2012**, *132*, 1191–1196. [CrossRef]
49. Guan, Y.G.; Hu, W.Z.; Xu, Y.P.; Yang, X.Z.; Ji, Y.R.; Feng, K.; Sa, R.G.W. Metabolomics and physiological analyses validates previous findings on the mechanism of response to wounding stress of different intensities in broccoli. *Food Res. Int.* **2020**, *140*, 110058. [CrossRef]
50. Wang, Y.; Ji, D.C.; Chen, T.; Li, B.Q.; Zhang, Z.Q.; Qin, G.Z.; Tian, S.P. Production, signaling, and scavenging mechanisms of reactive oxygen species in fruit-pathogen interactions. *Int. J. Mol. Sci.* **2019**, *20*, 2994. [CrossRef]
51. Zhu, J.K. Abiotic stress signaling and responses in plants. *Cell* **2016**, *167*, 313–324. [CrossRef] [PubMed]
52. Zuo, W.F.; Lu, L.; Su, M.Y.; Zhang, J.; Li, Y.Y.; Huang, S.L.; Li, C.X.; Wang, N.; Zhang, Z.Y.; Chen, X.S. Analysis of differentially expressed genes and differentially abundant metabolites associated with the browning of Meihong red-fleshed apple fruit. *Postharvest Biol. Technol.* **2021**, *174*, 111437. [CrossRef]
53. He, Q.; Xu, Y.H.; Wang, Y.X.; Zhang, N.N.; Zhang, L.G. Metabolic profiling and transcriptomic data providing critical flavonoid biosynthesis mechanisms disclose color differences of purple heading Chinese cabbages (*Brassica rapa* L.). *LWT-Food Sci. Technol.* **2022**, *168*, 113885. [CrossRef]
54. Bailly, C.; Benamar, A.; Corbineau, F.; Come, D. Changes in malondialdehyde content and in superoxide dismutase, catalase and glutathione reductase activities in sunflower seeds as related to deterioration during accelerated aging. *Physiol. Plant.* **2010**, *97*, 104–110. [CrossRef]
55. Zhang, J.; Yuan, L.; Liu, W.; Lin, Q.; Wang, Z.; Guan, W. Effects of UV-C on antioxidant capacity, antioxidant enzyme activity and color of fresh-cut red cabbage during storage. *Int. J. Food Sci. Technol.* **2016**, *52*, 626–634. [CrossRef]
56. Mori, T.; Terai, H.; Yamauchi, N.; Suzuki, Y. Effects of postharvest ethanol vapor treatment on the ascorbate-glutathione cycle in broccoli florets. *Postharvest Biol. Technol.* **2009**, *52*, 134–136. [CrossRef]
57. Barrett, D.M.; Beaulieu, J.C.; Shewfelt, R. Color, flavor, texture, and nutritional quality of fresh-cut fruits and vegetables: Desirable levels, instrumental and sensory measurement, and the effects of processing. *Crit. Rev. Food Sci.* **2010**, *50*, 369–389. [CrossRef] [PubMed]
58. Guan, Y.G.; Hu, W.Z.; Xu, Y.P.; Sarengaowa; Ji, Y.R.; Yang, X.Z.; Feng, K. Proteomic analysis validates previous findings on wounding-responsive plant hormone signaling and primary metabolism contributing to the biosynthesis of secondary metabolites based on metabolomic analysis in harvested broccoli (*Brassica oleracea* L. var. italica). *Food Res. Int.* **2021**, *145*, 110388.
59. Torres-Contreras, A.M.; Senes-Guerrero, C.; Pacheco, A.; Gonzalez-Agüero, M.; Ramos-Parra, P.A.; Cisneros-Zevallos, L.; Jacobo-Velazquez, D.A. Genes differentially expressed in broccoli as an early and late response to wounding stress. *Postharvest Biol. Technol.* **2018**, *145*, 172–182. [CrossRef]
60. Martinez-Hernandez, G.B.; Artés-Hernández, F.; Gómez, P.A.; Formica, A.C.; Artés, F. Combination of electrolysed water, UV-C, and super atmospheric O_2; packaging for improving fresh-cut broccoli quality. *Postharvest Biol. Technol.* **2013**, *76*, 125–134. [CrossRef]
61. Li, H.; Tian, S.P.; Qin, G.Z. NADPH oxidase is crucial for the cellular redox homeostasis in fungal pathogen *Botrytis cinerea*. *Mol. Plant Microbe Interact.* **2019**, *32*, 1508–1516. [CrossRef] [PubMed]
62. Song, J.L.; Huang, H.; Hao, Y.W.; Song, S.W.; Zhang, Y.T.; Su, W.; Liu, H.C. Nutritional quality, mineral and antioxidant content in lettuce afected by interaction of light intensity and nutrient solution concentration. *Sci. Rep.* **2020**, *10*, 2796. [CrossRef]

63. Singh, S.; Dey, S.S.; Kumar, R.; Bhatia, R.; Ghemeray, H.; Behera, T.K. Genetic analysis and interaction among CUPRAC, FRAP, phytochemical and phenotypic traits in cauliflower (*Brassica oleracea* var. botrytis L.). *Int. J. Chem. Stud.* **2019**, *7*, 1484–1494.
64. Purkiewicz, A.; Ciborska, J.; Tanska, M.; Narwojsz, A.; Starowicz, M.; Przybylowicz, K.E.; Sawicki, T. The impact of the method extraction and different carrot variety on the carotenoid profile, total phenolic content and antioxidant properties of juices. *Plants* **2020**, *9*, 1759. [CrossRef]
65. Reyes, L.F.; Villarreal, J.E.; Cisneros-Zevallos, L. The increase in antioxidant capacity after wounding depending on the type of fruit or vegetable tissue. *Food Chem.* **2007**, *101*, 1254–1262.
66. Surjadinata, B.B.; Cisneros-Zevallos, L. Biosynthesis of phenolic antioxidants in carrot tissue increases with wounding intensity. *Food Chem.* **2012**, *134*, 615–624. [CrossRef] [PubMed]
67. Vina, S.Z.; Chaves, A.R. Antioxidant responses in minimally processed celery during refrigerated storage. *Food Chem.* **2006**, *94*, 68–74. [CrossRef]
68. Zhan, L.J.; Li, Y.; Hu, J.Q.; Pang, L.Y.; Fan, H.P. Browning inhibition and quality preservation of fresh-cut romaine lettuce exposed to high intensity light. *Innov. Food Sci. Emerg. Technol.* **2012**, *14*, 70–76. [CrossRef]
69. Oms-Oliu, G.; Aguilo-Aguayo, I.; Martin-Belloso, O.; Soliva-Fortuny, R. Effects of pulsed light treatments on quality and antioxidant properties of fresh-cut mushrooms (*Agaricus bisporus*). *Postharvest Biol. Technol.* **2010**, *56*, 216–222. [CrossRef]
70. Berno, N.D.; Tezotto-Uliana, J.V.; dos Santos Dias, C.T.; Kluge, R.A. Storage temperature and type of cut affect the biochemical and physiological characteristics of fresh-cut purple onions. *Postharvest Biol. Technol.* **2014**, *93*, 91–96. [CrossRef]
71. Robles-Sanchez, M.R.; Alejandra Rojas-Gratue, M.; Odriozola-Serrano, I.; Gonzalez-Aguilar, G.; Martin-Belloso, O. Influence of alginate-based edible coating as carrier of antibrowning agents on bioactive compounds and antioxidant activity in fresh-cut kent mangoes. *LWT-Food Sci. Technol.* **2013**, *50*, 240–246. [CrossRef]
72. Torres-Contreras, A.M.; Nair, V.; Cisneros-Zevallos, L.; Jacobo-Velazquez, D.A. Plants as biofactories: Stress-induced production of chlorogenic acid isomers in potato tubers as affected by wounding intensity and storage time. *Ind. Crops Prod.* **2014**, *62*, 61–66. [CrossRef]

horticulturae

MDPI

Article

Fatty Acids and Minerals as Markers Useful to Classify Hass Avocado Quality: Ripening Patterns, Internal Disorders, and Sensory Quality

Pablo Rodríguez [1,*], Iris Soto [1], Jairo Villamizar [1,2] and Alexander Rebolledo [2]

[1] Research Unit ITAV: Innovaciones Tecnológicas para Agregar Valor a Recursos Agrícolas, Centro de Investigación La Selva, Corporación Colombiana de Investigación Agropecuaria Agrosavia, Km. 7, vía Rionegro—Las Palmas, Sector Llanogrande, Rionegro 054048, Antioquia, Colombia
[2] Centro de Investigación Palmira, Corporación Colombiana de Investigación Agropecuaria-Agrosavia, Diagonal a la Intersección de la Carrera 36A con Calle 23, Palmira 763531, Valle del Cauca, Colombia
* Correspondence: prodriguezf@agrosavia.co; Tel.: +57-317-658-2802

Abstract: Hass avocado quality varies by origin, season, and production practices. However, there is a lack of methodologies to guarantee that fruit reaching the market has consistent quality. The aim of this work was to identify predictive markers for quality management. Fruit samples produced under different nutrient management, elevation, date-to-harvest, and growth cycle conditions were analyzed. Dry matter, oil content, internal disorders, sensory attributes, minerals, and fatty acids were evaluated as quality variables. The results highlighted soil and weather differences among orchards. Nutrient management practices based on index balancing in some samples increased both productivity and fruit size. High variability was observed in the dry matter related to the age of the fruit at harvest. Ripening heterogeneity was very large in low-elevation orchards where the fruit was picked relatively early. High flesh mineral contents delayed fruit ripening. At low growing temperatures, more oleic and linoleic acids were present in fruits. The sensory texture and taste descriptors were affected by the fruit age and related to the flesh composition. Logistic, PLS-DA, and biplot models effectively represented the variabilities in the ripening pattern, composition, and sensory profile of avocado fruits and allowed the samples to be grouped according to the internal fruit quality.

Keywords: ripening heterogeneity; Hass avocado; quality predictive markers; internal disorders; fruit composition

Citation: Rodríguez, P.; Soto, I.; Villamizar, J.; Rebolledo, A. Fatty Acids and Minerals as Markers Useful to Classify Hass Avocado Quality: Ripening Patterns, Internal Disorders, and Sensory Quality. *Horticulturae* **2023**, *9*, 460. https://doi.org/10.3390/horticulturae9040460

Academic Editors: Wenzhong Hu, Tian Zhong and Xiuxiu Sun

Received: 2 March 2023
Revised: 28 March 2023
Accepted: 30 March 2023
Published: 4 April 2023

1. Introduction

Hass avocado consumption is increasing globally due to the demand for healthy and exotic flavors. Avocado consumption is led by the US with 4–7 kg per-capita consumption, followed by Canada, EU27 West, and EU27+UK [1]. This consumption trend can increase with a rise in Asian countries. Hass avocado-producing countries should conduct more research to achieve profitable productivity and internal quality that allows them to compete with fruit in destination markets. Effective nutrient management and agroecological characterization are essential for high-quality fruit and increased productivity in avocado orchards [2–4].

The marketing of 'Hass' avocados is limited by postharvest disorders, causing vast losses when the fruit is rejected for not meeting quality standards. Challenges include ripening heterogeneity, internal disorders, and sensory quality. Previous research has relied on univariate approaches to address the quality of Hass avocados and flesh mineral composition [5–7]. However, this approach limits the understanding of a phenomenon that is multivariate.

Heterogeneity in ripening has been focused on as a main logistical factor [8] but has not been associated with an increase in internal problems when avocado fruits require longer durations to ripen. Although fruit can ripen without internal disorders, such fruit does not always have high sensory quality [9].

Studies performed in the sensory area have implemented acceptance tests mainly with a focus on postharvest or breeding [10]. Other studies have proposed the content of fatty acids as a marker of origin [11]. However, metabolism during fruit growth limits the application of such methods to countries with highly contrasting production conditions, such as certain regions in Chile, Mexico, and Colombia. In addition, few works have sought to determine the relationships between fatty-acid profiles and the sensory quality of avocado fruit.

Multivariate studies are needed to address avocado quality, including internal and sensory aspects, to ensure consistency across destination markets. Agronomic management programs should guarantee productivity without compromising fruit quality. Harvest indices such as dry matter and oil content do not distinguish high or low postharvest quality. Nondestructive techniques often follow a univariate approach and cannot identify regions with differentiated fruit qualities. Multivariate studies can help determine how to manage regions according to market demands. Therefore, the aim of this work was to develop a methodology for evaluating the ripening pattern, internal disorders, flesh composition, and sensory quality of Hass avocados and to determine predictive quality markers to manage it.

2. Materials and Methods

2.1. Field Conditions

In the municipalities of Morales (orchard R) and El Tambo (orchard N) in Cauca, Colombia, fifteen commercial fruit-bearing 15-year-old 'Hass' avocado trees grafted on seed rootstock (possibly of the Antillean ecological group) were selected. Another previously analyzed, orchard EB located in Antioquia, Colombia, was used as a quality reference orchard because it produces fruit with a high internal quality consistently during the different studied harvest seasons [9].

In the same year in each orchard, trees with different flowering intensities are often present. In this work, trees were selected due to their high flowering level, vigor, and good phytosanitary status. For the R and N orchards, the study was conducted over two harvest cycles, the first corresponding to the intermediate flowering period with anthesis in August 2020 and the second corresponding to the main flowering period with anthesis in March 2021 (Figure 1). All the orchards had a density of 204 trees/Ha.

Climatic conditions were monitored during two consecutive years with a weather station (Watch Dog 2900ET; Spectrum Technologies, Plainfield, IL, USA); the following weather variables were recorded: the daily wind speed, rainfall, relative humidity, sunshine, and maximum, minimum, and average air temperatures. The recorded climatic variables were used to determine the water balance and the high and low rain seasons according to the daily reference evapotranspiration (ETo) based on the FAO (Food and Agriculture Organization) Penman–Monteith method [12].

The nutrient status and management were diagnosed and oriented by the Balanced Indices of Kenworthy (BIK) [13] for the orchards located in the department of Cauca in August 2020 (Figure 1). Nutrient removal with a 20-ton target yield was used to calculate the fertilization plan from the data obtained by Salazar-García and Lazcano-Ferrat [3]. The optimum doses of the nutrients were calculated from the soil fertility, the amount of nutrient removal for each element with a planned yield, and the fertilizer efficiency. The annual nutrient units (g) applied per tree were N (504.9), P_2O_5 (336.6), K_2O (656.9), CaO (22.9), MgO (48.2), and B (4.2). Soil-applied fertilizer was distributed evenly around the drip line of the tree, and fertilizer was applied monthly. The orchard was managed under standard practices by the farmer with the exception of trees fertilized with the plan management based on the BIK. The nutrient management by the farmer was higher, with applications per tree every two months in the following doses during the same year (g): N (236.6), P_2O_5 (530), K_2O (350), CaO (76.5), MgO (14).

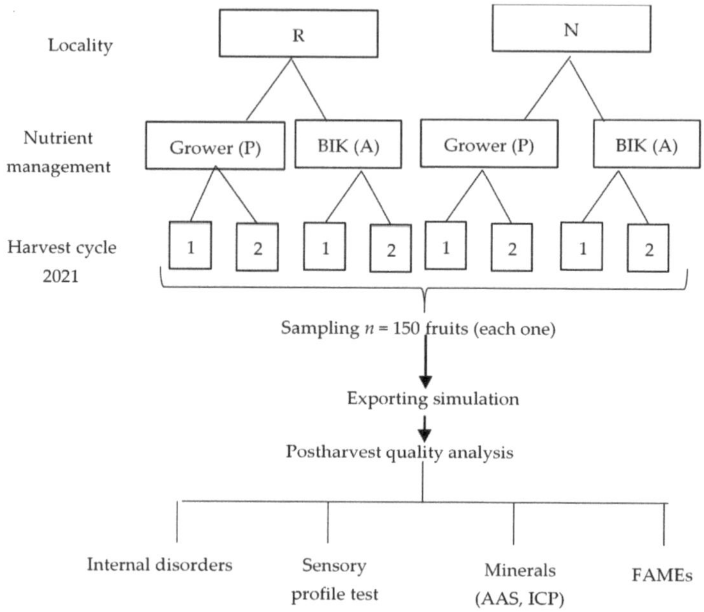

Figure 1. Schematic representation of the experimental design and postharvest quality analysis. Locality (orchard): R, N. BIK: Balanced Indices of Kenworthy. P: nutrient management of the farmer. A: Nutrient management based on the BIK. Harvest cycle: 1. Harvest from the intermediate flowering period. 2. Harvest from the main flowering period. FAMEs: fatty acid methyl esters. AAS: Atomic absorption spectroscopy. ICP: Inductively Coupled Plasma.

Soil chemical analyses of the top 45 cm layer were performed at the beginning of the experiment, and the results were used to calculate the nutrient supply capacity of the soil. To determine the plant nutrient status, fifty leaves from the vegetative growth were collected from trees at a height of 1.5 m every two months during the two consecutive years of study and washed, oven-dried to a constant mass, and sent to the laboratory at the "Corporación Colombiana de Investigación Agropecuaria" AGROSAVIA for mineral nutrient analysis.

The yield (kilogram/tree) and the number of fruits per tree were determined at the time of harvest during the two flowering cycles. The masses of a hundred randomly selected fruits from trees with high crop loads were collected and used to calculate the total number of fruits per tree. The harvested fruit were then classified according to the Codex Alimentarius standard (Codex Stand 197, [14]) with the following packing carton sizes (range in grams per fruit): size 10 (364–462), size 12 (300–371), size 14 (258–313), size 16 (227–274), size 18 (203–243), size 20 (184–217), size 22 (165–196), size 24 (151–175), size 26 (144–157), size 28 (134–147), size 30 (123–137), and size 32 (80–123).

2.2. *Postharvest Quality*

2.2.1. Sampling

In each orchard, a sample of 150 fruits per nutrient management strategy was taken for each harvest cycle (Figure 1). As described before, one sample was taken from the EB orchard for comparison with the R and N orchard results. Ten fruits per tree were harvested when they reached the dry matter (DM) value at commercial maturity employed by the grower according to the practices of the packing houses. As the DM varies by the production region, the fruit development time was established for each orchard. R1 (176 days), N1 (162 days), R2 and N2 (173 days), and EB (226 days). The fruits were harvested manually with previously disinfected pruning shears, leaving a 5 mm peduncle [15]. Fruit was harvested between 8:00 and 10:00 h from the outer 0.5 m of the canopy, between 1.5 and 3.0 m above ground, and at random from each quadrant of the tree. The DM and oil content (OC) were analyzed according to the AOAC methods (American Organization of Analytical Chemists, No. 934.01 and 963.15, respectively) [16].

2.2.2. Storage and Ripening

The samples were disinfected following the packing house practices and the methods reported by Hofman et al. (2003) [5]. The fruits were immersed in a prochloraz solution of 0.05% w/v (Mirage 45 ADAMA ANDINA, Pavullo, Italy) for 30 s. Then, the samples were dried at room temperature and packed into corrugated cardboard boxes. The fruits were stored for 2 weeks, simulating European market export conditions (at 5 °C/90% relative humidity (RH), Memmert HPP 110, Schwabach, Germany).

After that, the fruit was ripened (20 °C/90% RH, Memmert HPP 110). Ripening ended when fruits reached a rating of 5 according to White et al. (2009), corresponding to the firmness range from 8 to 12 N, as measured in the fruit's equatorial zone using a TA-XT Plus texture analyzer (Texture Technologies Corp., New York City, NY, USA) equipped with a 30-kg load cell [17]. Each day, two measurements were performed on unpeeled fruit with a stainless-steel compression plate (P/75.75-mm diameter).

2.2.3. Some Fruit Postharvest Quality Analysis

- Internal defects or disorders

Internal defects were evaluated by cutting fruits longitudinally in half, and the damage was identified as described in the international quality manual [17]. The internal disorder incidence (damage frequency in the sample) and severity (percentage of the flesh surface area affected) were evaluated. The internal disorders evaluated were stem end rot, body rot, vascular browning, flesh discoloration, and flesh bruising.

- Sensory analysis

The method used involved a sensory profile developed through a multidimensional approach [18]. Fruits without internal damage were cut longitudinally in two halves, and the seeds were removed. At the equatorial fruit zone, a cube (2 cm^3 each) was taken and placed into a 30-cm^3 plastic dish labeled with a random three-digit number. The test was then performed by a trained panel of 10 judges (male and female, 25 to 55 years old). The panelists consisted of personnel from the La Selva Agrosavia researcher center (Rionegro, Antioquia, Colombia). Five training sessions were performed to familiarize panelists with avocado quality attributes to identify the sensory intensity of descriptors and evaluate the sensory scale following the ISO standard [18]. The judges were selected based on their sensory acuity for basic characteristics (basic tastes, odors, and textures) and their ability to discriminate among products.

For treatment (see Figure 1), 32 ripened Hass avocado fruits were evaluated. The descriptors were selected based on previous works [9] and preliminary tests to verify the consensus of the panel about the Hass avocado sensory descriptors corresponding to the quality profile. The descriptors evaluated by panelists were fresh appearance, green color, yellow color, herbal smell, sweet taste, acidic taste, bitter taste, coconut sensation, nutty sensation, astringent sensation, oily sensation, hard texture, fibrous texture, lumpy texture, flux texture, and general quality. In general, those descriptors represent appearance, smell, taste, texture, and general quality attributes evaluated on a scale from 0 to 5. Panelists were also instructed to take a bite of cracker and drink some water to rinse their palate between each sample [19].

The sensory analysis was performed in the laboratory at the postharvest and agrofood processing pilot plant facilities (Agrosavia La Selva Research Center, Rionegro, Antioquia Colombia).

- Flesh mineral content

The analysis was performed following the procedure published by Escobar et al. [20]. Briefly, the flesh was dried at 70 °C in a forced convection oven (Memmert UF 110) until a constant weight was attained. Then, the samples were hydrolyzed, and the mineral content (K, Ca, Mg, S, and Na) was analyzed with an inductively coupled plasma (ICP) emission spectrometer (Thermo Scientific iCAP 6500 duo, Waltham, MA, USA). Atomic absorption spectroscopy (AAS, Agilent Technologies FS 240) was used to analyze the P, Fe, Cu, Mn, and Zn contents. The B content was determined in a spectrophotometer UV—VIS (Thermo Scientific Spectronic Genesys 10 S) at a wavelength of 430 nm following the methods of McKean [21].

- Fatty acid methyl esters (FAMEs)

The FAME profile was identified and quantified according to AOAC 996.06 [16]. The extracted total fat (TF) was methylated to fatty acid methyl esters using BF3 in methanol. FAMEs were quantitatively measured by gas chromatography and a flame ionization detector (GC-FID) (Thermo Trace 1300) against the C11:0 internal standard.

2.3. Statistical Analysis

The homoscedasticity and normality of the obtained data were analyzed to determine whether parametric or nonparametric tests should be applied. The fruit-ripening heterogeneity was analyzed using the nonparametric Kruskal–Wallis test. After that, Dunn's multiple-comparisons test was applied with a Bonferroni correction ($p < 0.05$). For the incidence of internal disorders, descriptive analyses of the qualitative data were used. Then, the relative frequency was analyzed by a k proportion test using a Chi-square test ($\alpha = 0.05$).

The disorder severity was analyzed by a Kruskal–Wallis test. The incidence of internal disorders was modeled through nonlinear regression using a logistic model (confidence interval 95%). Multivariate data of minerals and FAME flesh composition were analyzed by unsupervised (PCA) and supervised (PLS-DA) analysis methods. Sensory product characterization was performed to identify which descriptors best discriminated a set of fruits according to the orchard-specific nutrient management program and the growing cycle. For each descriptor, an ANOVA model was applied to check whether the scores given by the assessors differed significantly (XLSTAT 2022.2.1 Addinsoft, Paris, France).

3. Results

3.1. Field Conditions

The experimental avocado zones of R and N in the department of Cauca are located at elevations of 1600 and 1708 m.a.s.l, respectively. The reference orchard EB is located in the department of Antioquia at an elevation of 2464 m.a.s.l. The variations in climatic conditions among the EB, R, and N orchards were registered (Figure 2). Significant differences of approximately 5 °C were observed in the maximum and minimum temperatures among the R, N, and EB orchards (Figure 2a,b).

The cumulative annual precipitation totals were 2060 mm, 1795 mm, and 1411 mm in the R, N, and EB study areas, respectively, with a differential monthly distribution pattern throughout the year. An average of 211 mm of precipitation was recorded in the high-rain seasons from January to May and from October to December, while 61 mm of precipitation was recorded in the low-rain season from June to September in both R and N. A variation was reported in the precipitation distribution pattern in the EB location. A low-rain season occurred in January, February, September, and December, with 42 mm of precipitation falling (Figure 2c).

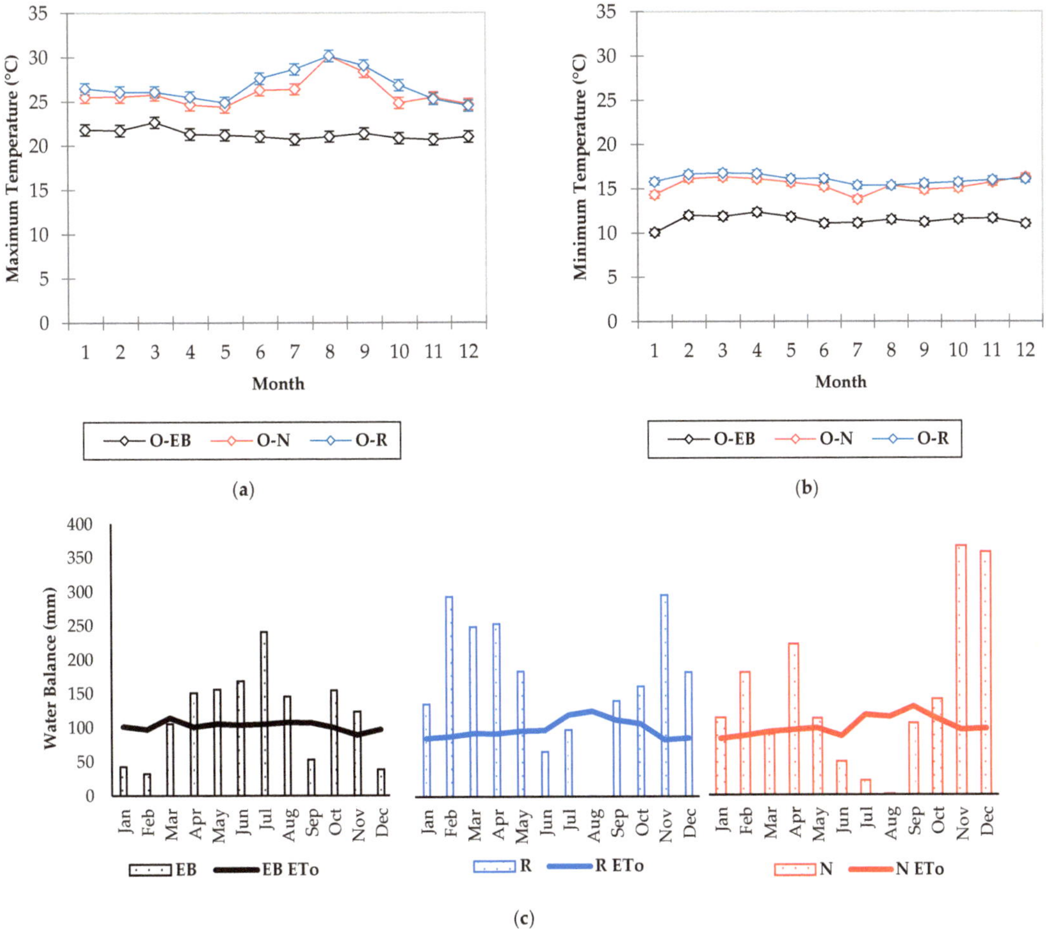

Figure 2. Climate conditions of the Hass avocado experimental zones corresponding to the following variables: (**a**) maximum temperature; (**b**) minimum temperature; and (**c**) evapotranspiration. O-EB: Orchard EB. O-N: Orchard N. O-R: Orchard R. ETo: evapotranspiration.

The soil characteristics in the experimental zones located in the Cauca region showed a sandy loam texture and pH conditions between moderate to hard acidic (Table 1). The electrical conductivity values indicated that these are salt-free soils, in addition to having high organic matter (OM) contents, indicating a high nitrogen availability.

The cation exchange capacity (CEC) showed values less than 10 meq/100 g, indicating the very poor ability of the soil to exchange nutrients such as calcium, magnesium, sodium, and potassium. These characteristics should be explained by the loamy-sandy texture of this kind of soil (Table 1). Some of these characteristics are common for the EB orchard, but the contents of mineral nutrients such as Ca, Mg, P, and minor elements were within the normal concentration ranges.

Two flowering periods were recorded in two consecutive years, characterized as the intermediate flowering period or cycle 1 (July September) and the main flowering period or cycle 2 (February-April); these flowering periods peaked with the largest proportion of flowers in anthesis in August and March, respectively.

Table 1. Ranges of chemical contents of the soils in the experimental avocado orchards.

Mineral	R Min–Max	N Min–Max	EB Min–Max
Depth (cm)	0.45	0.45	0.45
pH	5.42 ± 0.10	5.50 ± 0.06	5.12 ± 0.05
EC (ds/m)	0.33 ± 0.10	0.35 ± 0.06	1.05 ± 0.11
OM (%)	12.36 ± 0.47	14.85 ± 0.88	12.53 ± 0.69
P *	4.69 ± 0.72	5.45 ± 0.70	37.79 ± 5.08
S *	67.03 ± 7.4	75.13 ± 7.9	30.52 ± 1.83
Mg **	0.97 ± 0.15	0.99 ± 0.04	1.64 ± 0.27
Ca **	2.87± 0.40	3.90 ± 0.5	4.04 ± 0.61
K **	0.62 ± 0.09	0.44 ± 0.12	0.62 ± 0.05
Na **	0.09 ± 0.003	0.08 ± 0.008	0.10 ± 0.001
CEC **	4.64 ± 1.00	5.29 ± 0.6	6.98 ± 0.82
B *	0.22 ± 0.04	0.26 ± 0.03	0.33 ± 0.06
Mn *	2.39 ± 0.30	2.19 ± 0.20	13.05 ± 0.88
Cu *	2.49 ± 0.90	0.69 ± 0.20	5.65 ± 0.41
Fe *	101.95 ± 2.72	80.20 ± 5.60	309.32 ± 22.23
Zn *	1.54 ± 0.20	3.76 ± 1.20	8.56 ± 1.35
Soil texture	SL	SL	SL

R: Orchard R. N: Orchard N. EB: O Orchard EB. EC: electric conductivity. OM: organic matter. CEC: cation exchange capacity. SL: sandy loam. * = mg/kg, ** = meq/100 g.

The harvest from the intermediate flowering period was recorded in April, and that from the main flowering period was recorded in November 2021. An increase in the per-tree yield was registered in the harvest from the main flowering period with a nutrient management program based on the BIK standards. The wide standard error recorded in all treatments reflects the variation in the per-tree crop yield due to variations in the flowering intensity (Figure 3).

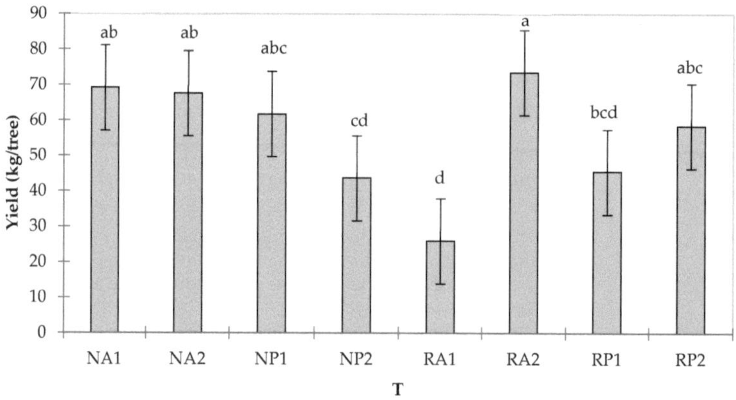

Figure 3. Yield (production average, kg/per tree) of Hass avocados in each orchard using 2 nutrient management practices during 2 production cycles. Different letters represent significant differences ($p < 0.05$) as determined using ANOVA and Tukey's post hoc test. T: treatments. Orchard: N and R. P: nutrient management of the farmer. A: Nutrient management based on the BIK standards. Harvest cycles: 1. Harvest from the intermediate flowering period. 2. Harvest from the main flowering period.

In accordance with the Codex Alimentarius standard [14], the fresh fruit weight at harvest showed an increase in the percentage of fruits from size 22 (165–196 g) in both the R and N orchards (Figure 4a) with the nutrient management program based on the BIK standards [13]. In the N orchard, no effect of nutrient management on improving the fruit weight from the intermediate flowering period was observed (Figure 4b).

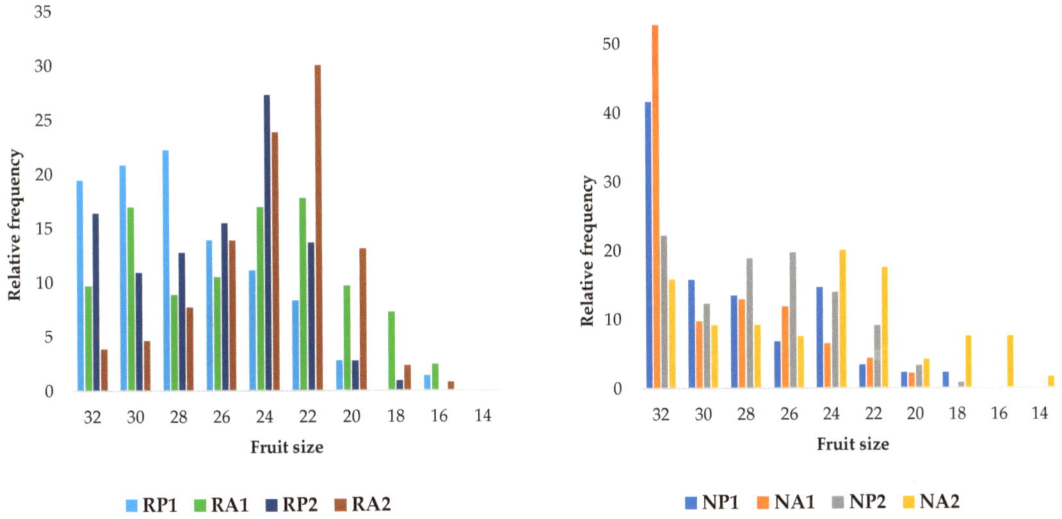

(**a**) (**b**)

Figure 4. Fruit harvest size classification results according to the Codex Stand 197, 1995. (**a**) R orchard and (**b**) N orchard. P: Nutrient management of the farmer. A: Nutrient management program based on the BIK standards. 1. Harvest from the intermediate flowering period. 2. Harvest from the main flowering period.

3.2. Postharvest Quality

In Figure 5, the Hass avocado ripening heterogeneity is shown. The harvest index (DM or fruit development time) has a greater impact on the ripening pattern of Hass avocados than either the nutrient management program or the growing season cycle does. The orchard located above 2400 m.a.s.l had shorter ripening ranges than the orchards situated at elevations below 1710 m.a.s.l. At high elevations, the development of fruit is slower than at low elevations because of the temperature differences (approximately 5 °C) among the N, R, and EB orchards (Figure 2a,b).

In terms of fruit development time, differences of at least 50 days were observed (Figure S1). In other words, the growers at low elevations harvested their fruit earlier than those at high elevations. Moreover, the mineral accumulations in the R and N orchards were larger than those in EB, as will be discussed later. Even though the EB orchard had a high range of variability at harvest (Figure S1), its softening pattern occurred faster. Therefore, other preharvest factors must influence the ripening heterogeneity pattern, and the need thus exists to include not only classical harvest indices but also other quality markers when predicting the robustness of fruit to exportation.

The incidence rates of internal disorders and the corresponding severity were higher in the N and R orchards than in the EB orchard (Figure S2a). Nutrient management A (BIK) reduced the incidence of disorders, but there was no significant difference from grower management (Table S1). Moreover, the disorder severity was lowest at the EB orchard (Table S2). This is contrary to the literature that reports that at high DM contents, the highest severity of flesh damage occurs. Therefore, these results confirm that internal damage appears when fruits take a long time to reach the RTE stage. For the N and R orchards, the mineral contents impeded the Hass avocado ripening pattern, as discussed above, and promoted the increased incidence and severity of internal disorders as the DTRE was lengthened (Figure S3).

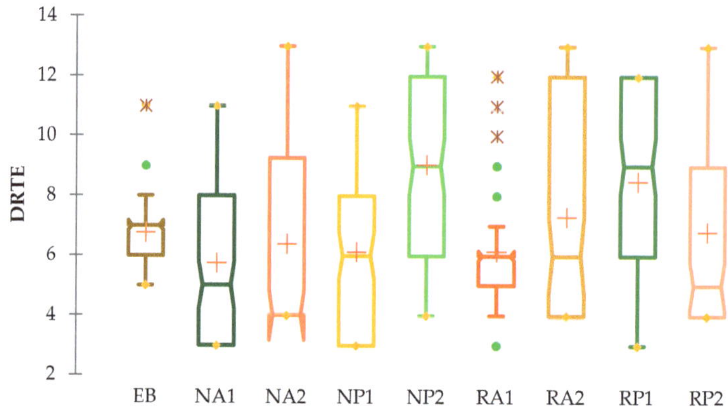

Figure 5. Ripening heterogeneity results of Hass avocados. Orchards: EB, N, and R. A: Nutrient management-based program on the Balanced Indices of Kenworthy (BIK) standards. P: Nutrient management of the farmer. 1. Harvest from the intermediate flowering period. 2. Harvest from the main flowering period. DRTE: Days to ready-to-eat. The crosses represent the mean values, the yellow dots indicate the minimum and maximum values, and the green dots and brown asterisks denote outliers.

Vascular browning was the main internal disorder found (37–76%), and the N and R orchards had the highest occurrence rates of this disorder (Figure S4a, Table S3). Stem end rot was the second most common internal disorder in the samples (6–46%), for which the EB orchard had the lowest incidence rate. Flesh discoloration (FD), flesh bruising (FB), flesh rots, and body rots were found in the samples with low relative incidence frequencies.

For both growth cycles, the percentages of internal damage severity at the R and N orchards were higher than 20% (Figures S5 and S6). The severity increased after the 8 DRTE (days to ready-to-eat) and, in many cases, exceeded the control limit and could affect the consumer acceptance of fruit in the international market [22] because of the disorder severity [23].

The principal component analysis (PCA) results of the flesh mineral and fatty acid contents are shown in Figure 6. The score plot shows that the orchards were effectively grouped according to their composition. In Figure 6b, the loading plot shows that the N orchard had the highest B, N, Zn, and S contents, except for four samples in growth cycle 1 with nutrient management of the grower, which showed low values of these minerals. The R orchard had more Mg, Ca, Cu, and Mg, especially during growth cycle 2. This could be the cause of a reduction in internal quality (increased internal disorders and reduced sensory quality), as will be discussed. The EB orchard had intermediate quantities of B, Fe, and Mn even though it had the highest fruit development time and OC.

The samples far from the EB grouping had low internal quality, high ripening heterogeneity, and even lower sensory quality, as will be shown later. Descriptive statistics of the flesh mineral composition for each orchard are presented in Table S4. These EB mineral contents could be used as a reference to determine Hass avocado heterogeneity and internal quality. Moreover, the results show that despite growing season 2 having low harvest indices, the mineral contents were high at this time. The PLS-DA analysis shows that the principal discriminant variables between orchards indicate that K, N, B, P, and S correspond to the first factor and B, Cu, S, N, and K correspond to the second factor (Figure 7).

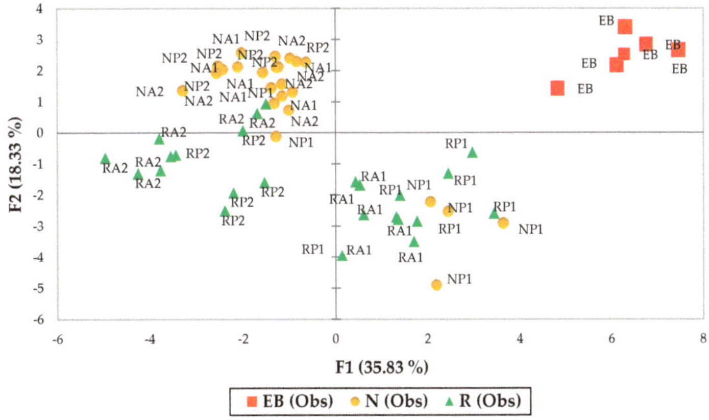

(**a**)

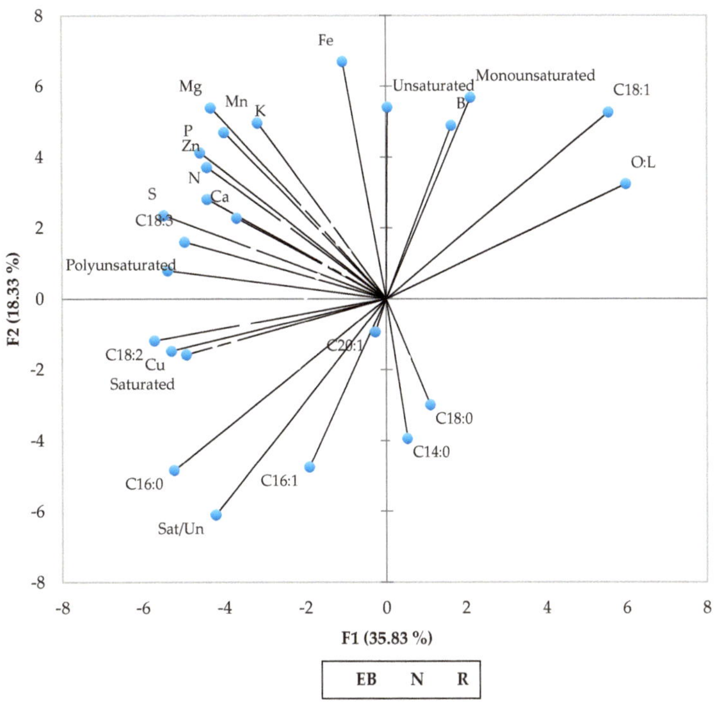

(**b**)

Figure 6. Principal component analysis (PCA) results of the mineral and FAME contents grouped by the samples: (**a**) Score plot; (**b**) Loading plots. Blue dots: active variables. Orchards: EB, N, and R. A: Nutrient management based on the Balanced Indices of Kenworthy (BIK). P: Nutrient management of the farmer. 1. Harvest from the intermediate flowering period. 2. Harvest from the main flowering period. Sat/Un (saturated and unsaturated relationship); C14:0 (myristic acid); C16:0 (palmitic acid), C16:1 (palmitoleic acid), C18:0 (stearic acid); C18:1 (oleic acid), C18:2 (linoleic acid), C18:3 (linolenic acid). C20:1 (arachidic acid), O:I (oleic and linoleic fatty acids relationship).

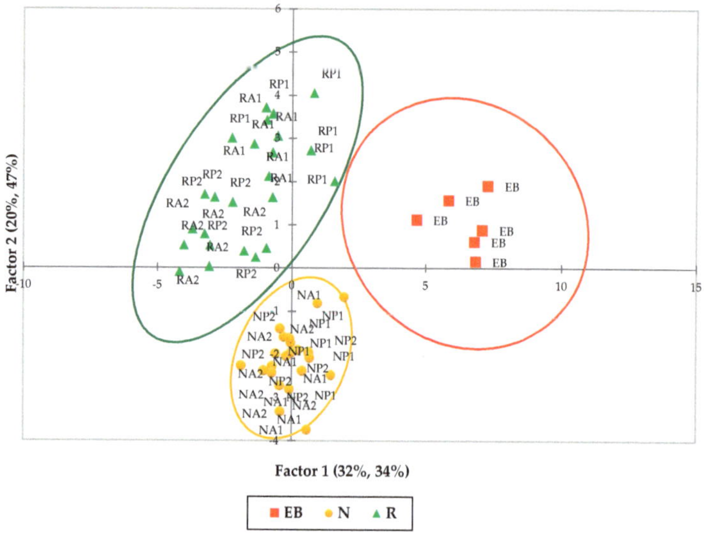

(**a**)

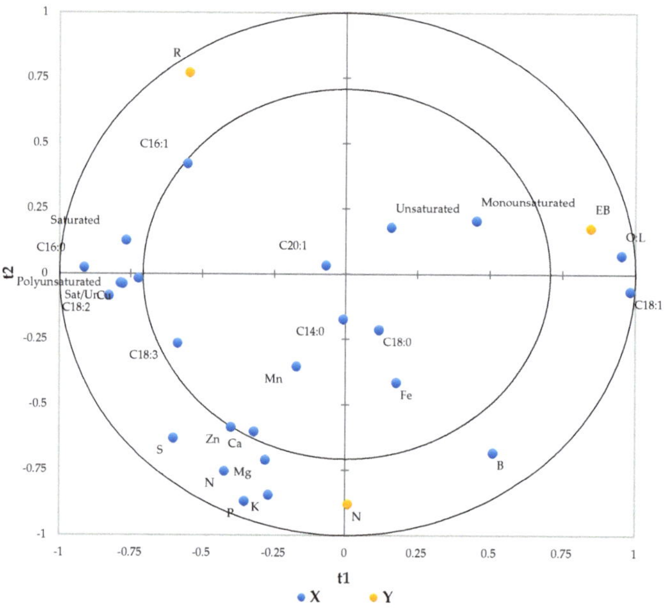

(**b**)

Figure 7. Partial least squares discriminant analysis (PLS-DA) results of the fatty acid and mineral contents grouped by orchard: Score plot (**a**); and loading plots (**b**). R^2 = 88.9 (3 factors). Orchards: EB, N, and R. A: Nutrient management based on the Balanced Indices of Kenworthy (BIK). P: Nutrient management of the farmer. 1. Harvest from the intermediate flowering period. 2. Harvest from the main flowering period. Sat/Un:(saturated and unsaturated relationship); C14:0 (myristic acid); C16:0 (palmitic acid), C16:1 (palmitoleic acid), C18:0 (stearic acid); C18:1 (oleic acid), C18:2 (linoleic acid), C18:3 (linolenic acid). C20:1 (arachidic acid), O:I (oleic and linoleic fatty acids relationship).

As a grouping factor, the localities or orchards were well classified with the PLS-DA analysis using fatty acid compositions as regressor variables. The fatty acid proportions found in each orchard are shown in Table 2. The orchard EB FAME proportion showed significant differences compared to the Cauca orchards (R and N). EB fruits had less saturated fat associated with the fatty acids C16:0 and C16:1, while the C18:1 proportion was significantly higher in the EB orchard. Therefore, the univariate approach confirms that the FAME content is a marker useful for classifying fruits, and as will be discussed, FAMEs have a great impact on sensory quality and are related to fruit age.

Table 2. Composition of fatty acids (% fatty acids on the total fatty acid methyl ester—FAMEs—content) in the three localities studied.

Fatty Acids	Orchards		
	R	N	EB
Saturated	27.531 ± 0.547 a	24.465 ± 0.489 b	19.017 ± 1.262 c
Monounsaturated	57.093 ± 1.207 b	57.062 ± 1.079 b	71.453 ± 2.787 a
Polyunsaturated	15.377 ± 0.482 a	13.901 ± 0.431 a	9.693 ± 1.114 b
Unsaturated	72.469 ± 1.443 b	70.963 ± 1.291 b	81.148 ± 3.333 a
Sat/Un	0.381 ± 0.008 a	0.349 ± 0.007 a	0.235 ± 0.018 b
C14:0	0.199 ± 0.046 a	0.204 ± 0.041 a	0.131 ± 0.107 a
C16:0	26.127 ± 0.327 a	24.174 ± 0.293 b	17.340 ± 0.756 c
C16:1	12.012 ± 0.243 a	10.103 ± 0.217 b	7.043 ± 0.561 c
C18:0	0.599 ± 0.118 a	0.866 ± 0.106 a	0.789 ± 0.274 a
C18:1	44.883 ± 0.548 c	49.498 ± 0.490 b	64.164 ± 1.265 a
C18:2	14.371 ± 0.337 a	13.436 ± 0.302 a	9.177 ± 0.779 b
C18:3	1.043 ± 0.035 a	1.008 ± 0.031 a	0.708 ± 0.080 b
C20:1	0.139 ± 0.015 a	0.168 ± 0.013 a	0.158 ± 0.034 a
O:L	3.225 ± 0.117 c	3.748 ± 0.105 b	7.096 ± 0.271 a

Orchards: R; N, and EB. Sat/Un (saturated and unsaturated relationship); C14:0 (myristic acid); C16:0 (palmitic acid), C16:1 (palmitoleic acid), C18:0 (stearic acid); C18:1 (oleic acid), C18:2 (linoleic acid), C18:3 (linolenic acid). C20:1 (arachidic acid), O:I (oleic and linoleic fatty acids relationship). The same letter in each row represents that there are no statistical differences between orchards (Tukey test, $p < 0.05$). Results are the average ($n = 40$) ± standard error.

Factors 1 and 2 explain 80% of the variance observed in the Y variables (orchards), and with three factors, the model has good robustness with regards to both prediction and stability (R^2 Calibration = 0.889, SEC = 0.106, RMSEC = 0.101, R^2 = 0.821 SECV = 0.122, RMSECV = 0.120). Therefore, the model could be useful for identifying whether new samples have a composition similar or not to the EB samples (colored in red and defined as an ellipse quality in Figure 7a).

The classification models found were:

F(EB) = (0.65 + 1.42 × 10^{-3} × Saturated) + (2.6 × 10^{-3} × monounsaturated) + (4.80 × 10^{-3} × Polyunsaturated) + (2.68 × 10^{-3} × Unsaturated) − (0.10 × Sat/Un) − (5.48 × 10^{-2} × C14:0) − (9.35 × 10^{-3} × C16:0) − 3.06 × 10^{-2} × C16:1) + (1.86 × 10^{-2} × C18:0) + (4.45 × 10^{-3} × C18:1) + (6.73 × 10^{-3} × C18:2) − 0.18 × C18:3) + (5.5042 × 10^{-2} × C20:1) + (3.89 × 10^{-2} × O:L) − (0.39 × N) − (0.42 × P) + (8.58 42 × 10^{-2} × K) − (1.87 × Ca) + (1.87 × Mg) − (5.12 × S) − (3.81 × 10^{-3} × Fe) − (2.86 × 10^{-2} × Cu) + (8.61 × 10^{-2} × Mn) − (1.62 × 10^{-2} × Zn) + (1.66 × 10^{-3} × B).

F(N) = −1.21 − (1.12 × 10^{-2} × Saturated) − (4.11 × 10^{-3} × monounsaturated) − (8.81 × 10^{-3} × Polyunsaturated) − (4.42 × 10^{-3} × Unsaturated) − (0.25 × Sat/Un) − (2.31 × 10^{-2} × C14:0) − (7.02 × 10^{-4} × C16:0) + (2.58 × 10^{-2} × C16:1) + (0.047 × C18:0) − (1.11 × 10^{-3} × C18:1) − (6.49 × 10^{-3} × C18:2) + (8.75 × 10^{-2} × C18:3) + (0.20 × C20:1) − (2.72 × 10^{-2} × O:L) + (0.63 × N) + (1.85 × P) + (0.37 × K) + (7.24 × Ca) − (1.11 × 10^{-2} × Mg) + (3.67 × S) + (0.014 × Fe) − (4.63 × 10^{-2} × Cu) − (7.47 × 10^{-2} × Mn) + (1.99 × 10^{-2} × Zn) + (1.028 × 10^{-2} × B)

F(R) = (1.56 + 9.77 × 10^{-3} × Saturated) + (1.54 × 10^{-3} × monounsaturated) + (4.01 × 10^{-3} × Polyunsaturated) + (1.74 × 10^{-3} × Unsaturated) + (0.35 × Sat/Un) + (7.79 × 10^{-2}

$\times$ C14:0) + (1.00 $\times$ 10^{-2}) $\times$ C16:0) + (4.76 $\times$ 10^{-3} $\times$ C16:1) − (6.54 $\times$ 10^{-2} $\times$ C18:0) − (3.35 $\times$ 10^{-3} $\times$ C18:1) − (2.40 $\times$ 10^{-4} $\times$ C18:2) + (9.22 $\times$ 10^{-2} $\times$ C18:3) − (0.26 $\times$ C20:1) − (1.18 $\times$ 10^{-2} $\times$ O:L) − (0.24 $\times$ N) − (1.43 $\times$ P) − (0.46 $\times$ K) − (5.37 $\times$ Ca) − (1.85 $\times$ Mg) + (1.46 $\times$ S) − (1.02 $\times$ 10^{-2} $\times$ Fe) + (7.49 $\times$ 10^{-2} $\times$ Cu) − (1.14 $\times$ 10^{-2} $\times$ Mn) − (3.69 $\times$ 10-3 $\times$ Zn) − (1.19 $\times$ 10^{-2} $\times$ B)

These models could allow us to predict the internal fruit quality and guarantee the mineral and fatty acids composition necessary to produce a robust fruit for the international supply chain of fresh fruit.

The quality of Hass avocado fruits must be qualified not only as the absence of internal disorders but also as an optimal sensory profile related to the origin (country, region, etc.), rootstock, management, and harvest practices, among other factors. A multivariate profile indicating the orchards, growing season, and nutrient management factors is shown in Figure 8.

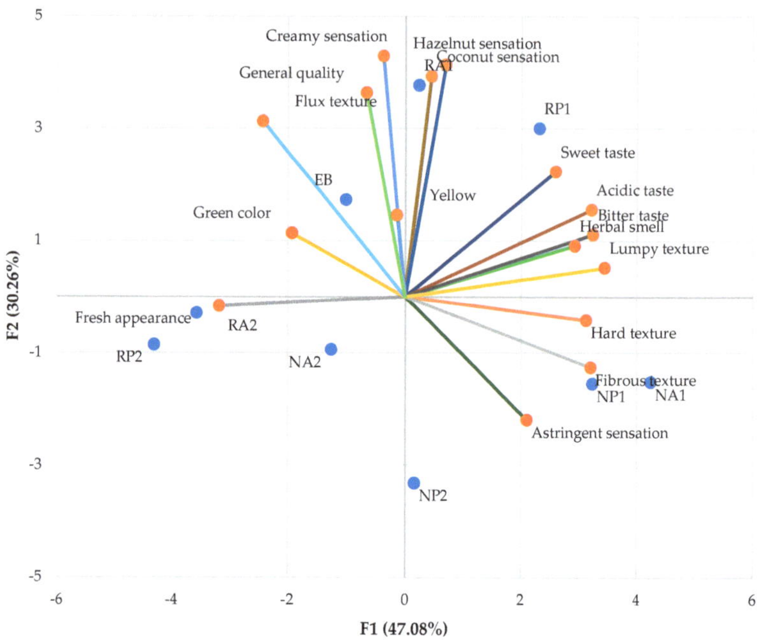

Figure 8. Sensory profile of fruits from different orchards under different nutrient management (P and A) and growth cycle (1 and 2) conditions. Orchards: EB, N, and R. A: Nutrient management based on the Balanced Indices of Kenworthy (BIK). P: Nutrient management of the farmer. 1. Harvest from the intermediate flowering period. 2. Harvest from the main flowering period.

The results show that the sensory descriptors related to high general quality are coconut sensation, hazelnut sensation, flux texture, and creamy sensation (oily sensation), and the means adjusted for these descriptors were high in EB, RA1, and RP1 tests (Figure 8).

Therefore, as discussed before, the composition has an impact on the overall Hass avocado fruit quality, and the "quality ellipse" established (Figure 7a) in the flesh mineral composition indicates that the oleic content and oleic: linoleic relationship are in accordance with the sensory quality profile. High mineral content is related to hard or firm, lumpy and fibrous textures, astringent sensations, and sweet, acid, and bitter tastes, effectively explaining how the sensory profile is related to the RP1, NA1, and NP1 samples. Growing cycle 2 had a low intensity in almost all of the descriptors, and the factor-2 fruits of the biplot showed a hard and fibrous texture with a high green color (Factor 2) and poor development of the discriminant descriptors related to a high sensory quality (Figure 8).

Moreover, the mineral composition and the internal and sensory quality of fruits are affected by the fatty acid composition. The EB orchard had a high oleic acid content and a strong O:L (oleic: linoleic) relationship (Figures 6 and 7). The R and N orchards had greater saturated fatty acid contents (myristic, palmitic, and stearic acids).

One difference between the R and N orchards was that unsaturated fatty acids were more abundant in fruits grown in the R orchard. The variability found in the fatty acid composition between orchards was well-represented by the PLS-DA analysis (Figure 7). These results explain the differences found in the sensory analysis. The samples with more unsaturated (EB and R) fatty acids (oleic, linoleic, linolenic, and palmitoleic acids) had the highest sensory quality (creamy sensation, flux texture, and hazelnut and coconut sensations).

4. Discussion

As a common characteristic in avocado-producing areas in Colombia, a bimodal rainfall distribution pattern is presented [24–28], as was registered in the studied avocado-producing area of Cauca. In this region, the high rainy season registered from January to May supplies the necessary water for nutrient mobility during the main flowering period. However, the intermediate flowering matches the low rainy season recorded from June to September. The flowering phase and initial fruit growth are highly demanding in terms of water and nutrients, and stress factors resulting from these two resources can increase the abscission of flowers and set fruits and can affect fruit growth [29–34]. Herein, in the experimental orchards in Cauca, the nutrient management program based on BIK supplied nutrients monthly, but the low water availability in the low rainy seasons potentially limited the transport of these nutrients to the plants during critical periods of high demand within the intermediate flowering period.

The soils in Cauca orchards showed a low capacity to exchange nutrients such as calcium, magnesium, sodium, and potassium. Previous reports showed a relationship between the nutritional condition of avocado trees and the availability of elements in the soil [35], and some characteristics, such as acidic pH values, low levels of organic matter, and high to very high levels of Cu, Fe, K, Ca, B, and Zn, lead to deficit concentrations of nutrients in the leaves [34]. BIK is a widely used tool for diagnosing the nutrimental status of Hass avocado trees [36,37].

In the methodology adopted herein, we considered an optimum standard value corresponding to each mineral nutrient in trees with yields exceeding 85 kg per tree and obtain the standard deviation for each nutrient in this population of trees. Using this tool to orientate nutrient management in the orchards of the Cauca region, yield increases of 55% and 26% in N and R orchards, respectively, were recorded in the harvest from the main flowering period. Additionally, improvements in fruit weight were recorded. The results shown herein are in accordance with the reports of Salazar-García and Lazcano-Ferrat, who oriented nutrient management based on the BIK methodology and increased fruit calibers with sizes greater than 170 g by 72% of the orchard yield, using specific fertilization with doses calculated for each nutrient throughout the growth cycle of the Hass avocado [4].

Ripening heterogeneity has been reported as a severe logistical problem for marketers and results in the fruits of inconsistent quality being delivered to consumers. This phenomenon is the result of the complex avocado fruit physiology and preharvest and postharvest factors [8]. Rapid softening during ripening has been associated with C7 sugars, organic acids, and the lauric acid, stearic acid, and oleic acid contents.

Therefore, avocado-producing countries with high variability face challenges due to the primary and secondary metabolites of Hass avocados produced during fruit development being affected by the climate, harvest practices, rootstock, etc. [8,38]. Moreover, the preharvest factors and harvest indices are variable in Colombian conditions [39]. In practice, the harvest begins in low-elevation regions at a DM value close to 23%. Our results agree with those of Fuentealba et al.: at low DM values, more ripening heterogeneity occurs [40].

This is because the biological age of the fruit does not guarantee the quantity of fatty acids or other metabolites that soften the flesh. Therefore, in Cauca orchards, growers must increase their development time to ensure adequate Hass avocado ripening.

The flesh mineral content has an effect on delaying the firmness of mesocarp softness because some elements are present and associated with the cell wall structure. Previously, nutrition has been reported to play an important role in postharvest Hass avocado quality. The effect on the internal quality of some minerals has been discussed separately, as briefly described below. High levels of calcium in avocado fruits have been associated with a delay in ripening due to the effects on respiration, the impedance of the ethylene peak, greater firmness after storage, and reduction in internal defects (stem end rot, body rot, vascular browning flesh discoloration, etc.) [41]. An excess of nitrogen fertilization induces the activation of cytokinins that inhibit the production of ethylene and ABA [42], making fruit more vulnerable to fungal pathogen attack [43] and more susceptible to postharvest rot and vascular browning.

Similar to cell wall-associated calcium, boron differences might be another factor contributing to the observed ripening heterogeneity because 95% of the flesh B content is present in the cell wall and acts as a bridge between the cell wall and membrane [8]. In contrast, B-deficient tissues have greater PPO activity and accumulate more phenolic compounds, the substrates for PPO-mediated browning reactions. Increasing the K supply beyond the minimum required value may lead to K-induced deficiencies of Ca and Mg, which could cause increased diffuse discoloration and body rot severity. Attention has been given to the balance between Mg, Ca, and K cations in the fruit, as this balance seems to be more strongly associated with avocado fruit quality than the Mg content alone [7].

According to Rooyen and Bower, copper is known to activate a group of oxidizing enzymes, such as polyphenol oxidase, monophenol oxidase, laccase, and systems of oxidizing ascorbic acid [39,44]. Copper acts as a catalyst for the enzymatic systems that lead to enzymatic browning and is thought to be interrelated with zinc in various oxidation-reduction reactions. Zinc is often coordinated with copper in the various oxidation-reduction reactions that lead to browning. For the other minerals, there is a lack of knowledge of their roles in affecting the internal quality of Hass avocado fruits.

The approaches of past studies aimed to find a simple relationship (linear or nonlinear) between mineral contents and internal disorders. However, our results show that no one mineral plays the main role in affecting the internal quality of Hass avocado fruits, and the associated variabilities could be well-represented in a multivariate approach. PCA and PLS-DA were very useful because the confidence of the ellipse of the high-quality group (EB) allowed the identification of multiple vectors (variables) that cause heterogeneity, internal quality, and sensory profile differences related to the flesh composition.

After the internal disorders, the other quality challenge of Hass avocado origins is obtaining a sensory quality that is satisfactory to consumer expectations. Therefore, high sensory quality could serve as a competitive strategy to improve the success of certain avocado origins in the international market, especially in emerging avocado markets. Studies have found that the harvest season has an effect on the sensory quality of avocado fruits.

Arpaia et al. reported that likeability increased with an advancing harvest date (DM content), and their research was performed with preference sensory tests [45]. Rodríguez et al. employed a sensory profile test and showed that the discriminant descriptors related to unmature fruit were green, bitter, acid, spicy taste, and sensations, whereas flux texture, oily taste, oily texture, and sweet taste were descriptors found in fruits at high DM contents [9].

The results of this work confirmed that harvest practices had an important impact on the Hass avocado fruit quality profile. The main issue under field conditions is that farmers do not have nondestructive equipment or lack knowledge regarding what markers to measure to sort fruit and guide the harvest decisions to guarantee the robustness and sensory quality of the fruit at the destination market [46,47]. Our results show that the sensory descriptors change among localities and growth cycles and could be explained as discussed above with regards to the flesh mineral and FAME composition.

There is no information that reports the role of the mineral contents on the sensory quality of Hass avocado fruits. However, the role of the fruit mineral composition on fruit softness and mesocarp discoloration in 'Pinkerton' avocado fruits was studied by Rooyen and Bower, who published mineral concentrations related to low, medium, and high risks of developing internal flesh disorders [39]. The results of this work contribute not only to reporting optimal mineral ranges for reducing internal damage but also to improving the sensory profile of Hass avocados.

The flesh–fatty acid relationship is influenced by the temperature of the localities or regions where the Hass avocado is produced. This phenomenon has been found in many Hass avocado-producing countries around the world. Some studies have reported that Hass avocados produced in regions with low temperatures had greater oil contents and higher proportions of oleic acid than those grown in high-temperature areas. In contrast, fruits grown in high-temperature regions had more palmitic acid [48,49]. Ferreyra et al. found that the fatty acid proportion changed with elevation in Chile [41].

The authors reported that high oleic acid contents were related to the maximum mean temperature. Therefore, different studies have postulated the fatty acid profile as a biomarker of fruit origin. However, high variabilities exist in avocado-producing countries such as México, Colombia, Chile, and Peru, thus making such classifications difficult. The EB locality has an oleic acid content similar to the cool regions or farms from South Africa [49], New Zealand [48], and Chile [41,49], whereas the low-elevation orchard has a FAME profile similar to producing localities in Peru; Spain [50]; and California, USA [51]. Fruits grown at the R orchard at both nutrient management levels had more palmitoleic acid than fruits of other international origins (low and high elevation or temperature) and were similar to the avocados produced in Florida, USA [52]. There is a lack of information relating the fatty acid profile to the sensory quality. In this work, it is clear that the sensory analysis was influenced by the proportion of fatty acids in the samples, especially with regards to unsaturated fats.

The mouth sensory texture perception could be associated with a change in the mesocarp tissue due to enzymatic (polygalacturonase and cellulase) action during fruit ripening and the mineral and fatty acid compositions. The parenchyma tissue contains oil bodies as a finely dispersed emulsion surrounded by a thin cellulose wall [53], whereas idioblasts are specialized cells containing large oil sacs covered by a thick cellular wall.

During Hass avocado ripening, the degradation of parenchyma cell walls and the release of some oil bodies in emulsion form occur [54]. Despite the importance of the role of fatty acids in Hass avocado composition and quality, few studies have reported the role of nutrient management in modulating fatty acids. This work contributes to identifying the quality markers related to nutrient management under field conditions. Our results show the importance of having a robust fruit flesh composition, and it is necessary to consider rootstock selection, nutrition management, and orchard location to guarantee consistent quality to consumers.

It is necessary to increase the knowledge on the development of nondestructive devices, NIR applications, or other technologies focused on analyzing the discriminant biomarkers found in this work to predict the integral quality of Hass avocados and their shelf life before export. These strategies could reduce postharvest losses and satisfy consumer expectations worldwide.

5. Conclusions

Nutrient BIK management increased the fruit sizes and yields without affecting the postharvest quality. Differences in growing temperatures among localities influenced the fruit compositions, especially the fatty acid composition and postharvest internal quality. The fruit produced at low temperatures had relatively high oleic acid contents, while fruits from orchards at high growing temperatures had more palmitic, palmitoleic, and stearic fatty acids and more internal disorders. Fruits harvested at more development days ripened faster and thus had reduced internal disorder incidence rates and severities.

The multivariate approach allowed us to find a robust model for representing and following the variability in the postharvest internal avocado fruit quality independently of the harvest season, locality, and nutrient management. The PLS-DA model could be a powerful tool for classifying the internal quality of fruits in avocado-producing countries. Therefore, in countries with highly variable growing conditions and production practices, the use of destructive analyses and multivariate data analyses could reduce the heterogeneity and allow the shipment of fruits with better internal quality to the international market.

Supplementary Materials: The following supporting information can be downloaded at https://www.mdpi.com/article/10.3390/horticulturae9040460/s1. Figure S1: Variability in the dry matter (DM) content of Hass avocado fruits during harvest cycles. Figure S2: Percentage of incidence (a) and severity (b) of internal disorders. Figure S3: Logistic regression analysis of internal disorder incidence. Figure S4: Incidence of internal disorders in avocado: relative frequency in vascular browning (a), stem end rot (b), flesh discoloration (c), flesh bruising (d), flesh rots (e), and body rots (f). Figure S5: Severity of internal disorders through Hass avocado ripening for growing cycle 1 of 2021. Figure S6: Severity of internal disorders through Hass avocado ripening for growing cycle 2 of 2021. Table S1: K proportion test results regarding the incidence of internal disorders (Marascuilo procedure). Table S2: Kruskal—Wallis two-tailed test results regarding disorder severity (multiple pairwise comparisons using Dunn's procedure with Bonferroni correction). Table S3: K proportion test results regarding the incidence of internal disorders (Marascuilo procedure comparison). Table S4: Descriptive statistics of the flesh mineral composition of Hass avocados at the RTE (ready-to-eat) stage.

Author Contributions: P.R.: Conceptualization, Methodology, Data curation, Writing—original draft, Formal analysis, Supervision, Project administration, Funding acquisition. I.S.: Methodology, Formal analysis, Writing—original draft. J.V.: Methodology, Writing—original draft. A.R.: Conceptualization, Methodology, Funding acquisition, Writing—original draft. All authors have read and agreed to the published version of the manuscript.

Funding: This research and the APC have been funded by the Sistema General de Regalías-SGR of the Department of Cauca (Colombia), under the Special Cooperation Agreement for Research No. 068 of 08/10/2018. The OCAD of the FCTeI designated the Corporación Colombiana de Investigación Agropecuaria-Agrosavia as the executing public entity of the project identified with the BPIN code 2018000100010 called "Development and validation of technologies to increase the productivity of the Hass Avocado crop in the Department of Cauca".

Institutional Review Board Statement: Not applicable.

Informed Consent Statement: Not applicable.

Data Availability Statement: The data that support the findings of this study are available from the corresponding author upon reasonable request.

Acknowledgments: The authors express their gratitude to the orchard growers, the support team, and the pilot plant of the La Selva and Palmira Research Centers (Agrosavia) for their support in the technical execution of this study and to the Department of Cauca (Colombia) for financial funding.

Conflicts of Interest: The authors declare no conflict of interest. The funders had no role in the design of the study; in the collection, analyses, or interpretation of data; in the writing of the manuscript; or in the decision to publish the results.

References

1. Anonymous. Avocado—World Statistics—Summer 2021 and Winter 2021–2022. *Frutitrop Online* **2022**, *283*, 98–110.
2. Selladurai, R.; Awachare, C.M. Nutrient Management for Avocado (*Persea Americana* Miller). *J. Plant Nutr.* **2020**, *43*, 138–147. [CrossRef]
3. Salazar, S.; Lazcano, I. Identifiying Fruit Mineral Removal Differences in Four Avocado Cultivars. *Better Crop. Int.* **2001**, *15*, 28–31.
4. Salazar, S.; Lazcano, I. La Fertilizacion En "Sitio Especifico" Incrementa Los Rendimientos Y El Tamaño De La Fruta Del Aguacate En Mexico. *Actas V Congr. Mund. Aguacate* **2003**, *35*, 373–379.

5. Hofman, P.J.; Vuthapanich, S.; Whiley, A.W.; Klieber, A.; Simons, D.H. Tree Yield and Fruit Minerals Concentrations Influence "Hass" Avocado Fruit Quality. *Sci. Hortic.* **2002**, *92*, 113–123. [CrossRef]
6. Hofman, P. (Ed.) *Optimising the Posharvest Qualities of Hass Avocado through Improved Calcium Nutrition*, 1st ed.; Horticultural Australia Ltd: Sydney, Australia, 2007; ISBN 0734114389.
7. Joyce, D.C. *Improved Fruit Quality and Robustness in Avocado Supply Chains (Mineral Nutrition)*, 7th ed.; Horticulture Innovation: Sydney, Australia, 2021; ISBN 9780734146854.
8. Hernández, I.; Fuentealba, C.; Olaeta, J.A.; Lurie, S.; Defilippi, B.G.; Campos-Vargas, R.; Pedreschi, R. Factors Associated with Postharvest Ripening Heterogeneity of "Hass" Avocados (*Persea Americana* Mill). *Fruits* **2016**, *71*, 259–268. [CrossRef]
9. Rodríguez-Fonseca, P.E.; Martínez-álvarez, O.L.; Ocampo-Arango, C.M.; Ardila-Castañeda, M.P.; Escobar-Calderón, J.V. Effect of Harvest Index and Storage Duration on Sensory Quality of Hass Avocado Grown in Colombia. *DYNA* **2021**, *88*, 131–139. [CrossRef]
10. Hausch, B.J.; Arpaia, M.L.; Campisi-Pinto, S.; Obenland, D.M. Sensory Characterization of Two California-Grown Avocado Varieties (Persea Americana Mill.) over the Harvest Season by Descriptive Analysis and Consumer Tests. *J. Food Sci. Wiley* **2021**, *86*, 4134–4147. [CrossRef]
11. Donetti, M.; Terry, L.A. Biochemical Markers Defining Growing Area and Ripening Stage of Imported Avocado Fruit Cv. Hass. *J. Food Compos. Anal.* **2014**, *34*, 90–98. [CrossRef]
12. Allen, R.G.; Pruitt, W.O.; Wright, J.L.; Howell, T.A.; Ventura, F.; Snyder, R.; Itenfisu, D.; Steduto, P.; Berengena, J.; Yrisarry, J.B.; et al. A Recommendation on Standardized Surface Resistance for Hourly Calculation of Reference ETo by the FAO56 Penman-Monteith Method. *Agric. Water Manag.* **2006**, *81*, 1–22. [CrossRef]
13. Kenworthy, A. Leaf Analysis as an Aid in Fertilizing Orchards. In *Soil Testing and Plant Analysis*; Walsh, L.M., Beaton, J.D., Eds.; Soil Science Society of America: Madison, WI, USA, 1973; pp. 381–392.
14. *FAO Codex Alimentarius CXS-197-1995*; Normas Internacionales de Los Alimentos. IFFO: London, UK, 1995; pp. 3–4.
15. Waissbluth, R.J.; Valenzuela, J. Determinación del porcentaje mínimo de materia seca para autorizar la cosecha de paltas cv. Hass para ser exportadas. In Proceedings of the VI World Avocado Congress, Viña del Mar, Chile, 12–16 November 2007; pp. 957–978.
16. AOAC. *Official Method of Analysis*, 18th ed.; AOAC: Washington, DC, USA, 2005.
17. White, A.; Woolf, A.; Hofman, P.; Arpaia, M.L. *The International Avocado Quality*; Postharves, White, A., Woolf, A., Research, P.F., Hofman, P., Fisheries, Q.P.I., Arpaia, M.L., Eds.; Plant & Food Research: Auckland, New Zealand, 2009.
18. *ISO (4121)*; Sensory Analysis-Guidelines for the Use of Quantitative Response Scales. Sensory Analysis. Technical Committee, I. 34/SC 12. ISO: Geneva, Switzerland, 2003; Volume 34.
19. Obenland, D.; Collin, S.; Sievert, J.; Negm, F.; Arpaia, M.L. Influence of Maturity and Ripening on Aroma Volatiles and Flavor in 'Hass' Avocado. *Postharvest Biol. Technol.* **2012**, *71*, 41–50. [CrossRef]
20. Escobar, J.V.; Cortés, M.; Correa, G.; Rondón, T.; Rodríguez, P. 'Hass' Avocado Internal Disorders under Simulated Export Conditions and Its Relationship with Flesh Mineral Content and Preharvest Variables. *Horticulturae* **2021**, *7*, 77. [CrossRef]
21. Mckean, S. Thinking with the Saint: The Miracle of Saint Januarius of Naples and Science in Early Modern Europe. *Early Sci. Med.* **1993**, *19*, 133–173. [CrossRef]
22. Gamble, J.; Harker, F.R.; Jaeger, S.R.; White, A.; Bava, C.; Beresford, M.; Stubbings, B.; Wohlers, M.; Hofman, P.J.; Marques, R.; et al. The Impact of Dry Matter, Ripeness and Internal Defects on Consumer Perceptions of Avocado Quality and Intentions to Purchase. *Postharvest Biol. Technol.* **2010**, *57*, 35–43. [CrossRef]
23. Matsui, T.; Kamata, T.; Koseki, S.; Koyama, K. Development of Automatic Detection Model for Stem-End Rots of 'Hass' Avocado Fruit Using X-Ray Imaging and Image Processing. *Postharvest Biol. Technol.* **2022**, *192*, 111996. [CrossRef]
24. Eslava, J. Some Particular Climatic Features of the Colombian Pacific Region. *Atmósfera* **1994**, *17*, 45–63.
25. Mejia, F.; Mesa, O.; Poveda, G.; Velez, J.; Hoyos, C.; Mantilla, R.; Barco, J.; Cuartas, A.; Montoya, M.; Botero, B. Distribución Espacial y Ciclos Anual y Semianual de La Precipitación En Colombia. *DYNA* **1999**, *127*, 7–26.
26. León, G.E.; Zea, J.A.; Eslava, J.A. Circulación General Del Tropico y La Zona de Confluencia Intertropical En Colombia. *Meteorol. Colomb.* **2000**, *1*, 31–38.
27. Poveda, G.; Vélez, J.I.; Mesa, O.J.; Cuartas, A.; Barco, J.; Mantilla, R.I.; Mejía, J.F.; Hoyos, C.D.; Ramírez, J.M.; Ceballos, L.I.; et al. Linking Long-Term Water Balances and Statistical Scaling to Estimate River Flows along the Drainage Network of Colombia. *J. Hydrol. Eng.* **2007**, *12*, 4–13. [CrossRef]
28. Poveda, G.; Álvarez, D.M.; Rueda, Ó.A. Hydro-Climatic Variability over the Andes of Colombia Associated with ENSO: A Review of Climatic Processes and Their Impact on One of the Earth's Most Important Biodiversity Hotspots. *Clim. Dyn.* **2011**, *36*, 2233–2249. [CrossRef]
29. Whiley, A.W.; Chapman, K.R.; Saranah, J.B. Water Loss by Floral Structures of Avocado (Persea Americana Cv. Fuerte) during Flowering. *Aust. J. Agric. Res.* **1988**, *39*, 457–467. [CrossRef]
30. Blanke, M.M.; Lovatt, C.J. Anatomy and Transpiration of the Avocado Inflorescence. *Ann. Bot.* **1993**, *71*, 543–547. [CrossRef]
31. Lahav, E.; Whiley, A.W.; Turner, D.W. Irrigation and Mineral Nutrition. In *Avocado: Botany, Production and Uses*; Schaffer, B., Wolstenholme, B.N., Whiley, A.W., Eds.; CABI: London, UK, 2013; pp. 301–341.
32. Schaffer, B.; Gil, P.; Mickelbart, M.V.; Whiley, A.W. Ecophysiology. In *Avocado: Botany, Production and Uses*; Schaffer, B., Wolstenholme, B.N., Whiley, A.W., Eds.; CABI: London, UK, 2013; Volume 1, pp. 168–199, ISBN 9788578110796.

33. Silber, A.; Naor, A.; Cohen, H.; Bar-Noy, Y.; Yechieli, N.; Levi, M.; Noy, M.; Peres, M.; Duari, D.; Narkis, K.; et al. Avocado Fertilization: Matching the Periodic Demand for Nutrients. *Sci. Hortic.* **2018**, *241*, 231–240. [CrossRef]
34. Cantuarias-Aviles, T.; da Silva, S.R.; Angolini, S.F.; do Amaral Brogio, B.; Baptista, E.G.; Micheletti, L.B. Water Status and Productivity of "Hass" Avocado Trees in Response to Supplemental Irrigation during Winter. *Pesqui. Agropecu. Bras.* **2019**, *54*, e00237. [CrossRef]
35. García-martínez, R. Yield, Quality and Post-Harvest Behavior of 'Hass' Avocado Fruits of Orchards with Different Fertilization. *Rev. Mex. Cienc. Agrícolas* **2021**, *12*, 205–218. [CrossRef]
36. Maldonado-Torres, R.; Álvarez-Sánchez, M.E.; Almaguer-Vargas, G.; Barrientos-Priego, A.F.; García-Mateos, R. Estándares Nutricionales Para Aguacatero 'HASS'. *Rev. Chapingo Ser. Hortic.* **2007**, *13*, 103–108. [CrossRef]
37. Mariscal, J.L.C.; Sánchez, M.E.Á.; Torres, R.M.; Gustavo, A.V. Aplicación de Abonos Orgánicos En El Rendimiento y Desarrollo Radicular En El Cultivo de Aguacate. *Rev. Mex. De Cienc. Agrícolas* **2020**, *11*, 263–274.
38. Medina-Carrillo, R.E.; Salazar-García, S.; Bonilla-Cárdenas, J.A.; Herrera-González, J.A.; Ibarra-Estrada, M.E.; Álvarez-Bravo, A. Secondary Metabolites and Lignin in 'Hass' Avocado Fruit Skin during Fruit Development in Three Producing Regions. *HortScience* **2017**, *52*, 852–858. [CrossRef]
39. Rooyen, V.Z.; Bower, J.P. The Role of Fruit Mineral Composition on Fruit Softness and Mesocarp Discolouration in "Pinkerton" Avocado (*Persea Americana* Mill.). *J. Hortic. Sci. Biotechnol.* **2005**, *80*, 793–799. [CrossRef]
40. Fuentealba, C.; Hernández, I.; Olaeta, J.A.; Defilippi, B.; Meneses, C.; Campos, R.; Lurie, S.; Carpentier, S.; Pedreschi, R. New Insights into the Heterogeneous Ripening in Hass Avocado via LC–MS/MS Proteomics. *Postharvest Biol. Technol.* **2017**, *132*, 51–61. [CrossRef]
41. Ferreyra, R.; Sellés, G.; Saavedra, J.; Ortiz, J.; Zúñiga, C.; Troncoso, C.; Rivera, S.A.; González-Agüero, M.; Defilippi, B.G. Identification of Pre-Harvest Factors That Affect Fatty Acid Profiles of Avocado Fruit (*Persea Americana* Mill.) Cv. 'Hass' at Harvest. *S. Afr. J. Bot.* **2016**, *104*, 15–20. [CrossRef]
42. Ferguson, I.; Volz, R.; Woolf, A. Preharvest Factors Affecting Physiological Disorders of Fruit. *Postharvest Biol. Technol.* **1999**, *15*, 255–262. [CrossRef]
43. Drori, N.; Kramer-Haimovich, H.; Rollins, J.; Dinoor, A.; Okon, Y.; Pines, O.; Prusky, D. External PH and Nitrogen Source Affect Secretion of Pectate Lyase by Colletotrichum Gloeosporioides. *Appl. Environ. Microbiol.* **2003**, *69*, 3258–3262. [CrossRef]
44. Van Rooyen, Z.; Bower, J.P. The Role of Fruit Mineral Composition, Phenolic Concentration and Polyphenol Oxidase Activity on Mesocarp Discolouration in 'Pinkerton'. *S. Afr. Avocado Grow. Assoc. Yearb.* **2003**, *26*, 72–75.
45. Arpaia, M.L.; Collin, S.; Sievert, J.; Obenland, D. Influence of Cold Storage Prior to and after Ripening on Quality Factors and Sensory Attributes of 'Hass' Avocados. *Postharvest Biol. Technol.* **2015**, *110*, 149–157. [CrossRef]
46. Rodriguez, P.; Henao, J.C.; Correa, G.; Aristizabal, A. Identification of Harvest Maturity Indicators for 'Hass' Avocado Adaptable to Field Conditions. *Horttechnology* **2018**, *28*, 815–821. [CrossRef]
47. Metlenkin, D.A.; Platov, Y.T.; Platova, R.A.; Zhirkova, E.V.; Teneva, O.T. Non-Destructive Identification of Defects and Classification of Hass Avocado Fruits with the Use of a Hyperspectral Image. *Agron. Res.* **2022**, *20*, 326–340. [CrossRef]
48. Rozenstvet, O.A.; Saksonov, S.V.; Filin, V.R.; Dembitsky, V.M. Seasonal Changes of Lipid Content in the Leaves of Some Ferns. *Physiol. Plant.* **2000**, *113*, 59–63. [CrossRef]
49. Kaiser, C.; Wolstenholme, B.N. Aspects of Delayed Harvest of "Haas" Avocado (*Persea Americana* Mill.) Fruit in a Cool Subtropical Climate. I.Fruit Lipid and Fatty Acid Accumulation. *J. Hortic. Sci.* **1994**, *69*, 437–445. [CrossRef]
50. Pedreschi, R.; Hollak, S.; Harkema, H.; Otma, E.; Robledo, P.; Westra, E.; Somhorst, D.; Ferreyra, R.; Defilippi, B.G. Impact of Postharvest Ripening Strategies on 'Hass' Avocado Fatty Acid Profiles. *South Afr. J. Bot.* **2016**, *103*, 32–35. [CrossRef]
51. Hausch, B.J.; Arpaia, M.L.; Kawagoe, Z.; Walse, S.; Obenland, D. Chemical Characterization of Two California-Grown Avocado Varieties (*Persea Americana* Mill.) over the Harvest Season with an Emphasis on Sensory-Directed Flavor Analysis. *J. Agric. Food Chem.* **2020**, *68*, 15301–15310. [CrossRef] [PubMed]
52. Pisani, C.; Ritenour, M.A.; Stover, E.; Plotto, A.; Alessandro, R.; Kuhn, D.N.; Schnell, R.J. Postharvest and Sensory Evaluation of Selected 'Hass' × 'Bacon' and 'Bacon' × 'Hass' Avocado Hybrids Grown in East-Central Florida. *HortScience* **2017**, *52*, 880–886. [CrossRef]
53. Yang, S.; Hallett, I.; Rebstock, R.; Oh, H.E.; Kam, R.; Woolf, A.B.; Wong, M. Cellular Changes in "Hass" Avocado Mesocarp During Cold-Pressed Oil Extraction. *J. Am. Oil Chem. Soc.* **2018**, *95*, 229–238. [CrossRef]
54. Cervantes-Paz, B.; Yahia, E.M. Avocado Oil: Production and Market Demand, Bioactive Components, Implications in Health, and Tendencies and Potential Uses. *Compr. Rev. Food Sci. Food Saf.* **2021**, *20*, 4120–4158. [CrossRef] [PubMed]

horticulturae

MDPI

Article

Could an Early Treatment with GA and BA Impact Prolonged Cold Storage and Shelf Life of Apricot?

Maja Milović [1], Žarko Kevrešan [2], Jasna Mastilović [3], Renata Kovač [2,*], Jelena Kalajdžić [1], Nenad Magazin [1], Aleksandra Bajić [2], Biserka Milić [1], Gordana Barać [1] and Zoran Keserović [1]

1 Department of Fruit Growing, Viticulture, Horticulture and Landscape Architecture, Faculty of Agriculture, University of Novi Sad, Dositej Obradović Sq. 8, 21000 Novi Sad, Serbia
2 Institute of Food Technology in Novi Sad, University of Novi Sad, Boul. Cara Lazara 1, 21000 Novi Sad, Serbia
3 Biosense Institute, University of Novi Sad, Dr Zorana Đinđića 1, 21000 Novi Sad, Serbia
* Correspondence: renata.kovac@fins.uns.ac.rs; Tel.: +381-21-485-3779

Abstract: Application of plant growth regulators (PGRs) in apricot orchards is a common practice with a goal of improving yield and/or quality of fruits at harvest. However, the question of whether such treatment alters postharvest properties is seldom answered. The effects of an early application of PGRs on postharvest changes on apricots were investigated on cultivar NS-4, grown on Myrobalan rootstock with blackthorn interstock in a 5-year-old orchard. PGR treatments included 50 and 100 ppm of benzyladenine (BA) and 200 ppm of gibberellic acid (GA_3), which were applied when the green ovary was surrounded by dying a sepal crown, at the stage where sepals beginning to fall. Apricots at the stage of commercial ripeness were used for the postharvest experiments. Analysis was performed at harvest, after 21 days of cold storage (at 1 ± 1 °C and 80 ± 10% RH), and after 3 days of shelf life (24 ± 2 °C). At harvest, significant differences were observed between treated and untreated fruits regarding flesh firmness, color, ethylene production and respiration rate, flavonoid, carotenoid and citric acid content, while application of BA100 changed TA and TSS. Prolonged cold storage reduced the initial differences in firmness, respiration rate, flavonoid and carotenoid contents, but new differences in fructose, malic and succinic acid contents began to appear. Shelf life reduced the difference in citric acid, but differences in TA, TSS, phenol and flavonoid content appeared. There is no difference in the sensory properties of treated and non-treated fruit after cold storage and shelf life.

Keywords: apricot; plant growth regulators; cold storage; shelf life

Citation: Milović, M.; Kevrešan, Ž.; Mastilović, J.; Kovač, R.; Kalajdžić, J.; Magazin, N.; Bajić, A.; Milić, B.; Barać, G.; Keserović, Z. Could an Early Treatment with GA and BA Impact Prolonged Cold Storage and Shelf Life of Apricot? *Horticulturae* **2022**, *8*, 1220. https://doi.org/10.3390/horticulturae8121220

Academic Editors: Wenzhong Hu, Yan Zhong and Xiuxiu Sun

Received: 15 November 2022
Accepted: 16 December 2022
Published: 19 December 2022

1. Introduction

In order to improve apricot production, many contemporary systems and techniques were applied in orchards that have resulted in the intensification of fruit production, such as the use of different rootstocks, interstocks, the increase of tree density, irrigation, pruning, thinning, fertilization, grassing, mulching, and weed control [1–4]. Despite numerous advances in apricot production, the main challenges remained unsolved: (1) frost in the early fruit development phases (flowering) leading to 50% variation in yield coefficient and, in some years, in total loss of apricot fruit [5]; and (2) the relatively short postharvest life of apricot fruit, limited by fruit softening, tissue breakdown and browning [6]. In attempts to improve the postharvest life of apricots, ethylene blockers [7], and different packing solutions (which include a modified and controlled atmosphere), application of calcium, salicylic acid [8], melatonin [9], putrescine [10], as well as use of edible coatings [11], and, recently, irradiation [12] have been frequently used.

Among frequently used methods aiming to improve apricot production is the application of plant growth regulators (PGRs). The process of fruit ripening is guided by the succession of different plant hormones, in terms of their concentration and time of appearance. Fruit development begins with high concentrations of auxins, brassinosteroids, jasmonates and cytokinins, followed by an increase in the concentration of gibberellic acid [13]. The ripening phase is characterized by a high concentration of abscisic acid, followed by small peaks of auxins and jasmonates, while the maximum production of ethylene is present during the transition of fruit from the ripening to senescence stage [13]. The idea behind the external application of PGRs is to alter the natural development of fruit with a goal to improve its production. That improvement caused by application of PGRs does not always imply only a higher yield, but in some cases also improves the quality of the fruit [14].

The application of PGRs regulates plant metabolism, which consequently affects flowering, fruit set and development, as well as fruit abscission, which ultimately reflects on fruit size, composition, and color [15–19]. Application of PGRs in early phases of fruit development is performed with different goals. Thus, gibberellic acid (GA) applied to fully bloomed plants affected fruit set and yield in apricot [20], while its applications 10 or 21 days before harvest affected thinning, fruit size, and firmness [21]. Cytokinin applied at the pit hardening stage had an effect on fruit physiological, organoleptic, and phytochemical properties [22]. In respect to benzyladenine (BA), this plant hormone belongs to the group of cytokinins, and is well-known for its role in plant growth and development, while the studies on its effect on fruit development were mostly focused to climacteric fruits, where it would slow down ripening, possibly through suppressing biosynthesis of ethylene [13]. It was previously proven that BA has an effect on fruit set, development, and shape. Namely, the combination of gibberellins A4 and A7 with the cytokinin BA improved fruit appearance in Red Delicious apples. This success resulted in its different uses on a wide variety of crops [15]. Application of GA_3 at 80% of full bloom affected fruit set, while its application 15 days prior to harvest had an effect on apricot shelf life at ambient temperature [23]. Noteworthy, some compounds with a similar structure to PGRs could have similar impact as PGRs on fruit storage [24].

However, studies examining an impact of PGR application on fruit quality after cold storage and shelf life are very rare, and mostly based on basic analysis of fruit, such as weight loss and spoilage [25]. Even rarer are the studies examining the relationship of an early PGR treatment with storage and shelf life on fruit quality and chemical composition [26]. Having in mind the frequent use of PGRs in practice today, and a lack of knowledge in terms of their effects on treated fruits, the results of the present study provide a glimpse and some interesting answers to the problems associated with PGR application and postharvest behavior of treated fruits. Therefore, this paper is expected to have a significant impact on today's practice in apricot production.

In our previous study, the effects of PGRs applied in the early stages of fruit development were investigated in apricot fruit after 15 days of cold storage followed by several days of shelf life [27]. The treatment had an impact on flesh firmness at harvest, but recorded differences were diminished after shelf life. No difference in titratable acidity (TA) and pH were observed. This study aimed to examine the impact of gibberellic acid (GA_3) and 6-benzyladenine (BA) on the physical characteristics and composition of apricot fruit after prolonged cold storage (i.e., 21 days) and shelf life.

2. Materials and Methods

2.1. Apricot Production and Preharvest Reatments

Apricot fruits (*Prunus armeniaca* L.) cv. NS-4 were obtained from trees grown at the Experimental field for fruit growing, Faculty of Agriculture, Novi Sad, located at Rimski Šančevi (45°3382′ N and 19°8445′ E, 86 m a.s.l.), Republic of Serbia. Cultivar NS-4 used Myrobalan seedlings (*P. cerasifera* Ehrh.) as a rootstock with blackthorn (*Prunus spinosa* L.) as an interstock. The orchard was established under black anti-hail nets and with drip irrigation. The lanes were covered with grass, while the space under the trees within each row was treated with herbicides. The standard agro-technical procedures were performed annually. The experiment was set up by using 5-year-old apricot trees grown at 4 × 2 m planting distance (1250 trees ha^{-1}). The one year trial was set up in a completely randomized design with six single trees used per treatment. Plant hormone treatments included 6-benzyladenine in two concentrations (50 mg L^{-1}—BA50 and 100 mg L^{-1}—BA100, Gerba 4LG—4% active ingredient 6-benzyladenine; L-Gobbi, Campo ligure, Italy), 200 mg L^{-1} gibberellin (GA_3, Gibberellin −1.8% active ingredient gibberellin; L-Gobbi, Campo ligure, Italy) and corresponding, non-treated control. PGRs were applied once by spraying to whole trees when the green ovaries were surrounded by dying sepal crown, just before the sepals begin to fall (stone fruit, principal grown stage 7: development of fruit, code 72, according to Meier [28]). Treatments were applied with a backpack sprayer (Stihl SR-420) and 4 L was used per treatment on six whole trees per treatment. In order to determine the effects on apricot fruits at harvest and after the prolonged storage period, fruit in commercial ripeness (I_{AD} 0.4–0.8), determined by DA-meter (TR Turoni, Bologna, Italy) were harvested [29].

For postharvest analysis, fruits were distributed in wooden crates 50 × 30 × 8 cm holding approximately 5 kg of apricots and placed in the cooling chamber at 1 ± 1 °C and 80 ± 10% RH for 21 days. Subsequently, the fruits were removed from cold storage and exposed for 3 days to shelf life at room temperature (24 ± 2 °C). In order to perform chemical analysis at each sampling period, quarters of 20 fruits was homogenized and immediately frozen in dry ice.

2.2. Fruit Color, Texture and Weight Loss

Fruit color was determined by performing two measurements on the opposite sides of the equatorial region, using CR-400 Chroma Meter (Konica-Minolta, Osaka, Japan). Flesh firmness was measured at the equatorial region of each fruit after peeling off a small circle of skin. A penetration test was performed with an 8 mm diameter stainless steel rounded cylinder probe with TA.XT Plus Texture Analyser (Stable Micro Systems, England, UK). Both analyses were performed on 20 whole apricots.

Weight loss was determined by daily measurement of fruit weight after harvest and after 21 days of cold storage at room temperature (24 ± 2 °C).

2.3. Ethylene Production and Respiration Rate

For determination of ethylene and respiration rate, approximately 250–300 g of fruits were placed in a 770 mL container, hermetically sealed with multilayer foil at 24 ± 2 °C. CO_2 measurement was performed by direct puncture of the sealed foil with a sampling needle of OXYBABY® 6.0 (WIT-Gasetechnik GmbH & Co KG T, Witten, Germany). Ethylene was analyzed from 2 mL of gas sampled by a plastic syringe and injected into a 10 mL headspace vial sealed with silicone septa. Ethylene was determined by gas chromatography (GC 7890, Agilent, Santa Clara, USA), equipped with a FID detector (Agilent, Santa Clara, USA) and auto sampler (COMBIPAL, CTC Analytics AG, Zwingen, Switzerland).

2.4. Chemical Analysis

Total soluble solids (TSS; %) were determined by a digital refractometer ATR-ST plus (Schmidt and Haensch, Berlin, Germany) on previously homogenized apricot samples at 20 °C. Titratable acidity (TA; g malic acid/100 g) was measured from 3 g of sample dissolved in 30 mL of deionized water. After homogenization, the sample was centrifuged (Centrifuge 5804R, Ependorf, Hamburg, Germany) at 13,776 g for 5 min, and 10 mL of supernatant was used for titration with 0.1 M NaOH.

Carotenoid content was analyzed according to Costache et al. [30], while phenol and flavonoid contents were extracted according to [31]. Phenol content was determined according to the Folin–Ciocalteu method [32], while the Pękal and Pyrzynska procedure [33] was used for flavonoids. A brief explanation of the methods used is described earlier in [6,27,29].

For determination of fructose, glucose, sucrose, citric, malic, and succinic acid content, the sample was extracted according to [34]. Separation was performed using HPLC Agilent

1200 series, (Agilent, Santa Clara, CA, USA). For sugar analysis, Zorbax Carbohydrate was used, (4.6 × 250 mm, 5 μm column; Agilent Technologies, Vienna, Austria), along with an evaporative light scattering detector (ELSD), while organic acids were analyzed on a NUCLEOGEL SUGAR 810 H (Macherey-Nagel, Dueren, Germany) column with diode array detector (DAD).

2.5. Sensory Evaluation

Sensory evaluation of apricot fruits after cold storage and shelf life was conducted by 12 trained panelists (6 women and 6 men), aged 20–65 years, according to the methodology adopted by Melgarejo et al. [35]. The panelists were asked to score visual appearance (tissue breakdown and browning) on halved fruits, and intensity of 5 fruit attributes (sweetness, acidity, apricot flavor, crispiness, and off flavor) on fruit slices. Evaluation was performed on a continuous scale ranging from 0 (lowest score) to 100 (highest score). The process was carried out at room temperature (20 °C) in individual cabins under white lighting. The evaluation was performed in two separated sessions within the same day. All participants received written information about the study, and they signed informed consent to participate.

2.6. Statistical Analysis

Obtained results were analyzed using two-way ANOVA. Duncan's multiple range test was used for testing the significance of differences between average values, while for weight loss, respiration, and ethylene, Tukey's HSD test was used. Principal component analysis (PCA) was performed on data collected at the same time and frequency. Statistical calculations were performed using TIBCO Data Science—Workbench (Statistica® 14.0.0).

3. Results

3.1. Weight Loss, Respiration Rate and Ethylene Production

Physiological parameters (weight loss, respiration rate, and ethylene production) were differently affected by the treatments to which were apricots subjected (Figure 1). Significant impact on weight loss at the end of shelf life was observed only between the control and BA100-treated fruits (Figure 1A,B).

Respiration rate was significantly decreased at the end of shelf life after the harvest in all treated apricots versus the control (Figure 1C), while after the cold storage, this effect was present only in BA100-treated fruits (Figure 1D).

After the harvest, ethylene production increased on the 5th day of shelf life, but only in the control when compared to the rest of the PGR-treated groups (Figure 1E). After cold storage, its production increased after the 2nd day of shelf life in all groups, but it became especially prominent on the 4th day in the control and BA50-treated fruits (Figure 1F).

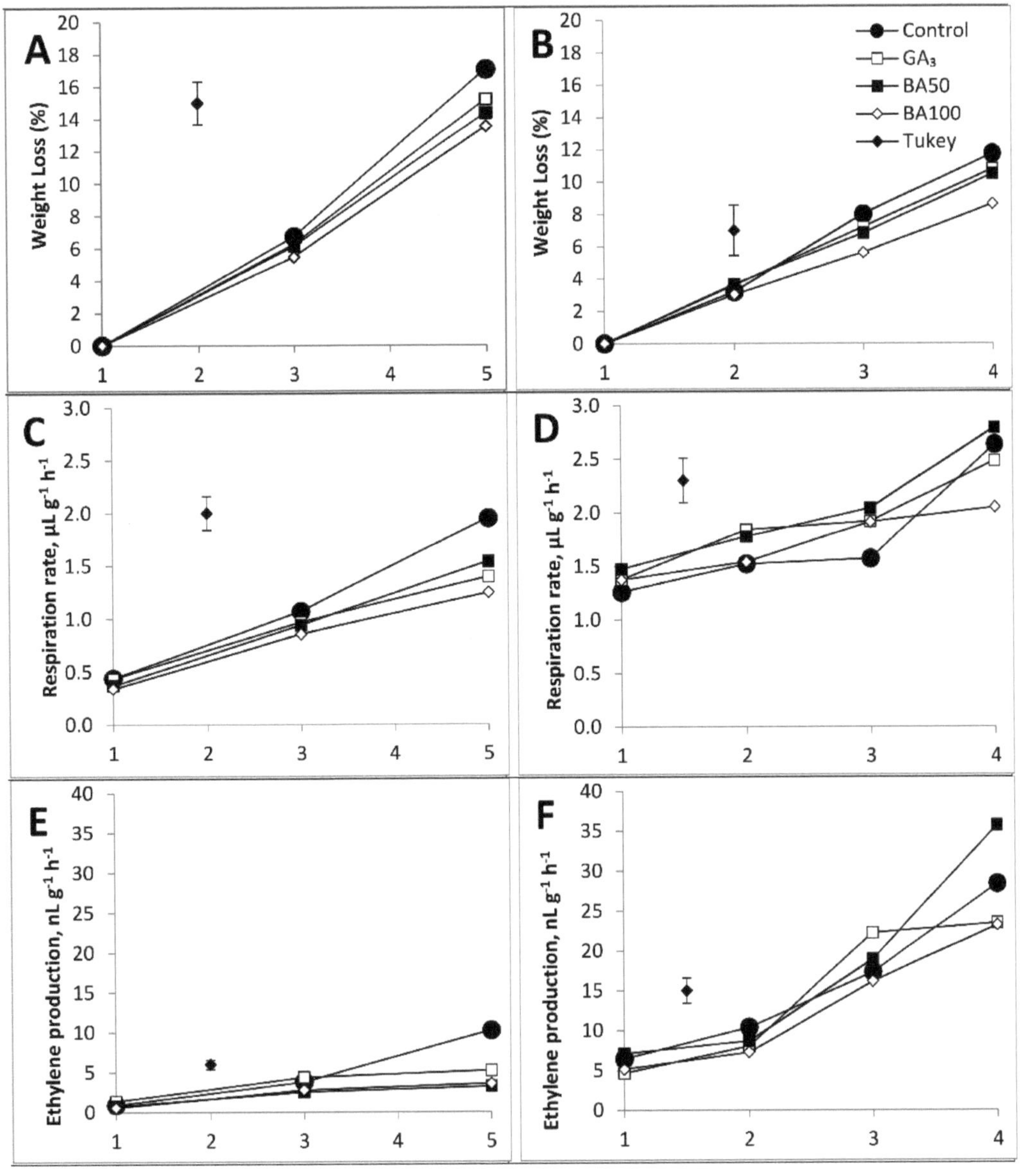

Figure 1. Weight loss, respiration rate, and ethylene production in PGR-treated apricots during shelf life. Fruit weight loss (**A**,**B**), respiration rate (**C**,**D**), and ethylene production (**E**,**F**) after harvest and cold storage, respectively. PGR treatment: control—untreated apricots; GA_3—200 mg L^{-1} of gibberellin A3; BA50—50 mg L^{-1} of 6-benzyladenine; BA100—100 mg L^{-1} of 6-benzyladenine. Differences higher than Tukey mark are significantly different ($p < 0.05$).

3.2. Physicochemical Properties

The PGR treatments showed different effects on physical properties and chemical composition on apricots at harvest, and after cold storage and shelf life (Table 1). In general, PGR treatments alone affected hue°, TA, and the contents of phenols, flavonoids, fructose, sucrose, and all organic acids in treated apricots. As expected, cold storage changed all analyzed parameters in all fruits, except the glucose content. Interaction between the PGRs and cold storage significantly affected TA, phenols, flavonoids, carotenoids, fructose, sucrose, malic, citric, and succinic acids, thereby suggesting that detected changes during cold storage are PGR-dependent.

Table 1. Effects of application of GA and BA on color, texture, quality and chemical composition of apricot fruit.

Treatment	Days	L *	Hue°	Flesh Firmness (N)	TA (%)	TSS %)	Phenols(mg kg^{-1} FW)	Flavonoids (mg kg^{-1} FW)	Carotenoids (mg kg^{-1} FW)	Fructose(g kg^{-1} FW)	Glucose(g kg^{-1} FW)	Sucrose(g kg^{-1} FW)	Malic Acid(g kg^{-1} FW)	Citric Acid(g kg^{-1} FW)	Succinic Acid(g kg^{-1} FW)
Control	0	62.2^{e}	77.5^{d}	21.2^{e}	1.21^{bc}	9.96^{c}	45.0^{a}	7.6^{b}	1.47^{a}	1.10^{a}	0.854	5.57^{d}	1.68^{abc}	1.98^{a}	0.63^{ab}
	21	59.2^{cd}	70.8^{bc}	6.9^{cd}	1.19^{ab}	10.98^{d}	54.7 abcd	8.0^{b}	2.85^{cde}	1.50^{c}	0.890	4.64^{bc}	1.58^{a}	1.95^{a}	0.65^{b}
	21 + 3	57.0^{c}	68.2^{abc}	2.9^{a}	1.36^{e}	10.78^{d}	49.4^{b}	5.6^{a}	3.21^{de}	1.52^{c}	0.897	4.36^{b}	1.64^{abc}	2.14^{b}	0.97^{d}
GA_3	0	60.1^{de}	71.1^{bc}	25.0^{f}	1.24^{bcd}	10.28^{c}	57.2^{bcd}	13.0^{d}	2.27^{bc}	1.16^{b}	0.866	5.74^{d}	1.68^{abc}	2.16^{bc}	0.59^{a}
	21	58.9^{cd}	65.6^{ab}	7.5^{cd}	1.23^{bcd}	12.10^{f}	53.0^{abc}	7.3^{b}	2.84^{cde}	1.26^{b}	0.848	4.23^{b}	1.72^{cd}	2.20^{bc}	0.76^{c}
	21 + 3	55.0^{ab}	64.0^{a}	1.1^{a}	1.56^{f}	12.74^{g}	64.4^{d}	7.6^{b}	3.33^{e}	1.65^{de}	0.799	4.24^{b}	1.83^{d}	2.26^{bcd}	0.76^{c}
BA50	0	61.0^{de}	70.8^{bc}	22.9^{ef}	1.27 bcde	9.86^{c}	54.1 abcd	9.9^{c}	2.25^{bc}	1.18^{ab}	0.904	5.68^{d}	1.74^{cd}	2.21^{bc}	0.59^{a}
	21	58.7^{cd}	64.8^{a}	5.6^{bc}	1.11^{a}	11.52^{c}	54.1 abcd	7.8^{b}	2.67^{cd}	1.63^{de}	0.846	4.24^{b}	1.60^{ab}	2.15^{bc}	0.77^{c}
	21 + 3	54.7^{a}	63.2^{a}	1.9^{a}	1.30^{cde}	12.56^{fg}	57.0^{bcd}	5.4^{a}	3.05^{de}	1.57^{d}	0.846	5.43^{d}	1.82^{d}	2.25^{bcd}	0.93^{d}
BA100	0	60.3^{de}	72.3^{c}	25.7^{f}	1.32^{de}	8.76^{a}	53.9 abcd	12.4^{d}	2.02^{b}	1.21^{b}	0.909	5.14^{cd}	1.69^{bc}	2.34^{de}	0.63^{ab}
	21	59.2^{cd}	66.7^{abc}	9.2^{d}	1.26^{bcd}	10.68^{d}	62.7^{cd}	8.6^{b}	2.34^{bc}	1.63^{de}	0.835	3.32^{a}	1.70^{bc}	2.27^{cd}	0.83^{e}
	21 + 3	55.9^{ab}	64.7^{a}	2.2^{a}	1.22^{bc}	12.34 cfg	74.3^{d}	8.0^{b}	3.26^{de}	1.71^{e}	0.838	4.56^{bc}	1.83^{d}	2.40^{e}	1.20^{f}
PGRs		NS	**	NS	**	NS	**	**	NS	**	NS	**	**	**	**
Storage		**	**	**	**	**	**	**	**	**	NS	**	**	**	**
PGRs x Storage		NS	NS	NS	**	NS	*	**	**	**	NS	**	*	NS	**

Values designated by the same letter were not significantly different ($p > 0.05$). Main factors and their interactions are presented, and their significance is annotated as follows: NS—not significant, *—significant at 0.05, and **—significant at 0.01; FW—fresh weight; PGR treatment: control—untreated apricots; GA_3—200 mg L^{-1} of gibberellin A3; BA50—50 mg L^{-1} of 6-benzyladenine; BA100—100 mg L^{-1} of 6-benzyladenine; storage: 0—at harvest; 21 day of cold storage; 21 + 3—21 day of cold storage followed by 3 days of shelf life.

3.2.1. Physical Properties

At harvest all PGR-treated fruits changed color towards red (hue°). The application of PGR did not affect the lightness of skin color, but since the hue° values were lower, the treated fruits appeared as more red, compared to the control (Table 1). Nevertheless, initial color differences between control and treated fruits faded away after cold storage and shelf life.

Flesh firmness of apricots after harvest was higher in GA_3- and BA100-treated fruits versus the control. After cold storage, fruit firmness was the highest in BA100-treated apricots and was only significant if compared to the second BA treatment. After shelf life, all differences between the treatments were diminished (Table 1).

3.2.2. Chemical Properties

After harvest, only BA100-treated apricots had higher TA and lower TSS than control apricots. Cold storage decreased TA only in BA50-treated apricots, but after shelf life, it increased again in all fruits, except in BA100-treated fruits (Table 1). In respect to TSS, cold storage increased this parameter only in GA_3- and BA50-treated fruits, but after shelf life, all treated apricots has significantly higher values of TSS compared to the control.

Applied PGRs also changed the chemical composition by increasing the contents of phenol, flavonoid, carotenoid, fructose, and citric acid.

At harvest, the highest phenol content was present in apricots treated with GA_3 and the lowest in the control (Table 1). After cold storage, no significant differences were observed among the experimental groups, but after shelf life, phenol content significantly increased in GA_3- and BA100-treated fruits versus the control. Similarly to phenols at harvest, flavonoids were the highest in GA_3-, followed by BA100-treated apricots, while the lowest content was recorded in the control (Table 1). Unlike the phenol content, which increased after cold storage and shelf life, flavonoid content had a decreasing trend. As a result, the lowest loss in these bioactive compounds was detected in GA_3- and BA100-treated apricots comparing to the rest of the groups. In respect to carotenoid content, all treated fruits had higher values than control fruit (Table 1). However, all recorded differences were eliminated after cold storage, and the slight increase in its content after shelf life did not make any difference among the treatments (Table 1).

In respect to sugar contents, only glucose did not change throughout the experiment (Table 1). Slightly higher fructose content was present at harvest in all treated fruits versus the control. During the cold storage, fructose content increased in all fruits, except in GA_3-treated fruits, but after shelf life, the highest increase in its content was detected in GA_3- and BA100-treated apricots. Sucrose level at harvest was similar for all fruits; it decreased during cold storage and after shelf life, with the highest content present in BA100-treated apricots versus the rest of the fruits.

In general, each of the examined organic acids did not show any decline toward the end of the experiment (Table 1). At harvest, contents of malic acid was similar among all fruits. After cold storage, GA_3- and BA100-treated apricots were characterized with a higher content of malic acid then the other two groups, while after shelf life, all treated apricots exhibited a significantly higher content of malic acid versus the control. Content of citric acid was higher in treated apricots at harvest as well as after cold storage versus the control, with BA100 group standing out due to its highest content of this acid. Shelf life increased the level of citric acid only in the control fruits. Succinic acid content was similar in all fruits at harvest. During cold storage, its content increased for all treated fruits. After shelf life, succinic acid increased in all apricots, with the highest content being recorded in BA100-, while the lowest in GA_3-treated apricots

3.3. Sensory Properties

According to the results of sensory analysis, cold storage affected all examined traits, except inappropriate taste and browning, while PGR treatments only had an impact on tissue breakdown (Table 2).

After cold storage, the only difference among the examined parameters was detected regarding aroma between BA50- and BA100-treated apricots. After shelf life, significant changes in sweetness, sourness, and aroma were recorded in BA100-treated fruits as well as in aroma in GA_3-treated apricots, when compared to initial estimation. After cold storage, both BA concentrations decreased tissue break down, but the only significant difference was noted when apricots from the GA_3 and BA100 groups were compared.

Following shelf life, only fruits treated with BA100 had significantly lower tissue breakdown when compared to the rest of the apricots. Still, tissue break down after shelf life was quite high for all treatments.

Table 2. Effects of application of GA and BA on sensory properties of apricot fruit.

Treatment	Days	Sweetness	Sourness	Aroma	Crispiness	Gumminess	Inappropriate Taste	Tissue Breakdown	Browning
Control	21	39.4 ^{ab}	34.5 ^{ab}	40.5 ^{ab}	35.7 ^{b}	39.9 ^{b}	3.6	20.0 ^{ab}	2.5
	21 + 3	52.5 ^{ab}	20.8 ^{ab}	60.9 ^{bc}	7.7 ^{a}	10.4 ^{a}	0.7	90.0 ^{d}	7.0
GA_3	21	40.7 ^{ab}	32.8 ^{ab}	40.5 ^{ab}	40.8 ^{b}	51.2 ^{b}	0.7	22.5 ^{b}	6.5
	21 + 3	58.4 ^{b}	15.3 ^{a}	72.3 ^{c}	0.7 ^{a}	7.1 ^{a}	0.7	97.2 ^{d}	3.3
BA50	21	54.3 ^{ab}	18.7 ^{ab}	56.8 ^{bc}	30.2 ^{b}	42.3 ^{b}	2.2	7.5 ^{ab}	3.0
	21 + 3	57.6 ^{b}	16.3 ^{a}	66.3 ^{c}	2.9 ^{a}	7.3 ^{a}	0.3	97.2 ^{d}	5.6
BA100	21	36.7 ^{a}	36.7 ^{b}	36.0 ^{a}	36.3 ^{b}	48.7 ^{b}	2.0	5.0 ^{a}	3.5
	21 + 3	56.0 ^{b}	15.9 ^{a}	68.4 ^{c}	7.7 ^{a}	6.9 ^{a}	1.4	72.5 ^{c}	7.0
PGRs		NS	NS	NS	NS	NS	NS	**	NS
Storage		**	**	**	**	**	NS	**	NS
PGRs x Storage		NS	NS	NS	NS	NS	NS	NS	NS

Values designated by the same letter were not significantly different ($p > 0.05$). Main factors and their interactions are presented, and their significance is annotated as follows: NS—not significant, **—significant at 0.01; PGR treatment: control—untreated apricots; GA_3—200 mg L^{-1} of gibberellin A3; BA50—50 mg L^{-1} of 6-benzyladenine; BA100—100 mg L^{-1} of 6-benzyladenine; storage: 0—at harvest; 21 day of cold storage; 21 + 3—21 day of cold storage followed by 3 days of shelf life.

3.4. PCA Analysis

Based on the PCA analysis (Figure 2), the first two factors explained more than 70% of variability in the case of color, texture, and all indicators of metabolic activities: TA, TSS, phenols, flavonoids, carotenoids, sugars, and organic acids sampled at three sampling points (Figure 2A). In the case of sensory descriptors assessed after cold storage and shelf life, the first two factors explain more than 90% (Figure 2B). It is noticeable that most of the variation is explained by Factor 1 (55.66% and 75.98%, respectively). Given the nature of the sample separation in both cases, it could be tentatively assumed that Factor 1 represents the sampling time, while Factor 2 may represent PGR treatments.

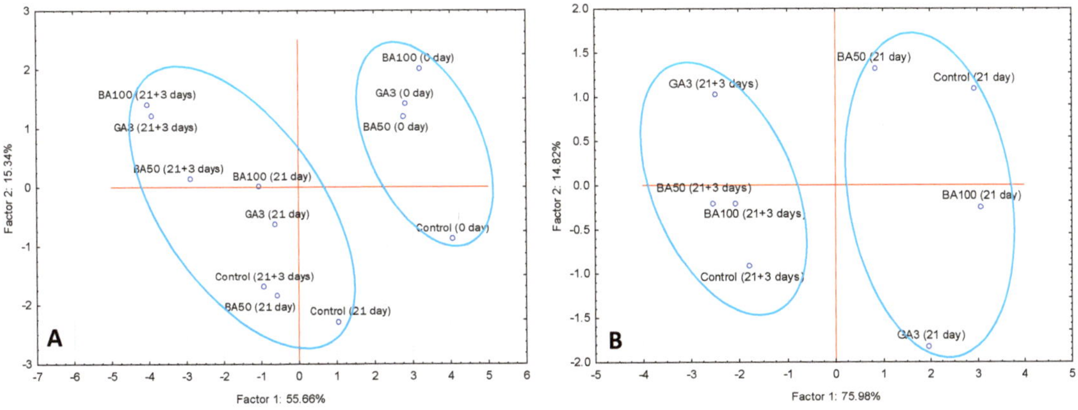

Figure 2. Principal component analysis (PCA) of apricot fruit separation at three sampling points based on color, texture, quality, and chemical composition (**A**) and two sampling points based on sensory characteristics (**B**). Treatments: GA_3—200 mg L^{-1} of gibberellin A3; BA50—50 mg L^{-1} of 6-benzyladenine; BA100—100 mg L^{-1} of 6-benzyladenine; storage: 0—at harvest; 21 day of cold storage; 21 + 3—21 day of cold storage followed by 3 days of shelf life. Circled treatments represent the samples belonging to the following groups: before and after cold storage (**A**) and after cold storage and after shelf life (**B**), respectively.

4. Discussion

Apricot perishability is one of the main causes limiting its storability and, consequently, market potential. This factor becomes even more important when apricots are exposed to ambient temperatures [36], and it is one of the key reasons why some studies have no records of apricot shelf life [37]. Despite this, there are reports on the decrease in physiological loss in apricot cv. Harcot, treated with GA_3 following eight days of exposure to ambient temperature [23]. Our results also indicate a change in weight loss, but a decrease in weight loss in cv. NS-4 was caused by BA100, rather than GA_3, compared to control (Figure 1A,B).

Application of PGRs often changes fruit texture and color at harvest. In the present study, PGRs changed fruit color to redder tones, and increased fruit firmness (Table 1). Similar changes in apricot color were reported after the treatment with 25, 50, 75, and 100 ppm of GA_3 [38]. Obtained results of the increase in flesh firmness are similar to the previous reports, in which apricot fruit firmness increased at harvest as a result of application of 10 and 15 ppm GA_3 [20], and also after application of 50 and 100 ppm BA and 50 and 100 ppm GA_3 [39]. In apricots, beside different concentrations of PGRs, the time of the application also affects flesh firmness, which is shown in the case of GA [40].

The changes in color and texture at harvest are accompanied with the changes in chemical composition of apricot fruits; i.e., TA, TSS, phenols, flavonoids, carotenoids fructose, sucrose, malic, citric, and succinic acid contents (Table 1). Although recorded changes are not uncommon for apricot fruit, their trends are not always similar. Contrary to our results, treatment with BA did not change TA [41], nor did the treatment with GA_3 [23]. On the other hand, the treatments with GA_3 and BA reduced TA [39]. Furthermore, treatment with different concentrations of BA (50, 100 and 150 ppm) did not change TSS [41], nor did the treatment with 50 and 100 ppm of GA in two seasons of Patterson apricot [40]. Still, the treatment with 75 and 100 ppm of GA increased TSS in both applied timings [38]. Regarding the increase in total phenol content, similar findings were reported after application of 10 and 15 ppm GA_3 [20], while other cytokinins (forchlorfenuron) had no effect in two apricot cultivars [22]. Glucose level was not changed by PGRs, which is consistent with reports where lower concentration of synthetic cytokinin did not change glucose, fructose, sucrose, or sorbitol contents [22]. Moreover, similar changes in quality, as a consequence of PGR treatment at harvest, were reported for litchi [18], mango [42], cherry [43], plum [26], and apricot [41]. Such different trends, despite of similar treatments, could be explained by several different aspects. Some of them may include the usage of different cultivars, different PGR concentrations and application time, but also the variations in climatic conditions should not be overlooked. Another important factor is the insufficiently defined ripening stage of apricot, which inevitably leads to different stages of fruit maturity at the moment of harvest, and contributes to inconsistent or contradictory results. Regarding this, the fruit selection may be resolved using a noninvasive method for determination of the ripening stage for each apricot individually [22].

Externally added PGRs interact with the naturally defined sequence of the occurrence of plant hormones and could lead to subtle changes in plant and fruit metabolism, as detected in this study and in previous reports. Fruit metabolism does not stop after its picking, thus it continues during the postharvest period, causing changes in almost all measured parameters (Figure 1, Tables 1 and 2). Alongside 21 days of cold storage, preharvest treatment of apricots with PGRs reflected on a number of observed parameters during cold storage: TA, fructose and sucrose contents, citric, malic, and succinic acid contents, bioactive compounds (phenols, flavonoids), and carotenoids.

The mentioned changes caused by PGR application, according to PCA analysis, are tentatively assigned as Factor 2, which could explain only 15.34% and 14.82% of variability, respectively (Figure 2A,B). Such small percentages coupled with relatively close proximity of samples after cold storage and shelf life pointed out a small impact, and thereby the overall small difference caused by the application of PGRs.

Considering our previous study in which the apricots were stored for 15 days [27], 21 day of cold storage did not eliminate the differences in weight loss. Furthermore, some parameters appeared to be affected by PGRs as the storage period extended; i.e., TA, fructose and sucrose, organic acids, bioactive compounds (phenols, flavonoids), and carotenoids.

According to some reports, application of PGRs could have a negative effect on apricot storability, reflected by a reduction of firmness in NAA-treated apricots after 20 days of cold storage [44]. This effect was not recorded in this study, since the PGRs alone or in combination with 21 day of cold storage did not affect flesh firmness (Table 1).

Subsequent shelf life, as expected, increased the respiration rate and ethylene production when compared to the values before cold storage; also it reduced lightness, increased malic acid content, and, in the case of GA_3 and BA100, increased TSS, phenol, flavonoid, and fructose contents. None of these differences could be observed by sensory analysis.

5. Conclusions

As expected, the application of PGR improved most of the quality traits and fruit composition of apricots at harvest. The significant differences between the treated and untreated fruits were observed at harvest in terms of their flesh firmness, color, ethylene production and respiration rate, flavonoid, carotenoid, and citric acid contents. Prolonged cold storage diminished the initial differences in firmness, respiration rate, flavonoid and carotenoid contents, but new differences in fructose, and malic and succinic acid contents began to appear. Noteworthy, shelf life reduced the difference in citric acid, but the differences in TA, TSS, phenol, and flavonoid contents were recorded. In terms of sensory properties, the treated and non-treated apricots did not show differences following the cold storage and shelf life. Despite the observed differences in chemical composition at the end of cold storage, fruit firmness, color, and other sensory traits did not differ between the treated and non-treated fruits. Differences between the treated and non-treated apricots were recorded after shelf life in respect to fruit color, more precisely, treated fruits had lower L*, which made them more dark than control fruits. Still, the recorded fluctuations in chemical composition needs another extensive and separate study, which will be focused on the underlying processes that defines the ripening and senescence, as well as the PGR treatment.

Author Contributions: Conceptualization and Experimental Design: J.M., Ž.K., B.M., N.M. and Z.K.; Experiment in the field: M.M., J.K. and G.B.; Postharvest experiment and laboratory analyses: Ž.K., A.B., G.B., M.M. and J.K.; Validation of laboratory methods R.K.; Results analysis and Data curation: J.M. and Ž.K.; Writing—Original Draft Preparation: Ž.K., J.M. and R.K.; Writing—Review & Editing: Ž.K., J.M. and R.K.; Supervision: J.M. and Ž.K. All authors have read and agreed to the published version of the manuscript.

Funding: This research was financed by the Ministry of Education, Science and Technological Development of Republic of Serbia [Contract numbers: 451-03-68/2022-14/200222] and project "The use of plant growth regulators and biostimulants for the improvement of fruit quality and storage ability" funded from 2016–2019, by Provincial Secretariat for Higher Education and Scientific Research, Autonomous Province of Vojvodina, Republic of Serbia.

Data Availability Statement: Not applicable.

Conflicts of Interest: The authors declare no conflict of interest.

References

1. Sansavini, S.; Giannerini, G.F. Advances in apricot growing and management. *Acta Hortic.* **1991**, *293*, 409–430. [CrossRef]
2. Bassi, D.; Bartolini, S.; Viti, R. Recent advances on environmental and physiological challenges in apricot growing. *Acta Hortic.* **2006**, *717*, 23–32. [CrossRef]
3. Moale, C.; Ghiurea, M.; Sîrbu, C.E.; Somoghi, R.; Cioroianu, T.M.; Faraon, V.A.; Lupu, C.; Trică, B.; Constantinescu-Aruxandei, D.; Oancea, F. Effects of Siliceous Natural Nanomaterials Applied in Combination with Foliar Fertilizers on Physiology, Yield and Fruit Quality of the Apricot and Peach Trees. *Plants* **2021**, *10*, 2395. [CrossRef]
4. Mendelné Pászti, E.; Bujdosó, G.; Mendel, Á. Vegetative Characteristics of Three Apricot Cultivars Grafted on Six Different Rootstocks. *Horticulturae* **2022**, *8*, 1004. [CrossRef]
5. Milatović, D.; Đurović, D.; Zec, G. Osetljivost sorti kajsije na zimski i pozni prolećni mraz. In *Zbornik Radova IV Savetovanja „Inovacije u Voćarstvu"*; Poljoprivredni fakultet Univerziteta u Beogradu: Belgrade, Serbia, 2013; pp. 239–247.
6. Mastilović, J.; Kevrešan, Ž.; Milović, M.; Kovač, R.; Milić, B.; Magazin, N.; Plavšić, D.; Kalajdžić, J. Effects of ripening stage and postharvest treatment on apricot (*Prunus armeniaca* L.) cv. NS4 delivered to the consumers. *J. Food Process. Preserv.* **2022**, *46*, e16399. [CrossRef]
7. Rebeaud, S.G.; Cioli, L.; Cotter, P.Y.; Christen, D. Cultivar, maturity at harvest and postharvest treatments influence softening of apricots. *Postharvest Biol. Technol.* **2023**, *195*, 112134. [CrossRef]
8. Batool, M.; Bashir, O.; Amin, T.; Wani, S.M.; Masoodi, F.A.; Jan, N.; Bhat, S.A.; Gul, A. Effect of oxalic acid and salicylic acid treatments on the post-harvest life of temperate grown apricot varieties (*Prunus armeniaca*) during controlled atmosphere storage. *Food Sci. Technol. Int.* **2022**, *28*, 557–569. [CrossRef]
9. Medina-Santamarina, J.; Zapata, P.J.; Valverde, J.M.; Valero, D.; Serrano, M.; Guillén, F. Melatonin treatment of apricot trees leads to maintenance of fruit quality attributes during storage at chilling and non-chilling temperatures. *Agronomy* **2021**, *11*, 917. [CrossRef]
10. Hosseinifarahi, M.; Mousavi, S.M.; Mohsen, R.A.D.I.; Jowkar, M.M.; Romanazzi, G. Postharvest application of hot water and putrescine treatments reduce brown rot and improve shelf life and quality of apricots. *Phytopathol. Mediterr.* **2020**, *59*, 319–329.
11. Algarni, E.H.; Elnaggar, I.A.; Abd El-wahed, A.E.W.N.; Taha, I.M.; Al-Jumayi, H.A.; Elhamamsy, S.M.; Mahmoud, S.F.; Fahmy, A. Effect of Chitosan Nanoparticles as Edible Coating on the Storability and Quality of Apricot Fruits. *Polymers* **2022**, *14*, 2227. [CrossRef]
12. Muzzaffar, S.; Bhat, M.M.; Wani, T.A.; Wani, I.A.; Masoodi, F.A. Postharvest biology and technology of apricot. In *Postharvest Biology and Technology of Temperate Fruits*; Mir, S.A., Shah, M.A., Mir, M.M., Eds.; Springer: Cham, Switzerland, 2018; pp. 201–222.
13. Ji, Y.; Xu, M.; Wang, A. Recent advances in the regulation of climacteric fruit ripening: Hormone, transcription factor and epigenetic modifications. *Front. Agric. Sci. Eng.* **2021**, *8*, 314–334.
14. Milić, B.; Tarlanović, J.; Keserović, Z.; Magazin, N.; Miodragović, M.; Popara, G. Bioregulators can improve fruit size, yield and plant growth of northern highbush blueberry (*Vaccinium corymbosum* L.). *Sci. Hortic.* **2018**, *235*, 214–220. [CrossRef]
15. Nickel, L. Plant Growth Regulators in Agriculture and Horticulture. In *Bioregulators for Crop Protection and Pest Control, Proceedings of the Developed from a Symposium Sponsored by the Division of Agrochemicals at the 205th National Meeting of the American Chemical Society, Denver, CO, USA, 28 march–2 April 1993*; Hedin, P.A., Ed.; American Chemical Society: Washington, DC, USA, 1993; pp. 1–14.
16. Abd El-Naby, S.K.M.; Mohamed, A.A.A.; El-Naggar, Y.I.M. Effect of melatonin, GA3 and NAA on vegetative growth, yield and quality of 'canino' apricot fruits. *Acta Sci. Pol. Hortorum Cultus* **2019**, *18*, 167–174. [CrossRef]
17. Mhetre, V.B.; Patel, V.B.; Singh, S.K.; Verma, M.K.; Mishra, G.P.; Dahuja, A.; Kumar, C. Effect of new generation bio-regulators on anthocyanins and berry quality of grape cv. Beauty Seedless. *Indian J. Agric. Sci.* **2021**, *91*, 920–923. [CrossRef]
18. Gupta, A.; Tripathi, V.K.; Shukla, J.K. Influence of GA3, Zinc and Boron on Fruit drop, Yield and Quality of Litchi (*Litchi chinensis* Sonn.). *Biol. Forum-Int. J.* **2022**, *14*, 1079–1083.

19. Laňar, L.; Scháňková, K.; Náměstek, J. Searching for new possibilities of bloom delay in apricots. *Acta Hortic.* **2022**, *1344*, 183–188. [CrossRef]
20. Devrari, N.; Negi, M.; Thakur, N. Studies on effect of gibberellic acid and naphthalene acetic acid spray on fruit set and yield of apricot (*Prunus armeniaca* L.). *Int. J. Agric. Sci.* **2017**, *7*, 59–64.
21. Southwick, S.M.; Yeager, J.T. Use of gibberellin formulations for improved fruit firmness and chemical thinning in 'Patterson' apricot. *ISHS Acta Hortic. X Int. Symp. Apric. Cult.* **1993**, *384*, 425–430. [CrossRef]
22. Roussos, P.A.; Ntanos, E.; Denaxa, N.K.; Tsafouros, A.; Bouali, I.; Nikolakakos, V.; Assimakopoulou, A. Auxin (triclopyr) and cytokinin (forchlorfenuron) differentially affect fruit physiological, organoleptic and phytochemical properties of two apricot cultivars. *Acta Physiol. Plant.* **2021**, *43*, 25. [CrossRef]
23. Lal, S.; Kumar, D.; Singh, D.B.; Ahmed, N.; Kumar, R.; Dar, G.A. Effect of pre-harvest application of calcium chloride and gibberellic acid on shelf-life and post-harvest quality of apricot (*Prunus armeniaca* L.) cv. Harcot. *J. Hortic. Sci.* **2011**, *6*, 46–51.
24. Kevrešan, Ž.; Milić, B.; Bajić, A.; Kovač, R.; Milović, M.; Kalajdžić, J.; Barać, G. Does application of naphthenic acids in early fruit development stage result in prolonged effect on cold storage and shelf life of apricot fruit? *Food Feed. Res.* **2022**, *49*, 139–153.
25. Garhwal, O.P.; Choudhary, M.R.; Bairwa, L.N.; Kumawat, K.L.; Kumar, P.; Basile, B.; Corrado, G.; Rouphael, Y.; Gora, J.S. Effects of Time of Pruning and Plant Bio-Regulators on the Growth, Yield, Fruit Quality, and Post-Harvest Losses of Ber (*Ziziphus mauritiana*). *Horticulturae* **2022**, *8*, 809.
26. Barać, G.; Mastilović, J.; Kevrešan, Ž.; Milić, B.; Kovač, R.; Milović, M.; Kalajdžić, J.; Bajić, A.; Magazon, N.; Keserović, Z. Effects of Plant Growth Regulators on Plum (*Prunus domestica* L.) Grown on Two Rootstocks at Harvest and at the Postharvest Period. *Horticulturae* **2022**, *8*, 621. [CrossRef]
27. Milić, B.; Mastilović, J.; Kevrešan, Ž.; Kovač, R.; Bajić, A.; Keserović, Z.; Magazin, N.; Milović, M.; Kalajdžić, J.; Barać, G. Consequences of NAA, BA and GA3 treatment in early fruit development phase on postharvest properties of apricot cv. NS4. *Acta Sci. Pol. Hortorum Cultus* **2022**, *21*, 49–59. [CrossRef]
28. Meier, U. *Growth Stages of Mono- and Dico-Tyledonous Plants*; Blackwell Wissenschafts-Verlag: Berlin, Germany, 1997.
29. Kovač, R.; Kevrešan, Ž.; Mastilović, J.; Magazin, N.; Milić, M.; Milović, M.; Bajić, A.; Kalajdžić, J.; Barać, G.; Keserović, Z. I_{AD} values of apricot (*Prunus armeniaca* L.) at harvest in relation to fruit quality and sensory properties during cold storage and shelf life. *N. Z. J. Crop Hortic. Sci.* **2022**, *50*, 205–222. [CrossRef]
30. Costache, M.A.; Campeanu, G.; Neata, G. Studies concerning the extraction of chlorophyll and total carotenoids from vegetables. *Rom. Biotechnol. Lett.* **2012**, *17*, 7702–7708.
31. Larrauri, J.A.; Rupérez, P.; Saura-Calixto, F. Mango peel fibers with antioxidant activity. *Z. Lebensm. Forsch. A* **1997**, *205*, 39–42. [CrossRef]
32. Singleton, V.; Orthofer, R.; Lamuela-Raventos, R.M. Analysis of total phenols and other oxidation substrates and antioxidants by means of folin-ciocalteu reagent. *Methods Enzymol.* **1999**, *299*, 152–178.
33. Pękal, A.; Pyrzynska, K. Evaluation of aluminium complexation reaction for flavonoid content assay. *Food Anal. Methods* **2014**, *7*, 1776–1782. [CrossRef]
34. Milenković, L.; Mastilović, J.; Kevrešan, Ž.; Bajić, A.; Gledić, A.; Stanojević, L.; Cvetković, D.; Šunić, L.; Ilić, S.Z. Effect of shading and grafting on yield and quality of tomato. *J. Sci. Food Agric.* **2020**, *100*, 623–633. [CrossRef] [PubMed]
35. Melgarejo, P.; Calín-Sánchez, Á.; Carbonell-Barrachina, Á.A.; Martínez-Nicolás, J.J.; Legua, P.; Martínez, R.; Hernández, F. Antioxidant activity, volatile composition and sensory profile of four new very-early apricots (*Prunus armeniaca* L.). *J. Sci. Food Agric.* **2014**, *94*, 85–94. [CrossRef]
36. Ishaq, S.; Rathore, H.A.; Masud, T.; Ali, S. Influence of postharvest calcium chloride application, ethylene absorbent and modified atmosphere on quality characteristics and shelf life of apricot (*Prunus armeniaca* L.) fruit during storage. *Pak. J. Nutr.* **2009**, *8*, 861–865. [CrossRef]
37. Nagy, N. Effect of preharvest applications of calcium, anti-ethylene compounds and their combinations on "Canino" apricot fruit quality and storability. *Zagazig J. Agric. Res.* **2018**, *45*, 1609–1631. [CrossRef]
38. Khajehyar, R.; Rahemi, M.; Fallahi, E. The impact of various rates and dates of gibberellic acid applications on fruit set in apricot. *Int. J. Fruit Sci.* **2015**, *15*, 324–338. [CrossRef]
39. Abdel-Mohsen, M.A.; Kamel, H.M. Fruit set and quality of 'canino' apricot fruits as affected by spraying with yeast, growth regulators and micronutrients. *J. Plant Prod.* **2015**, *6*, 1431–1441. [CrossRef]
40. Southwick, S.M.; Yeager, J.T.; Weis, K.G. Use of gibberellins on 'Patterson' apricot (*Prunus armeniaca*) to reduce hand thinning and improve fruit size and firmness: Effects over three seasons. *J. Hortic. Sci.* **1997**, *72*, 645–652. [CrossRef]
41. Canli, F.A.; Sahin, M.; Temurtas, N.; Pektas, M. Improving fruit quality of apricot by means of preharvest benzyladenine and benzyladenine plus gibberellin applications. *Horttechnology* **2014**, *24*, 424–427. [CrossRef]
42. Bezerra, J.B.N.; Jesus, P.R.R.D.; Souza, I.D.; Bezerra, W.C.; Martins, G.C.S.B.; Ribeiro, V.G. Plant regulators on the growth, quality and production of 'Tommy Atkins' mango fruits. *Rev. Bras. Frutic.* **2021**, *43*, e546. [CrossRef]
43. Webster, A.D.; Spencer, J.E.; Dover, C.; Atkinson, C.J. The influence of sprays of gibberellic acid (GA3) and aminoethoxyvinylglycine (AVG) on fruit abscission, fruit ripening and quality of two sweet cherry cultivars. *Acta Hortic.* **2006**, *727*, 467–472. [CrossRef]
44. Mesa, K.; Reginato, G.; Contador, L.; Infante, R. Prohexadione Calcium and Naphthalene Acetic Acid Sprays Improve Fruit Size and Maintain Fruit Quality of 'Castlebrite' Apricot. *Eur. J. Hortic. Sci.* **2012**, *77*, 115.

horticulturae

Article

Application of Cinnamaldehyde Solid Lipid Nanoparticles in Strawberry Preservation

Shangjian Li [1,2], Jiajia Chen [1,2], Yuntong Liu [1,2], Qinhua Zheng [1,2], Weijian Tan [1,2], Xiaolin Feng [1,2], Kexin Feng [1,2] and Wenzhong Hu [1,*]

1 School of Pharmacy and Food Science, Zhuhai College of Science and Technology, Zhuhai 519040, China
2 College of Life Science, Jilin University, Changchun 130000, China
* Correspondence: huwenzhongtd@sina.com

Abstract: Strawberries are a popular food. However, the growth and reproduction of microorganisms on the surface of strawberries change their quality and may cause food poisoning. We compared the effects of solid lipid nanoparticles containing cinnamaldehyde (SLN-CA) and unencapsulated cinnamaldehyde on the freshness of strawberries stored for seven days. The impacts of SLN-CA at different concentrations on strawberry firmness, weight loss, rate of fruit rot, and sensory quality were investigated at 25 °C. Superoxide dismutase (*SOD*) and catalase activities and malonaldehyde (*MDA*) and vitamin C contents of strawberry cell homogenates were measured during storage. The experimental results showed that SLN-CA treatment can effectively reduce the probability of decay in strawberries without causing excessive weight loss. SLN-CA can reduce softening, maintain a high level of *SOD* activity in cells, reduce the accumulation of *MDA* and consumption of organic acids, and improve the sensory characteristics of strawberries and thereby their shelf life. Therefore, SLN-CA is a promising preservation method to increase the shelf life and safety of strawberries.

Keywords: postharvest; solid lipid nanoparticles; *Fragaria*; fruit storage

Citation: Li, S.; Chen, J.; Liu, Y.; Zheng, Q.; Tan, W.; Feng, X.; Feng, K.; Hu, W. Application of Cinnamaldehyde Solid Lipid Nanoparticles in Strawberry Preservation. *Horticulturae* **2023**, *9*, 607. https://doi.org/10.3390/horticulturae9050607

Academic Editor: Anita Sønsteby

Received: 17 March 2023
Revised: 15 May 2023
Accepted: 19 May 2023
Published: 22 May 2023

1. Introduction

Strawberry (*Fragaria* × *ananassa* Duch.) belongs to the family Rosaceae. The fruit is deliciously sweet, aromatic, and red in colour and has a unique flavour. The pulp is rich in dietary fibres and minerals, whereas the skin is rich in anthocyanins and contains many biologically active components, including polyphenolic compounds, flavonoids, vitamins, pectin, and organic acids, which are highly nutritious [1–5]. Strawberries do not have strong skin, and the cell wall strength is relatively low, which easily causes mechanical damage during production and transportation. Mechanical damage causes damage to the exterior of the strawberry, resulting in the leakage of cellular fluid. The cytosol contains a large amount of nutrients that greatly facilitate the infection and growth of microorganisms. The microorganisms' growth is the main cause of post-harvest rot and spoilage in strawberries [6–8]. In addition, fresh strawberries have high respiration and fast metabolic rates, which further accelerate fruit decay [7]. Mildew, softening, and rotting of strawberries can have serious adverse effects during storage, transportation, and sale [9,10]. Therefore, strawberries are considered a fragile commodity with a very limited shelf life, typically 6–9 days at 25 °C, and an estimated loss rate of up to 40% from the harvest and storage to table process [11]. Rapid cooling after picking and storing strawberries at a low temperature of 0–4 °C is an effective and common method to maintain the fruit quality. [12]. However, the process of an unbroken cold chain, from picking to the supermarket, is cost-intensive in terms of material and personnel [13].

Cinnamaldehyde (CA), the main component of essential oils extracted from natural cinnamon plants, has good antibacterial activity (MIC:1.88 mM) [14]. CA has been found to inhibit the growth of various bacteria, such as *Escherichia coli*, *Salmonella*, *Bacillus subtilis*, and

Pseudomonas aeruginosa. CA has been recognised as "Generally Recognized as Safe" (GRAS) by the US Food and Drug Administration (FDA) and Flavor and Extract Manufacturers Association. Additionally, the European Commission allows the use of CAs in food. [15]. CA is a natural, green, safe, and highly sensitive antibacterial preservative with great development potential and good application prospects in the food industry [16,17]. The application of these natural compounds in the food industry may be a potential alternative; however, the cost of their application and other issues such as their strong flavour and potential toxicity limit their use in food conservation [18].

Owing to its pungent odour, CA is difficult to apply directly to fruits and vegetables. Moreover, a high concentration of CA has a burning effect on the surface of products and shortens their shelf life. Additionally, high volatilisation and poor stability of CA reduce its long-term antibacterial effects. A practical solution to overcome these limitations is to encapsulate CA in solid lipid nanoparticles to form a nanocomplex.

Superoxide dismutase (*SOD*) and catalase (CAT) are important antioxidant enzymes in plant cells [19], which eliminate harmful free radicals both inside and outside the cells, maintain cellular metabolic homeostasis, and protect cells from oxidative damage [20]. *SOD* primarily clears superoxide free radicals produced within cells, whereas CAT clears hydrogen peroxide molecules inside and outside the cells. Together, they maintain the oxidation-reduction balance within the cells and protect strawberry cells from oxidative damage. Malonaldehyde (*MDA*) is a toxic metabolite primarily used as a marker of oxidative stress to measure the degree of oxidative damage to biological molecules such as cell membranes and enzymes [21].

Solid lipid nanoparticles (SLN) are nanoscale drug carriers developed in the 1980s and 1990s [22]. SLN typically consist of three fundamental components: solid lipids, surfactants, and water. Lipid compounds are generally recognised as safe (GRAS) and have been shown to possess biocompatibility and biodegradability since they are naturally present in living organisms. Solid lipid nanoparticles can encapsulate drugs, thereby protecting them from degradation [23], decomposition, or inactivation. Sustained drug release can be achieved by adjusting factors such as the particle size, composition, and surface structure [24]. Owing to the high stability, simple preparation, high drug loading, excellent biocompatibility, and biodegradability of SLN, they are one of the current research hotspots [25].

Therefore, the aim of this study was to investigate the effect of solid lipid nanoparticles containing cinnamaldehyde (SLN-CA) on postharvest strawberry preservation and to evaluate fruit firmness, colour, *SOD*, *MDA*, CAT, cell membrane integrity, and sensory evaluation results after SLC-CA treatment to provide efficient, long-lasting, natural, and pollution-free preservatives for future agricultural and horticultural industries.

2. Materials and Methods

2.1. *Strawberry*

Strawberries (cultivar Sweet Charlie) were harvested in December from a strawberry garden in the Jinwan District of Zhuhai City, Guangdong Province, China. Strawberries of similar size, maturity, and shapes, with no mechanical damage, weighing 18–22 g, and with approximately 80% red fruit surface were selected. Fresh strawberries were immediately sent to the laboratory at the School of Pharmacy and Food Science, Zhuhai Institute of Science and Technology, Jinwan District, Zhuhai, China. The experiment started when the samples arrived at the laboratory and lasted for seven days.

2.2. *Preparation of SLN-CA*

SLN-CA was prepared by ultrahigh-pressure homogenisation (UHPH) [26]. Ultrapure water was stirred on a magnetic stirrer at 65 °C, and 1.5% (w/v) Poloxamer 188 (BR, CAS: 9003-11-6, Shanghai Yuanye Bio-Technology Co., Ltd., Shanghai, China) and 1.5% (w/v) Lecithin High Potency (Shengqing Bio Ltd., Xi'an, China) were added as the aqueous phase. We added 1.5% (w/v) Tween-80 (PC, CAS: 9005-65-6, Shanghai Yuanye BioTechnology Co., Ltd., Shanghai, China), 1.5% (w/v) monostearin (PC, CAS: 9005-65-6,

Shanghai Yuanye BioTechnology Co., Ltd., Shanghai, China), and 4% (*v*/*v*) CA (AR, 95%, CAS: 104-55-2, Shanghai Yuanye Biotechnology Co., Ltd., Shanghai, China) to the oil phase and stirred using a magnetic stirrer at 65 °C. The aqueous phase was gradually added to the oil phase under constant stirring, followed by emulsification and homogenisation using a microfluidizer. Finally, the homogenised nanoemulsion was rapidly cooled and solidified into nanoparticles.

2.3. Handling Strawberries and Preservation

To evaluate the quality of strawberries, 240 strawberries were divided into eight groups for weight loss experiments. Additionally, 672 strawberries were randomly divided into eight groups, out of which seven groups underwent treatment with different concentrations of SLN-CA (16.6, 8.3, 4.6, 2.08, and 1.04 µL/mL), CA (6.6 and 1.04 µL/mL), and ultrapure water. For each treatment, 84 fruits were placed in a large storage box.

2.4. Decay Index of Strawberry Spoilage

The freshness preservation ability of the SLN-CA solution for strawberries was evaluated by analysing the decay incidence and severity during postharvest storage of fruits [27]. Sixty strawberries were selected from the treated pile and divided into three groups, and the occurrence and severity of spoilage in each group were recorded daily for seven consecutive days.

The decay index was graded into 5 levels. (0 = absence of symptoms; 1 = 1–25% of injured area; 2 = 26–50%; 3 = 51–75%; and 4 = 76–100%) [28].

$$X = \frac{n_1 + n_2 + n_3 + n_4}{n} \times 100$$

$$S = \frac{1n_1 + 2n_2 + 3n_3 + 4n_4}{4n} \times 100$$

where X is Corruption rate (%); S is Severity (%); n_1–n_4 are the number of strawberries in decay classes 1–4; n is the total number of strawberries evaluated in each replicate experiment.

2.5. Quality Evaluation

2.5.1. Weight Loss

The weight loss of each fruit during storage was determined by daily weighing and expressed as a percentage of weight loss relative to the initial weight according to the following formula:

$$W = \frac{(m_0 - m_n)}{m_0} \times 100\%$$

where W is the weight loss rate (%) of the strawberry samples; m_0 is the weight of the strawberries before storage, and m_n is the weight on day n.

2.5.2. Firmness

Firmness was measured using a texture analyser (Bosin, TA. TOUCH, Shanghai, China) with a 6 mm cylindrical probe and mode selection: texture profile analysis. The test was conducted at a pre-test speed of 2 mm/s, test speed of 1 mm/s, and post-test speed of 1 mm/s. The target mode was set to a displacement (mm) value of 5 mm. Each experimental group was further divided into five parallel groups. Changes in skin and flesh firmness after treatment with preservation liquid during storage were analysed.

2.5.3. Colour Variation and Visual Appearance

After seven days of storage, the strawberries were removed from the storage environment and equilibrated at room temperature. Colour was measured at 25 °C using a colorimeter (KONICA MINOLTA, CR-10 Plus, Tokyo, Japan). Brightness (ΔL*) and colour saturation (Chroma, ΔC*) were selected for colour index analysis. Three biological replicates were used for each experiment.

2.6. Biochemical Index

2.6.1. Preparation of Strawberry Homogenate

Strawberries were chopped into pieces and added to 1:1 ultrapure water. This mixture was homogenised in a homogenisation bag through beating.

2.6.2. Total Soluble Solids (TSS)

An ABBE refractometer (way-2W, INESA, Shanghai, China) was used to measure the TSS. The measured values were corrected according to the temperature coefficient of the refractive index, and measurements were recorded daily during storage [29].

2.6.3. Titratable Acid (TA)

The acid-base titration method with phenolphthalein was used as an indicator of TA. Tissue TA content was calculated according to the NaOH titrant consumption and expressed as a mass fraction (%) according to the following formula:

$$Titratable\ acid = \frac{V \times c \times (V_1 - V_0) \times f}{V_s \times m} \times 100\%$$

where V is the total volume of sample extract in mL; V_s is the volume of filtrate taken during titration in mL; c is the concentration of NaOH titrant in mol/L; V_1 is the volume of NaOH solution consumed for titration of filtrate in mL; V_0 is the volume of NaOH solution consumed for titration of distilled water in mL; m is the mass of sample in g; f is the conversion factor in g/mmol; Citric acid 0.064 was chosen as the f for this experiment.

2.6.4. SOD Activity

The strawberry homogenate was added to 8 times the volume of phosphate buffer (phosphate buffer: 0.1 mol/L pH 7–7.4). The mixture was centrifuged at 6000 rpm at 4 °C and the supernatant extracted. *SOD activity* was detected using an *SOD* assay kit (Nanjing Jiancheng Institute of Biological Engineering, Nanjing, China) at 450 nm using a UV spectrophotometer (UV-2600, Shimadzu, Kyoto, Japan). The *SOD inhibition* rate was obtained to calculate *SOD activity* (U) using the following formula:

$$SOD\ activity = \frac{SOD\ Inhibition \times 12}{50\%} \times P \div \frac{M}{V}$$

where P is the dilution; M is the strawberry weight in g; V is the volume of the added PBS in mL.

2.6.5. MDA

The extraction and reaction were performed using a plant *MDA* test kit (Nanjing Jiancheng Institute of Biological Engineering, Nanjing, China), and the results were detected using a UV spectrophotometer at 532 nm. *MDA content* (nmol/g) was calculated using the following formula:

$$MDA\ content = \frac{A_E - A_z}{A_S - A_Z} \times C \div \frac{M}{V}$$

where C is the concentration of the standard (10 nmol/mL); M is the weight of the plant tissue (g); V is the total amount of added extract (mL); A_E is the absorbance of the sample; A_S is the absorbance of the standard; A_z is the absorbance of the blank.

2.6.6. Vitamin C (Vc)

A standard curve for Vc was established using spectrophotometry, and measurements were recorded daily during storage. Each treatment was repeated three times.

2.6.7. CAT

The decomposition of H_2O_2 by CAT can be rapidly aborted by adding ammonium molybdate. The remaining H_2O_2 interacted with ammonium molybdate to produce a yellowish complex, which was measured at 405 nm, and CAT activity was calculated. CAT viability was assessed using a CAT assay kit (Nanjing Jiancheng Institute of Biological Engineering, Nanjing, China) (visible light).

2.7. Sensory Evaluation

The nine-point preference scale method for the organoleptic evaluation of strawberries is universally applicable for evaluating product acceptability and preference and is widely used in many fields for a wide range of products, such as food and beverages, cosmetics, and household products. Strawberries stored for seven days were evaluated for shape, colour, appearance, scent, and overall acceptability based on a nine-point preference scale [30,31] from 1 (dislike very much) to 9 (like very much). The hedonic evaluation scale was as follows: 1, dislike very much; 2, dislike a lot; 3, dislike moderately; 4, dislike slightly; 5, indifferent; 6, like slightly; 7, like moderately; 8, like a lot; and 9, like very much. Ten practitioners (five males and five females, aged 19–45 years) in food and pharmaceutical science and technology-related disciplines served as evaluators. The evaluators were trained for the sensory evaluation related to strawberries before the experiment. Sensory evaluation of the strawberries was performed on a separate test bench in a separate sensory laboratory. The assessors evaluated the appearance, colour, odour, shape, and overall acceptability of the strawberries at room temperature (approximately 25 °C). This method was used to determine the acceptability of SLN-CA-treated strawberries after seven days of storage and to verify the effect of SLN-CA on the sensory properties of the fruits.

3. Results and Discussion

3.1. Decay Index of Strawberry Spoilage

Decay of fruits and vegetables is one of the most important indicators for evaluating preservatives [28]. Regarding decay incidence (Figure 1a), treatment with 16.6 μL/mL of CA resulted in complete berry rotting on the third day, whereas treatment with 1.04 μL/mL of CA led to complete rotting on the fifth day. However, the treatment of strawberries with 1.05, 2.05, and 4.16 μL/mL SLN-CA positively reduced the rate of fruit rot compared to the that in the control. Even after treating strawberries with unencapsulated CA, the low-concentration treatment group showed much more severe decay than the blank control group. Strawberries treated with high concentrations of CA and SLN-CA did not exhibit mould, and their decay was due to ulceration of the strawberry peel caused by corrosion because of the high concentrations of CA [32,33]. Regarding the severity (Figure 1b), the same result of decay rate was observed. Pure CA treatment caused a large area of strawberry rot. In contrast, 2.05 and 4.16 μL/mL SLN-CA treatment could effectively alleviate the strawberry rot problem.

Strawberries treated with 8.3–1.04 μL/mL SLN-CA showed a reduction in decay incidence and severity relative to the blank control group. Among them, 1.05 μL/mL freshness solution-treated strawberries showed decay on the surface during the experiment, and the multiplication of the decay mainly caused their decay. The same effective concentration of SLN-CA freshness solution can release CA slowly so that strawberries are not initially exposed to a high concentration of essential oil. Therefore, strawberries treated with low concentrations of preservation solutions have insufficient effective concentrations to inhibit decay growth later. In contrast, strawberries treated with high concentrations of unencapsulated essential oil experienced severe rot due to the sudden release and high concentrations.

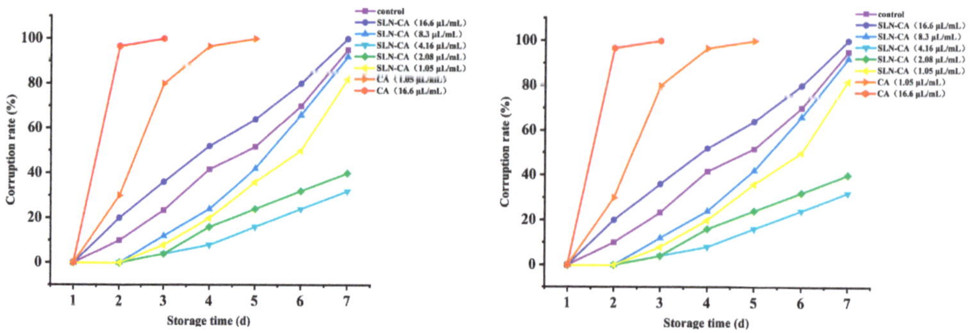

(**a**) (**b**)

Figure 1. Effects of different treatments on incidence (**a**) and severity (**b**) of the postharvest decay in fresh strawberries during a seven-day storage period at 25 °C and 60% humidity.

3.2. Weight Loss

The change in weight during storage is an important indicator of strawberry quality. Weight loss during storage results from respiration and water loss due to transpiration [34]. The results showed that strawberries treated with a high concentration (16.6 μL/mL) of pure CA started to lose weight severely on the second day, and the weight loss rate was as high as 42.21 ± 10.39% on the third day. On the third day, the whole fruit was rotten; therefore, no further data were recorded for this group. In addition, strawberries treated with a low concentration (1.05 μL/mL) of pure CA showed a weight loss of up to 42.88 ± 5.81% and whole fruit decay on the fifth day (Table 1). The weight loss rate of strawberries treated with SLN-CA was lower than that of strawberries treated with pure CA, and no serious rotting occurred within seven days. Moreover, treatment with 8.3–2.08 μL/mL SLN-CA did not cause significant weight loss in strawberry fruits compared to the control group ($p \geq 0.05$). The overall weight loss of the 8.3–2.08 μL/mL SLN-CA treatments was near 21.44%. The highest concentration (16.6 μL/mL) of preservation solution resulted in a slightly higher weight loss compared to the other concentration groups. The weight loss rate was maintained at 28.98 ± 3.11% on the seventh day. Strawberries treated with a 1.05 μL/mL preservation solution had a better preservation effect than other concentration groups ($p < 0.05$), and the weight loss rate was maintained at 16.72 ± 1.33% at the seventh day (Figure 2). SLN-CA can form a stable complex system that forms a protective film on the surface of strawberries, reducing evaporation and oxidation, preventing microbial contamination, and thus reducing the weight loss and decay incidence of strawberries. However, unencapsulated CA can burn fruit cells [31]. This conclusion was also reflected in a study on the application of thyme essential oil for the preservation of freshly cut apples [33]. The photographs of strawberries in Figure 3 show that SLN-CA treatment is effective in alleviating rotting in strawberries compared to the control group.

3.3. Firmness

Fruit firmness is an important indicator of shelf life and an important aspect for evaluating quality and commodity value [35]. The initial firmness of the strawberries in the experiment was 115 ± 7.62 gf. The firmness test results for each test group, as shown in Figure 3, revealed that the strawberries treated with 2.08 μL/mL SLN-CA had the best peel firmness, with 117 gf (Figure 4a). There were no significant differences ($p > 0.05$) in the flesh firmness of strawberries treated with 4.16, 2.08, and 1.05 μL/mL SLN-CA for seven days with 72 gf compared to that in the blank control group. The good firmness is due to the SLN-CA treatment, which reduces the likelihood of strawberries being infected with pathogenic fungi such as *Rhizopus* spp. and *Botrytis cinerea* and reduces the enzymatic digestion of strawberry cells by microorganisms. Strawberries contain a large amount of water, and decay can grow on the surface and inside the fruit, consuming nutrients and water. This can cause strawberries to become soft and sticky, eventually losing their crispness and flavour. Additionally, decay can secrete acids and enzymes, which can accelerate the decay and spoilage of strawberries. Therefore, decay on strawberries can lead to a decrease in firmness. These enzymatic reactions cause the softening of strawberries [36]. Notably, after seven days of storage, the firmness of the pure CA treatment group was lower compared to that of the blank control group (Figure 4b) because excessive essential oil treatment poisoned the strawberries, causing cell destruction and accelerating the aging of the strawberry tissue, resulting in a decrease in firmness [18].

Table 1. Results of the effect of different treatments on the weight of strawberries before and after storage and weight loss rate. Different letters above the columns indicate significant differences among the treatments based on Duncan's multiple-range test ($p < 0.05$); D means the strawberry is completely rotten and nothing is measured.

Processing Means	Weight before (g)	Weight afterwards (g)	Weight Loss Rate (%)
CA (1.05 μL/mL)	12.14 ± 2.05 [a]	D	D
CA (16.6 μL/mL)	12.92 ± 1.83 [a]	D	D
SLN-CA (1.05 μL/mL)	12.38 ± 2.23 [a]	10.317 ± 2.2 [b]	16.72 ± 1.33 [a]
SLN-CA (2.08 μL/mL)	12.45 ± 2.19 [a]	9.83 ± 2.14 [b]	21.04 ± 2.22 [b]
SLN-CA (4.16 μL/mL)	12.63 ± 2.62 [a]	10.01 ± 2.49 [b]	20.72 ± 4.95 [b]
SLN-CA (8.32 μL/mL)	12.76 ± 1.95 [a]	9.93 ± 1.88 [b]	22.17 ± 3.36 [b]
SLN-CA (16.6 μL/mL)	12.29 ± 1.18 [a]	8.57 ± 1.12 [a]	28.98 ± 3.11 [c]
Control	12.18 ± 1.65 [a]	9.51 ± 1.584 [b]	21.98 ± 4.33 [b]

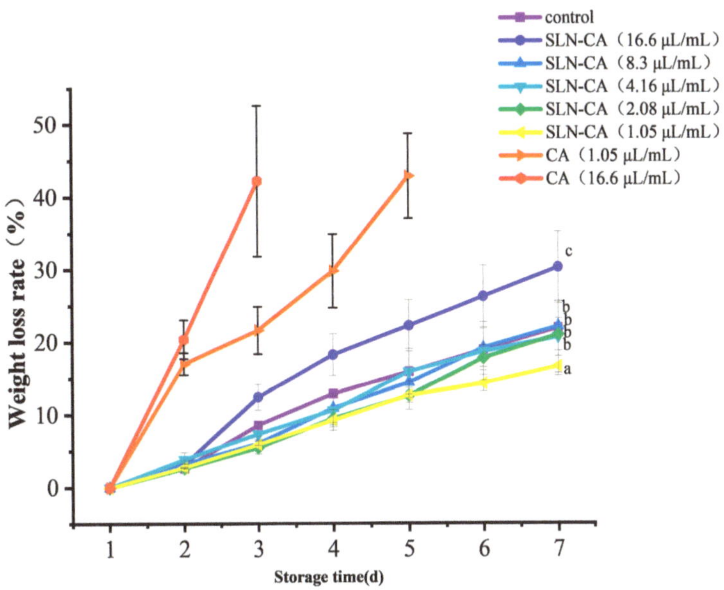

Figure 2. The change curves of the weight loss rate of strawberries treated with different concentrations of SLN-CA preservation solution and CA for seven days. Error bars show the standard deviation of the means (thirty strawberry repeats); different letters above the columns indicate significant differences among the treatments based on Duncan's multiple-range test ($p < 0.05$).

Figure 3. Pictures of strawberries stored at 25 °C and 60% humidity for seven days after different treatments.

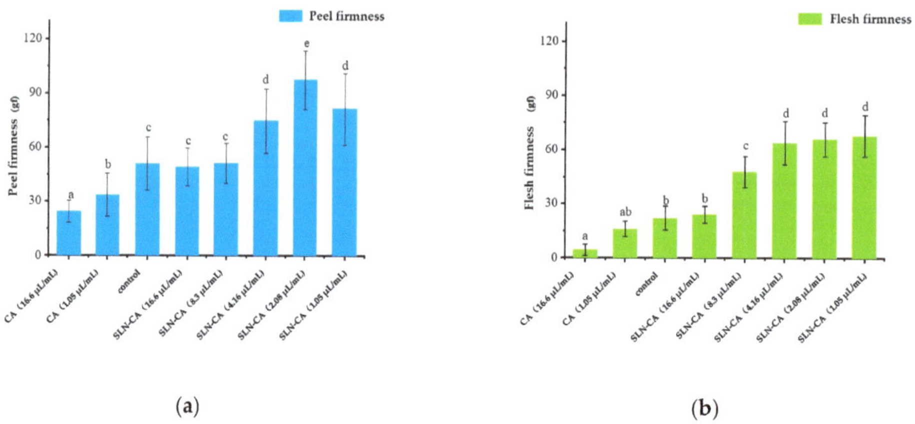

Figure 4. Strawberry peel firmness (**a**) and strawberry flesh firmness (**b**) in different treatments and blank control during seven days. Different letters above the columns indicate significant differences among the treatments based on Duncan's multiple-range test ($p < 0.05$).

3.4. Colour Variation and Visual Appearance

Colour and appearance are important indicators of consumer perceptions of fruit and vegetable quality. The colour changes in fresh-cut apples in the different treatment groups are shown in Figure 4. ΔL^* and ΔC^* were chosen as the colour indices. The ΔL^* value indicates the brightness of the sample, which is one of the indicative parameters of surface darkening due to enzymatic browning or pigment aggregation during storage; the lower the value, the more serious the browning. The ΔC^* value indicates the colour saturation of the sample; the higher the value, the brighter the quality. After seven days of storage, the treated strawberries were brighter than the blank controls ($p < 0.05$). However, the 16.6 μL/mL SLN-CA- and pure CA-treated groups were less bright compared to the blank control (Figure 5a) because the excessive essential oil burned the strawberry skin and poisoned the strawberries, causing a severe decrease in brightness [18,31,37]. The ΔC^* value indicated the colour saturation of the samples. The colour saturation results after seven days of storage were similar to the brightness results. Strawberries treated with 2.08 and 1.05 μL/mL SLN-CA had higher ΔC^* compared to that of the blank control and other treatments ($p < 0.05$). Treatment with pure CA significantly reduced ΔC^* in strawberries (Figure 5b). High concentrations of CA significantly increased the degree of browning in strawberries, probably because the high concentration of CA disrupts the strawberry cells, causing a loss of separation between enzymes and substrates regionally distributed in the strawberry tissue and accelerating enzymatic browning. High concentrations of CA are toxic to fruits and vegetables and have been reported in various fruits and vegetables. Adding rosemary, thyme, and oregano CA to lettuce, coleslaw, and chopped cabbage reduced the product appearance scores to unacceptable levels throughout storage, and all the products lost their commercial value after 8 days of storage [38]. In addition, the slow release of cinnamaldehyde contained in the SLN-CA played a role in avoiding the toxicity of high concentrations of CA in strawberry cells. Therefore, the brightness and saturation of the SLN-CA treatment group were significantly higher than those of the CA group at the same essential oil concentrations.

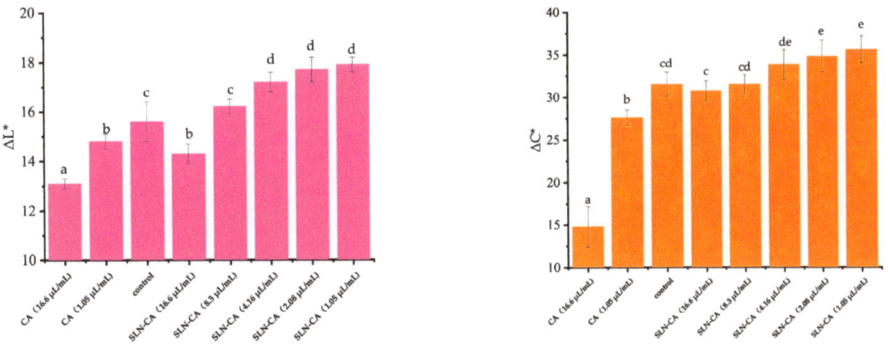

(a) (b)

Figure 5. Strawberry lightness (**a**) and strawberry colour saturation (**b**) in different treatments and blank control during seven days. Different letters above the columns indicate significant differences among the treatments based on Duncan's multiple-range test ($p < 0.05$).

3.5. Vc, TSS, and TA

TSS, including sugars, acids, vitamins, and minerals, has a crucial influence on strawberry quality. The decomposition of TSS and the accumulation of acids by microbial metabolism are the reasons for the rapid decline in the quality of strawberries in the middle and late stages of postharvest storage. The TSS of strawberries treated with SLN-CA was higher compared to that of the blank treatment group (Figure 6a) ($p < 0.05$), possibly because of two factors. First, after treatment with SLN-CA, some microorganisms on the surface of the strawberries were inhibited, protecting the integrity of the cells and allowing the enzymes and substrates distributed regionally in the strawberry tissues to remain separate and not digested enzymatically. Second, the lipid nanoparticles were coated on the surface of the strawberry, forming a solid lipid-cling film, creating an aerated environment with low oxygen and high carbon dioxide and reducing the metabolic rate in strawberry cells. In the CA-treated group, the essential oil was not encapsulated and acted at a high initial release concentration, scorching the cells and decreasing the TSS. Organic acids are essential components of strawberry flavour. Changes in acidity are mainly influenced by the metabolic rate, particularly the respiration rate. Respiration depletes organic acids; therefore, the acidity decreases during storage, which is the leading cause of fruit ageing [39]. The Vc contents of the different treatment groups after seven days of storage are shown in Figure 6b. The decrease in TA content in the 1.05 μL/mL SLN-CA-treated group may be due to two factors: First, the concentration was too low to inhibit microbial growth, resulting in acid accumulation via microbial metabolism. Second, microbial multiplication disrupts the integrity of strawberry cells, causing a loss of separation between enzymes and substrates regionally distributed in the tissue and accelerating the breakdown of TA. In contrast, the higher TA content in the pure CA-treated group may be due to the high concentration of CA that disrupts the integrity of strawberry cells [38]. The Vc content of the different treatment groups after seven days of storage is shown in Figure 6c, and the results were similar to those of TSS and TA.

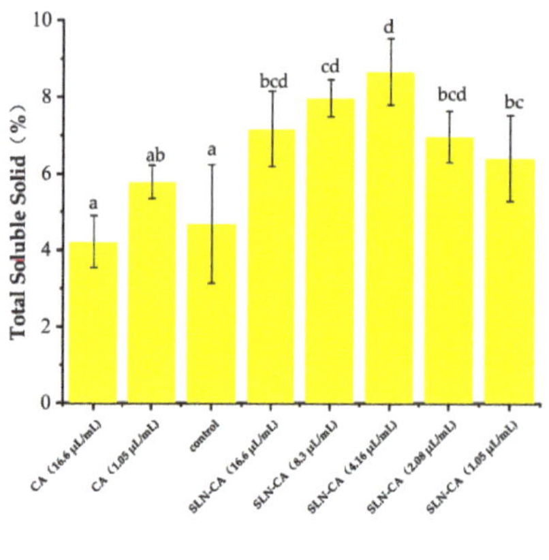

(**a**)

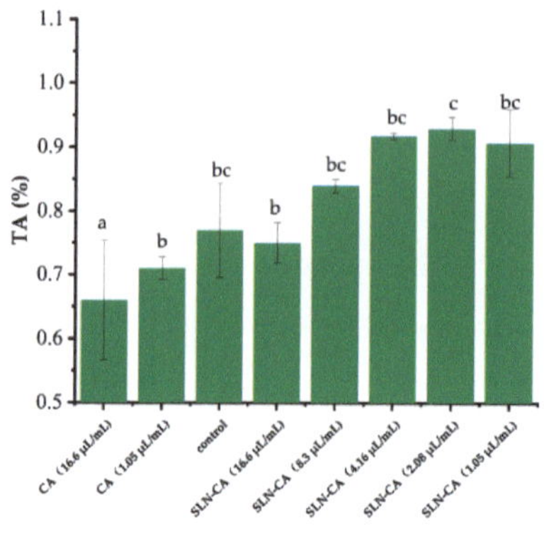

(**b**)

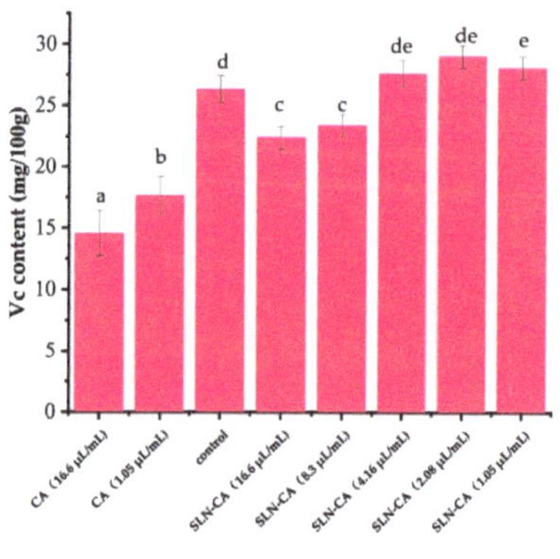

(**c**)

Figure 6. Result of Total soluble solids (**a**), *Titratable acid* (**b**), and Vitamin C content (**c**) in different treatments and blank control during seven days. Different letters above the columns indicate significant differences among the treatments based on Duncan's multiple-range test ($p < 0.05$).

3.6. SOD, MDA, and CAT

The effects of the different treatments on the *SOD*, *MDA*, and Vc of strawberries after seven days of storage are shown in Table 2.

Antioxidant enzymes play important roles in the suppression of oxidative stress. Strawberries constantly produce substances that are harmful to cells, such as H_2O_2. To reduce cell damage, strawberries must produce more antioxidant enzymes to remove harmful substances. Free-radical-induced oxidative stress plays an important role in fruit senescence.

SOD is the main reactive oxygen species scavenging enzyme, which catalyses the conversion of O^{2-} to H_2O_2 and O_2, keeping free radicals at a low level and avoiding membrane damage. The first line of defence for scavenging reactive oxygen species in plant cells is important for maintaining plant cell activity [40]. Strawberries treated with 1.05 μL/mL of SLN-CA for seven days had the highest *SOD* activity, reaching 214.77 ± 0.65 U/g, which was higher compared to that of the blank control ($p < 0.05$) and 1.05 μL/mL CA-treated group. CA-treated strawberries with 16.6 μL/mL had 197.86 ± 1.83 U/g after seven days of storage, which was lower compared to that of the blank control ($p < 0.05$). Similarly, CAT activity in SLN-CA-treated strawberries was significantly higher compared to that in the immediate treatment groups with unencapsulated CA. Strawberries treated with 2.08 μL/mL SLN-CA showed higher CAT activity compared to that with the other groups ($p < 0.05$).

Table 2. Results of the effect of different treatments on *SOD*, *MDA*, and Vc of strawberries before and after storage and weight loss rate. Different letters above the columns indicate significant differences among the treatments based on Duncan's multiple-range test ($p < 0.05$).

Processing Means	*SOD* (U/g)	CAT (U/g)	*MDA* (nmol/g)
CA (1.05 μL/mL)	207.25 ± 1.62^{a}	88.21 ± 7.62^{bc}	577.75 ± 20.68^{de}
CA (16.6 μL/mL)	197.86 ± 1.83^{b}	64.98 ± 8.69^{a}	598.63 ± 17.56^{e}
SLN-CA (1.05 μL/mL)	214.77 ± 0.65^{e}	92.91 ± 5.11^{bc}	546.83 ± 11.43^{c}
SLN-CA (2.08 μL/mL)	212.28 ± 1.08^{cd}	98.6 ± 6.87^{b}	463.07 ± 6.92^{a}
SLN-CA (4.16 μL/mL)	213.02 ± 1.34^{de}	94.6 ± 6.42^{bc}	523.23 ± 6.51^{b}
SLN-CA (8.32 μL/mL)	210.87 ± 0.86^{c}	88.45 ± 3.08^{bc}	565.93 ± 8.66^{cd}
SLN-CA (16.6 μL/mL)	202.08 ± 1.06^{b}	82.65 ± 7.16^{b}	580.69 ± 8.93^{de}
Control	210.42 ± 0.39^{c}	94.56 ± 5.44^{bc}	558.48 ± 7.65^{cd}

MDA is the final product of membrane lipid peroxidation, and SLN-CA treatment reduced the *MDA* content in strawberry cells, demonstrating that SLN-CA treatment could effectively reduce the peroxidation of strawberry cell membranes and further supporting that SLN-CA treatment can effectively maintain the integrity of cells and tissues and the fresh quality of strawberry fruit. The *MDA* concentration in strawberries treated with SLN-CA was significantly lower compared to that in strawberries naturally treated with unencapsulated CA. The lowest *MDA* concentration was observed in 2.08 μL/mL SLN-CA-treated strawberries ($p < 0.05$).

The effect of the different CA treatments on *SOD*, CAT, and *MDA* in strawberries could be due to the effective reduction in the growth of microorganisms. The microorganisms on the surface of the strawberry draw nutrients from the strawberry cells and destroy them, leading to their decay. However, unencapsulated CA treatment at high concentrations burned strawberry cells, destroying their cellular integrity.

3.7. Sensory Evaluation

The pungent smell and taste of essential oils can affect the organoleptic properties of food [41]. So, an organoleptic evaluation was carried out on strawberry fruit on the first and seventh days of storage at 25 °C. Strawberry fruit colour, scent, appearance, shape, and overall acceptability were rated on a nine-point preference scale ranging from 1 (highly disliked) to 9 (extremely liked). On day one (Figure 7a), shape, colour, appearance, scent, and overall acceptability in the 4.16, 2.08, and 1.05 μL/mL SLN-CA-treated groups did not significantly differ compared to the blank control group ($p > 0.05$). Unwrapped CA has a pungent odour that affects the sensory odour of strawberries. The 16.6 μL/mL CA-treated group had an odour score of only four and an overall acceptability score of five. In comparison, the 1.05 μL/mL CA-treated group had an odour score of only seven and an overall acceptability score of seven, indicating that the unencapsulated CA greatly affected the sensory odour quality of the strawberries. After seven days (Figure 7b), the results were recorded as 0 and not plotted in the graph because the 16.6 and 1.05 μL/mL CA-treated groups were severely decayed at four and six days of storage, respectively. Therefore, the significance of the sensory evaluation was lost. The highest sensory evaluation score between 7 and 8 and between 6 and 7 was overserved for 2.08 and 4.16 μL/mL SLN-CA-treated groups, respectively. These groups also performed better than the blank control group. However, the SLN-CA treatment of more than 8.3 μL/mL negatively affected the sensory odour of strawberries, and the score dropped between 1 and 4. These results are consistent with those of other studies, indicating that high concentrations of essential oils have a negative impact on the overall acceptability of food. For example, when evaluating the sensory characteristics of freshly cut sweet melons with lemongrass, the addition of 0.7% lemongrass to the coating formulation significantly reduced the sensory score [42].

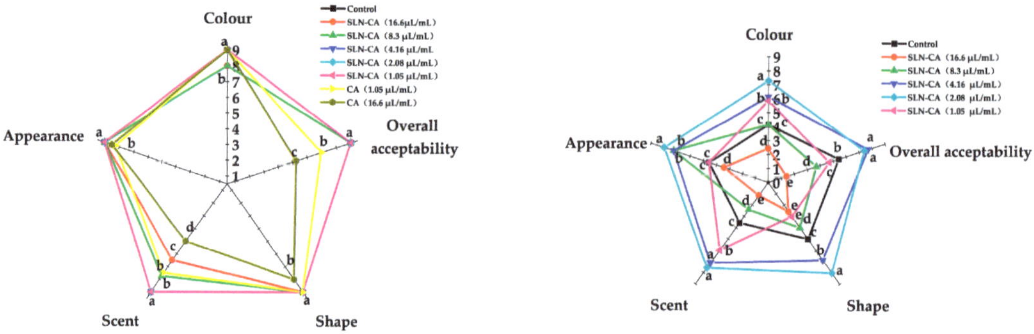

(**a**) (**b**)

Figure 7. Sensory characteristics of strawberries on the first (**a**) and seventh day (**b**) of storage at 25 °C. In the control group of first (**a**), the SLN-CA score for a concentration range of 1.05–4.16 μL/mL was 9. The nine-point hedonic evaluation scale was used as follows: 9, like very much to 1, dislike very much. For each sample, the means designated by different letters are significantly different ($p < 0.05$).

4. Conclusions

One of the most critical indicators for evaluating preservatives is their ability to prevent mould and decay of fruits and vegetables. The positive effect of SLN-CA on freshness was also confirmed by tests on fresh strawberries, in which SLN-CA treatment significantly reduced the incidence and severity of spoilage. As a potential preservative, SLN-CA can also reduce the *MDA* content of cells, increase *SOD* and CAT activity, and prevent cell damage. SLN-CA, compared to CA, avoids the toxic effects of high CA concentrations on strawberry cells, maintains better colour and appearance, and reduces the adverse sensory effects of CA on strawberries. Conclusively, this study tested the potential application of SLN-CA in the preservation of strawberries to provide efficient, long-lasting, natural, and pollution-free preservatives for future agricultural and horticultural industries.

Supplementary Materials: The following supporting information can be downloaded at: https://www.mdpi.com/article/10.3390/horticulturae9050607/s1.

Author Contributions: Conceptualization, J.C.; Data curation, S.L. and X.F.; Formal analysis, J.C.; Funding acquisition, W.H.; Investigation, Q.Z.; Methodology, J.C. and S.L.; Project administration, W.H.; Resources, W.H.; Software, S.L., X.F., W.T. and Y.L.; Supervision, W.H.; Validation, J.C., S.L. and K.F.; Visualization, S.L., Q.Z. and K.F.; Writing—original draft, J.C. and S.L.; Writing—review and editing, J.C., S.L., Q.Z. and W.T. All authors have read and agreed to the published version of the manuscript.

Funding: This research received no external funding.

Data Availability Statement: The data presented in this study are contained within the article or Supplementary Material.

Acknowledgments: The support from Laboratory, School of Pharmacy and Food Science, Zhuhai College of Science and Technology, is appreciated. This study was supported by the 'Thirteenth Five-Year Plan' for National key research and development program (Grant No. 2016YFD0400903).

Conflicts of Interest: The authors declare no conflict of interest.

References

1. Kallio, H.P.; Hakala, M.A.; Pelkkikangas, A.-M.; Lapveteläinen, A. Sugars and acids of strawberry varieties. *Eur. Food Res. Technol.* **2000**, *212*, 81–85. [CrossRef]
2. Tulipani, S.; Marzban, G.; Herndl, A.; Laimer, M.; Mezzetti, B.; Battino, M.A. Influence of environmental and genetic factors on health-related compounds in strawberry. *Food Chem.* **2011**, *124*, 906–913. [CrossRef]
3. Nunes, M.C.d.N.; Brecht, J.K.; Morais, A.M.M.B.; Sargent, S.A. Possible Influences of Water Loss and Polyphenol Oxidase Activity on Anthocyanin Content and Discoloration in Fresh Ripe Strawberry (cv. Oso Grande) During Storage at 1 °C. *J. Food Sci.* **2005**, *70*, 56. [CrossRef]
4. Aday, M.S.; Buyukcan, M.B.; Caner, C. Maintaining the quality of strawberries by combined effect of aqueous chlorine dioxide with modified atmosphere packaging. *J. Food Process. Preservat.* **2013**, *37*, 568–581. [CrossRef]
5. Fierascu, R.C.; Temocico, G.; Fierascu, I.; Ortan, A.; Babeanu, N.E. Fragaria genus: Chemical composition and biological activities. *Molecules* **2020**, *25*, 498. [CrossRef]
6. Jesmin, S.; Al-Jubayer, A.; Eusuf, S.b.; Kamal, A.H.M.A.; Islam, J.M.M.; Ferdoush, F.; Kabir, S.E.; Khan, M.A. Gamma Radiation Treated Chitosan Solution for Strawberry Preservation: Physico-Chemical Properties and Sensory Evaluation. *Int. Lett. Nat. Sci.* **2016**, *60*, 30–37. [CrossRef]
7. Jiang, Y.; Yu, L.J.; Hu, Y.; Zhu, Z.; Zhuang, C.; Zhao, Y.; Zhong, Y. Electrostatic spraying of chitosan coating with different deacetylation degree for strawberry preservation. *Int. J. Biol. Macromol.* **2019**, *139*, 1232–1238. [CrossRef]
8. Basu, A.; Nguyen, A.; Betts, N.M.; Lyons, T.J. Strawberry As a Functional Food: An Evidence-Based Review. *Crit. Rev. Food Sci. Nutrit.* **2014**, *54*, 790–806. [CrossRef]
9. Nunes, M.C.d.N.; Morais, A.M.M.B.; Brecht, J.K.; Sargent, S.A.; Bartz, J.A. Prompt Cooling Reduces Incidence and Severity of Decay Caused by Botrytis cinerea and Rhizopus stolonifer in Strawberry. *Hortechnology* **2005**, *15*, 153–156. [CrossRef]
10. Ceredi, G.; Antoniacci, L.; Montuschi, C.; De Paoli, E.; Mari, M.; Gengotti, S. Ten years of field trials on grey mold control on strawberries. *Acta Hortic.* **2009**, *842*, 327–330. [CrossRef]
11. Duran, M.; Aday, M.S.; Zorba, N.N.D.; Temizkan, R.; Büyükcan, M.B.; Caner, C. Potential of antimicrobial active packaging 'containing natamycin, nisin, pomegranate and grape seed extract in chitosan coating' to extend shelf life of fresh strawberry. *Food Bioprod. Process.* **2016**, *98*, 354–363. [CrossRef]
12. Tahir, H.E.; Xiaobo, Z.; Jiyong, S.; Mahunu, G.K.; Zhai, X.; Mariod, A.A. Quality and postharvest-shelf life of cold-stored strawberry fruit as affected by gum arabic (*Acacia senegal*) edible coating. *J. Food Biochem.* **2018**, *42*, e12527. [CrossRef]
13. Nunes, M.; Emond, J.; Brecht, J. Quality of strawberries as affected by temperature abuse during ground, in-flight and retail handling operations. In Proceedings of the International Conference on Quality in Chains; An Integrated View on Fruit and Vegetable Quality 604, Wageningen, The Netherlands, 6–9 July 2003; pp. 239–246.
14. Doyle, A.A.; Stephens, J.C. A review of cinnamaldehyde and its derivatives as antibacterial agents. *Fitoterapia* **2019**, *139*, 104405. [CrossRef] [PubMed]
15. Friedman, M. Chemistry, Antimicrobial Mechanisms, and Antibiotic Activities of Cinnamaldehyde against Pathogenic Bacteria in Animal Feeds and Human Foods. *J. Agric. Food Chem.* **2017**, *65*, 10406–10423. [CrossRef]
16. Chen, J.; Li, Y.; Shi, W.; Zheng, H.; Wang, L.; Li, L. Release of Cinnamaldehyde and Thymol from PLA/Tilapia Fish Gelatin-Sodium Alginate Bilayer Films to Liquid and Solid Food Simulants, and Japanese Sea Bass: A Comparative Study. *Molecules* **2021**, *26*, 7140. [CrossRef]
17. Abdullah, S.Z.; Tong, W.Y.; Leong, C.R.; Ab Rashid, S.; Rozman, N.A.S.; Mohammad Hamid, N.H.; Karim, S.; Tumin, N.D.; Muda, S.A.; Yacob, L.S. Development of cinnamaldehyde loaded-alginate based film for food packaging. *Asia Pacific J. Mol. Biol. Biotechnol.* **2021**, *10*, 1150. [CrossRef]
18. Sánchez-González, L.; Vargas, M.; González-Martínez, C.; Chiralt, A.; Chafer, M. Use of essential oils in bioactive edible coatings: A review. *Food Eng. Rev.* **2011**, *3*, 1–16. [CrossRef]
19. Xiong, S.; Wang, Y.-D.; Chen, Y.; Gao, M.; Zhao, Y.-X.; Wu, L. Effects of Drought Stress and Rehydration on Physiological and Biochemical Properties of Four Oak Species in China. *Plants* **2022**, *11*, 679. [CrossRef]
20. Huang, Y.; Sun, C.; Guan, X.; Lian, S.; Li, B.-h.; Wang, C.-x. Butylated Hydroxytoluene Induced Resistance Against Botryosphaeria dothidea in Apple Fruit. *Front. Microbiol.* **2021**, *11*, 599062. [CrossRef]
21. Yue, M.; Jiang, L.; Zhang, N.; Zhang, L.; Liu, Y.; Wang, Y.; Li, M.; Lin, Y.; Zhang, Y.; Zhang, Y.; et al. Importance of FaWRKY71 in Strawberry (*Fragaria* × *ananassa*) Fruit Ripening. *Int. J. Mol. Sci.* **2022**, *23*, 12483. [CrossRef]
22. Banerjee, A.; Qi, J.; Gogoi, R.; Wong, J.; Mitragotri, S. Role of nanoparticle size, shape and surface chemistry in oral drug delivery. *J. Control. Rel. Off. J. Controll. Rel. Soc.* **2016**, *238*, 176–185. [CrossRef]
23. Wang, C.; Lv, J.; Xue, C.-c.; Li, J.; Liu, Y.; Xu, D.; Jiang, Y.; Jiang, S.; Zhu, M.; Yang, Y.; et al. Novel role of COX6c in the regulation of oxidative phosphorylation and diseases. *Cell Death Discov.* **2022**, *8*, 336. [CrossRef] [PubMed]
24. Ngwuluka, N.C.; Kotak, D.J.; Devarajan, P.V. Design and Characterization of Metformin-Loaded Solid Lipid Nanoparticles for Colon Cancer. *AAPS PharmSciTech.* **2017**, *18*, 358–368. [CrossRef] [PubMed]
25. Hosseini, S.F.; Zandi, M.; Rezaei, M.; Farahmandghavi, F. Two-step method for encapsulation of oregano essential oil in chitosan nanoparticles: Preparation, characterization and in vitro release study. *Carbohydr. Polym.* **2013**, *95*, 50–56. [CrossRef] [PubMed]
26. Brandl, M.; Bachmann, D.; Drechsler, M.; Bauer, K.H. Liposome Preparation by a New High Pressure Homogenizer Gaulin Micron Lab 40. *Drug Dev. Ind. Pharm.* **1990**, *16*, 2167–2191. [CrossRef]

27. Dhakshinamoorthy, D.; Sundaresan, S.; Iyadurai, A.; Subramanian, K.S.; Janavi, G.J.; Paliyath, G.; Subramanian, J. Hexanal Vapor Induced Resistance against Major Postharvest Pathogens of Banana (*Musa acuminata* L.). *Plant Pathol. J.* **2020**, *36*, 133–147. [CrossRef]
28. Lin, Y.; Huang, R.; Sun, X.; Yu, X.; Xiao, Y.; Wang, L.; Hu, W.; Zhong, T. The p-Anisaldehyde/β-cyclodextrin inclusion complexes as fumigation agent for control of postharvest decay and quality of strawberry. *Food Control* **2021**, *130*, 108346. [CrossRef]
29. Mohammed, M.; Sallam, A.; Alqahtani, N.; Munir, M. The Combined Effects of Precision-Controlled Temperature and Relative Humidity on Artificial Ripening and Quality of Date Fruit. *Foods* **2021**, *10*, 2636. [CrossRef]
30. Peryam, D.R.; Pilgrim, F.J. Hedonic scale method of measuring food preferences. *Food Technol.* **1957**, *11*, 9–14.
31. Hu, W.; Feng, K.; Xiu, Z.; Jiang, A.; Lao, Y. Thyme oil alginate-based edible coatings inhibit growth of pathogenic microorganisms spoiling fresh-cut cantaloupe. *Food Biosci.* **2019**, *32*, 100467.
32. Perdones, Á.; Sánchez-González, L.; Chiralt, A.; Vargas, M. Effect of chitosan–lemon essential oil coatings on storage-keeping quality of strawberry. *Postharvest Biol. Technol.* **2012**, *70*, 32–41. [CrossRef]
33. Hu, W.; Feng, K.; Xiu, Z.; Jiang, A.; Lao, Y. Efficacy of thyme oil-alginate-based coating in reducing foodborne pathogens on fresh-cut apples. *Int. J. Food Sci. Technol.* **2019**, *54*, 3128–3137.
34. Dhital, R.; Mora, N.; Watson, D.G.; Kohli, P.; Choudhary, R. Efficacy of limonene nano coatings on post-harvest shelf life of strawberries. *LWT* **2018**, *97*, 124–134. [CrossRef]
35. Asghari, M.; Hasanlooe, A.R. Methyl jasmonate effectively enhanced some defense enzymes activity and Total Antioxidant content in harvested "Sabrosa" strawberry fruit. *Food Sci. Nutrit.* **2016**, *4*, 377–383. [CrossRef] [PubMed]
36. Feliziani, E.; Romanazzi, G. Postharvest decay of strawberry fruit: Etiology, epidemiology, and disease management. *J. Berry Res.* **2016**, *6*, 47–63. [CrossRef]
37. Moreira, M.R.; Cassani, L.; Martín-Belloso, O.; Soliva-Fortuny, R. Effects of polysaccharide-based edible coatings enriched with dietary fiber on quality attributes of fresh-cut apples. *J. Food Sci. Technol.* **2015**, *52*, 7795–7805. [CrossRef]
38. Scollard, J.; Francis, G.A.; O'Beirne, D. Some conventional and latent anti-listerial effects of essential oils, herbs, carrot and cabbage in fresh-cut vegetable systems. *Postharvest Biol. Technol.* **2013**, *77*, 87–93. [CrossRef]
39. Khalifa, I.; Barakat, H.; El-Mansy, H.A.; Soliman, S.A. Improving the shelf-life stability of apple and strawberry fruits applying chitosan-incorporated olive oil processing residues coating. *Food Packag. Shelf Life* **2016**, *9*, 10–19. [CrossRef]
40. Hasanuzzaman, M.; Fotopoulos, V.; Nahar, K.; Fujita, M. *Reactive Oxygen, Nitrogen and Sulfur Species in Plants: Production, Metabolism, Signaling and Defense Mechanisms*; John Wiley & Sons: Hoboken, NJ, USA, 2019.
41. Ji, H.; Kim, H.; Ryu, J.-H. Antimicrobial Activities of Gaseous Essential Oils against Xerophilic Mold (*Penicillium corylophilum*). In Proceedings of the IAFP 2018 Annual Meeting, Salt Lake City, UT, USA, 8–11 July 2018; pp. 49–54.
42. Raybaudi-Massilia, R.M.; Mosqueda-Melgar, J.; Martín-Belloso, O. Edible alginate-based coating as carrier of antimicrobials to improve shelf-life and safety of fresh-cut melon. *Int. J. Food Microbiol.* **2008**, *121*, 313–327. [CrossRef]

MDPI AG
Grosspeteranlage 5
4052 Basel
Switzerland
Tel.: +41 61 683 77 34

Horticulturae Editorial Office
E-mail: horticulturae@mdpi.com
www.mdpi.com/journal/horticulturae

www.ingramcontent.com/pod-product-compliance
Lightning Source LLC
Chambersburg PA
CBHW041403210926
51850CB00013B/3166
* 9 7 8 3 7 2 5 8 2 4 2 3 6 *